WHERE TO WATCH BIRDS IN NORTHWEST ENGLAND

LANCASHIRE, CHESHIRE, MERSEYSIDE AND GREATER MANCHESTER

STEPHEN DUNSTAN, IAN McKERCHAR AND JANE TURNER

H E L M
LONDON • OXFORD • NEW YORK • NEW DELHI • SYDNEY

HELM
Bloomsbury Publishing Plc
50 Bedford Square, London, WC1B 3DP, UK
Bloomsbury Publishing Ireland Limited,
29 Earlsfort Terrace, Dublin 2, D02 AY28, Ireland

BLOOMSBURY, HELM and the Helm logo are
trademarks of Bloomsbury Publishing Plc

First published in the United Kingdom 2026

Copyright © Stephen Dunstan, Ian McKerchar and Jane Turner, 2026

Stephen Dunstan, Ian McKerchar and Jane Turner have asserted
their right under the Copyright, Designs and Patents Act, 1988,
to be identified as Authors of this work.

For legal purposes the Acknowledgements on p. 6
constitute an extension of this copyright page.

All rights reserved. No part of this publication may be: i) reproduced or transmitted in any form, electronic or mechanical, including photocopying, recording or by means of any information storage or retrieval system without prior permission in writing from the publishers; or ii) used or reproduced in any way for the training, development or operation of artificial intelligence (AI) technologies, including generative AI technologies. The rights holders expressly reserve this publication from the text and data mining exception as per Article 4(3) of the Digital Single Market Directive (EU) 2019/790.

Bloomsbury Publishing Plc does not have any control over, or responsibility for, any third-party websites referred to or in this book. All internet addresses given in this book were correct at the time of going to press. The authors and publisher regret any inconvenience caused if addresses have changed or sites have ceased to exist, but can accept no responsibility for any such changes.

A catalogue record for this book is available from the British Library.
Library of Congress Cataloguing-in-Publication data has been applied for.

ISBN: PB: 978–1–3994–1532–3; ePDF: 978–1–3994–1534–7;
ePub: 978–1–3994–1533–0

2 4 6 8 10 9 7 5 3 1

Typeset in the UK by Mark Heslington
Maps by Brian Southern and L Wright Design

Printed and bound in Great Britain by Clays Ltd, Elcograf S.p.A.

To find out more about our authors and books visit
www.bloomsbury.com and sign up for our newsletters.

For product safety related questions
contact productsafety@bloomsbury.com.

Cover photographs.
Front: Black-necked Grebes in breeding plumage at Woolston Eyes –
Peter Sutton, Cheshire and Wirral Ornithological Society; Willow Tit at
Carr House Green Common – Paul Ellis, Fylde Bird Club.

Back: Bittern, Ashley Cooper/Getty; Wigeon, Carl Mckie/Shutterstock; Twite,
Erni/Shutterstock. Spine: Leach's Petrel, Agami Photo Agency/Shutterstock.

CONTENTS

Acknowledgements	6
Introduction	7
Map of the region	9
How to use this book	10
A note on bird recording in Northwest England	13
Glossary of species names	14
Glossary of terms and acronyms	16

CHESHIRE — 19

1	Rostherne Mere	20
2	Tatton Park and Knutsford Moor	24
3	Alderley Woods	28
4	Rudheath Sand Quarries	30
5	Chelford Sand Quarries	32
6	Macclesfield Forest, Langley Reservoirs and Teggs Nose Country	36
7	The Dane Valley	41
8	The Sandbach Flashes	43
9	Marbury Country Park and Witton Lime Beds	47
10	Woolston Eyes	52
11	Moore Nature Reserve	55
12	Frodsham Marsh and The Weaver Bend	58
13	Hale Head	63
14	Delamere Forest	67
15	Beeston Castle and Peckforton Hills	71
16	The River Dee	74
17	Burton Mere Wetlands	77
18	Inner Dee Marshes to Heswall including Parkgate	81
19	The Outer Dee: Heswall to West Kirby	86
20	Hilbre Bird Observatory	88
21	Red Rocks, Hoylake and Meols Shore	93
22	Hoylake Carrs and Gilroy Road Nature Park	97
23	Leasowe Lighthouse and North Wirral Country Park	100
24	New Brighton and Wallasey	103
25	New Ferry Shore and Port Sunlight River Park	106
26	Woodlands on the Wirral	108

LANCASHIRE — 111

North Lancashire — *113*

27	Leighton Moss RSPB	113
28	Carnforth Marsh (Morecambe Bay RSPB)	117
29	Hest Bank	118
30	Jenny Brown's Point	120
31	Dockacres/Pine Lake Waters, Carnforth	122

Contents

32	Heysham Area including Middleton	125
33	Morecambe Promenade	128
34	Lune Estuary	130
35	Cockersand (including Bank End)	134

The Fylde Coast — 137

36	Pilling Marsh/Cocker's Dyke	137
37	Over Wyre Fields	139
38	Ribble Estuary North Shore	141
39	Marton Mere	145
40	Wyre Estuary	148
41	Fleetwood Area	152
42	Blackpool Promenade and Coastal Sites	156

East Lancashire — 160

43	Stocks Reservoir	160
44	Dunsop and Langden Valleys	162
45	Rishton Reservoir	165
46	Foulridge Reservoir	167
47	Pendle Hill	169
48	Fishmoor Reservoir	171
49	Alston Reservoirs and Alston Wetland	173

Preston/Garstang Area — 175

50	Brockholes LWT Reserve	175
51	Preston Dock	177
52	Grimsargh Reservoirs	179
53	Abbeystead, Barnarcre and Garstang Lakes and Reservoirs	180
54	Myerscough Quarry	182

Southwest Lancashire and North Merseyside — 185

55	Seaforth Nature Reserve (permit only)	185
56	Crosby Marina	187
57	Mere Sands Wood LWT Reserve	189
58	Marshside RSPB Reserve	191
59	Hesketh Out Marsh RSPB Reserve and Banks Marsh NNR	194
60	Sefton Coast	197
61	Rimrose Valley	201
62	Martin Mere WWT	202
63	Lunt Meadows LWT Reserve	205

Chorley Area and West Pennine Moors — 208

64	West Pennine Moors	208
65	Anglezarke and Rivington Reservoirs	213
66	Yarrow Valley Park	214
67	Croston and Mawdesley Moss	216

GREATER MANCHESTER — 217

68	Wigan Flashes	219
69	Abram Flashes	222

70	Pennington Flash	224
71	Hope Carr	227
72	The Horwich Moors	229
73	Rumworth Lodge	234
74	Elton Reservoir	235
75	Cutacre Country Park	238
76	Little Woolden Moss LWT Reserve	239
77	Carrington Moss	242
78	Chorlton Water Park	245
79	Sale Water Park	247
80	Heaton Park Reservoir (private site)	248
81	Watergrove Reservoir	249
82	Light Hazzles, Whiteholme and Warland Reservoirs	251
83	Hollingworth Lake	252
84	Stalybridge Country Park, Brushes Valley	254
85	Dove Stone	256
86	Audenshaw Reservoirs (permit only)	258
87	Etherow Country Park	259

Seawatching in the Northwest	**261**
Top sites for accessibility	**264**
Top sites for public transport	**265**
Thirty species to see in Northwest England	**266**
Online resources and contacts	**272**
Bibliography	**275**
Northwest England bird list	**276**
Index to species	**281**

ACKNOWLEDGEMENTS

We would like to thank the authors of the previous editions in the Helm series covering broadly the same area as this book – *Where to Watch Birds in Cumbria, Lancashire and Cheshire* (Guest and Hutcheson, 1997) and *Where to Watch Birds in North West England and the Isle of Man* (Conlin, Cullen, Marsh, Reid, Sharpe, Smith and Williams, 2008). Although bird distributions can change rapidly, with some sites disappearing and others being created, this book is firmly built upon the foundations laid by its predecessors.

We would also like to thank the volunteer officers of Cheshire and Wirral Ornithological Society, Lancashire and Cheshire Fauna Society and the Greater Manchester Bird Recording Group for assistance provided by assessment of records and maintenance of the county lists that underpin the combined list for the area in this book.

Stephen Dunstan would like to thank Chris Batty for assistance with queries on Fylde sites, Neil Southworth for input on Chorley sites, Steve Martin for proactive support on West Pennine Moors material, Dave Bickerton for suggestions on east Lancashire and Steve White for input on south-west Lancashire and North Merseyside sites. Stephen would also like to thank the wider group of people who have worked with him on producing the *Lancashire Bird Report* over the last 30 years, whose knowledge and enthusiasm are hopefully reflected in how the county is conveyed in this work.

Jane Turner would like to thank the following dedicated local patch-workers who have improved these entries: Greg Baker for Marbury and the Witton Flashes, Steve Barber for the Chelford Sand Quarries, Bill Bellamy for Rostherne, Allan Conlin for Leasowe Lighthouse and the North Wirral Country Park, Dave Craven for Hale Head, Andrew Goodwin for the Sandbach flashes, Graham Jones for Burton Mere Wetlands, Bill Morton for Frodsham Marsh, and Steve Williams for Hilbre, as well as all those birders who have submitted their records to the County Society or via Bird Track.

Ian McKerchar would like to thank Andy Makin for information on the Scout Road watchpoint of the Horwich Moors.

INTRODUCTION

Whilst north-west England may bring to mind the cities of Liverpool and Manchester, or a number of former mill towns and seaside tourist resorts for varied clientele, the birdlife of the area is very diverse for a largely urban environment. The major estuaries of the Dee and Ribble are in Cheshire and Lancashire respectively, and the southern half of Morecambe Bay is also in Lancashire. Nationally important reedbeds thrive on the RSPB reserves at Leighton Moss in Lancashire and Burton Mere on the Cheshire/Flintshire border and also in Greater Manchester at the Wigan and Leigh Flashes and Pennington Flash. One of the Wildfowl and Wetland Trust's national flagship reserves is at Martin Mere near Burscough, hosting thousands of wildfowl including Pink-footed Geese and Whooper Swans in winter. There is an accredited bird observatory on Hilbre Island, offshore from West Kirby, and a bird observatory–standard setup in the shadow of the nuclear power stations at Heysham.

Large numbers of waders and wildfowl take refuge on the Dee, Mersey and Ribble estuaries as well as the Lancashire part of Morecambe Bay. Pintail and Wigeon occur in internationally important numbers, whilst Pink-footed Geese are a key sound of the winter as they move from coastal roosts to inland feeding grounds. Whooper Swans are popular winter visitors, and spring migrants can be seen throughout the region. Shorebirds, whilst increasingly pressured by human activity, still form big flocks as detailed in the monthly Wetland Bird Survey counts. More recent colonists/visitors to these areas include egrets, Spoonbills, Mediterranean Gulls and less conspicuous species like Water Pipit. Avocets, once exotic visitors, are now established breeders on the margins of our coasts. Cetti's Warblers are now a familiar sound, and less often sight, across the lowland parts of the area.

The coasts of Cheshire and Lancashire can offer rewarding seawatching. In winter this includes large flocks of Common Scoter and regular gatherings of Red-throated Divers. At migration times auks and skuas move through whilst in summer Storm Petrels feature in gales, and even in calm conditions terns, Gannets and Manx Shearwaters pass offshore. Autumn brings Long-tailed Skuas, Sabine's Gulls and Grey Phalaropes, especially during westerly storms. Above all, though, Leach's Petrel is the sought-after seabird most associated with north-west England. Any birder lucky enough to experience a classic Leach's Petrel passage off Cheshire or Lancashire will always remember it. Seabirds are clearly few and far between in landlocked Greater Manchester, but demonstrating the unpredictability of birding the county it has a White-billed Diver on its species list while Cheshire and Lancashire do not.

Several carefully managed reedbed sites are maintained by conservation organisations. As well as the long-established Leighton Moss nature reserve other satellite habitat to support this has been developed in the Silverdale area, and there are other similar habitats at Burton Mere and an extension to Martin Mere Wildfowl and Wetland Trust and the relatively new Lancashire Wildlife Trust reserve at Lunt Meadows. Bitterns and Bearded Tits breed at several sites in the region, whilst Marsh Harriers have prospered so much it is easy to forget how rare they once were. Climate change has seen Cetti's Warblers colonise these and other sites, as well as three species of egret often roosting in reedbeds.

There are a number of sites along the Cheshire and Lancashire coast where

Introduction

regular monitoring of seabird and passerine passage occurs. In spring and autumn coastal sites, whilst less rewarding than east-coast sites, produce migrant passerines. Visible migration of pipits, wagtails and thrushes can be as rewarding at inland sites as well as on the coast. Nationally important tern colonies are found within the docks at Seaforth and Common Terns also nest on pontoons at Preston Dock, with significant Lesser Black-backed Gull and Black-headed Gull colonies both on the coast and well inland at sites including Tarnbrook Fell and Belmont Reservoir.

Inland reservoirs, whilst generally artificial sites, can be an important focus of birding activity. At migration time they can attract Swallows and martins, passage terns (including Black Tern), Little Gulls and shorebird species along banks or feeding on any exposed mud. Audenshaw Reservoirs in Manchester and Stocks Reservoir near Slaidburn have attracted an extensive list of scarce and rare visitors over the years. Some reservoirs also attract large gull roosts, Fishmoor in Blackburn and Lower Rivington near Chorley often attracting scare species including Glaucous, Iceland and increasingly Caspian Gulls. Landfill sites and increasingly recycling facilities like Redgate in Gorton give opportunities to study gulls during the day.

Upland birding in the area can be more challenging, particularly as the once almost guaranteed prize of trips of Dotterel is less predictable, but it can bring rewards. The interaction between Hen Harriers and Red Grouse is contentious, but both are highlights of the heather moors of Lancashire and Greater Manchester. Ring Ouzel, Wheatear, Pied Flycatcher and Redstart are other main features of the upland summer whilst in winter flocks of Snow Buntings may still be chanced upon on the tops.

A lot has changed in birding in the 15 years since the publication of *Where to Watch Birds In North West England & the Isle of Man*, which covered some of the same geography as this new guide. Few people now carry pagers, and local bird news has been democratised by social media apps. Camera advances mean that excellent images are no longer the preserve of an elite of photographers. Thermal imaging devices mean that you can see birds in the dark if you wish. Sound recording improvements allow birds to be identified that weren't seen by the observer – and in some cases there wasn't even an observer! Phone and computer technology is beginning to change how birds are identified.

What has not changed is the value of knowledge of where and when to watch birds. This book is a guide to what can be seen in a part of England that whilst generally fairly urban offers some fantastic birding opportunities. Readers will be acutely aware of the ongoing climate change challenge, and the ethical dilemmas in using carbon fuels to travel to see birds. The relatively compact area covered by Cheshire, Greater Manchester and Lancashire and the good public transport infrastructure, particularly around Liverpool and Manchester, makes greener birding achievable and this book reflects that with information on how to get to sites by bus, tram and train.

The list of different wild bird species seen in Cheshire, Lancashire and Greater Manchester combined has surpassed the 400 mark in the last couple of years. This reflects the tremendous variety of habitats that still occur in the area, and the potential for diverse birding experiences to be had in a relatively urban and highly populated environment. Many amazing bird spectacles can be enjoyed in north-west England. Huge flocks of Knot roosting on Heysham Harbour wall, thousands of Common Scoter off Blackpool and Southport, the Whooper Swan feed at Martin Mere, Starling murmurations at several sites, spring seabird passage along our coasts, terns and Little Gulls descending on inland waters, raptors quartering the Dee and Ribble marshes are just some examples. We hope this book helps inspire you to get out in the area and enjoy some of its fantastic birding.

Map of the north-west region covered by this book, following the historic county lines used by bird clubs and recorders.

HOW TO USE THIS BOOK

Sites are organised by county: Cheshire and Wirral first, then Lancashire and North Merseyside, and finally Greater Manchester. The entries generally follow the format listed below, simplified for some less productive or subsidiary sites.

This book does not include a separate section for Merseyside, as bird recording in the north-west still follows the historic county boundaries of Cheshire and Lancashire rather than modern administrative divisions. Bird clubs, recorders, and reporting structures are long-established and have proved resilient to the changes imposed by local government reorganisation. In Cheshire, bird recording is coordinated by the Cheshire and Wirral Ornithological Society (CAWOS), whilst in Lancashire and North Merseyside it is overseen by the Lancashire and Cheshire Fauna Society. Both societies produce annual reports and maintain consistent recording frameworks that align with the wider ornithological community and the bird news services, ensuring that everyone works to the same, stable set of boundaries.

SITE NAME
The generally used name of the site as referred to by the landowner or management organisation. Other names used generally as an alternative by birdwatchers or locals will be mentioned in the main body of the text where this may help understanding.

STREET ADDRESS
Where there is an official site entrance this will be the street on which this is located. In other cases, it will generally be the address of a convenient access point, particularly where there is a public transport stop or a car parking option.

ORDNANCE SURVEY (OS) MAP REFERENCE
This is the traditional grid reference, and consists of the relevant grid square (two letters) and the first three numbers of the easting and northing coordinates respectively. The British Trust for Ornithology use OS grid references in survey work and many birders still also use them to record sighting locations. As well as paper maps you can, with an appropriate subscription, input these coordinates at explore.osmaps.com to see the exact location on screen.

WEBSITE
Where a location has an official website the address of this will be given. In some cases where there is no official website and a clearly authoritative unofficial website this will be mentioned.

WHAT3WORDS APP LOCATION
This widely used app divides the world into three-metre squares and gives each of these a unique combination of three words to pinpoint the exact location. The app is available free on mobile phones and can give navigation from your current location to the relevant birding site.

PHONE
Where this exists it will be the main phone number (often a landline but sometimes a mobile number) for the nature reserve or site manager.

EMAIL
This is the current email address at the time of publication that will connect you directly with the person or team most responsible for the site and its birds.

OPENING TIMES
Where relevant the times (and in some cases days) access for birdwatching is granted at the time of publication. In some cases this will require payment of an entrance fee or membership of the managing charity. A very small number of sites are permit only; where this is the case it is clearly stated.

PARKING
Details of where to leave your car when visiting the site. This may be an official car park or parking near an access point. Where charges apply, this will be noted.

PERMITS
Some sites restrict access to certain areas for conservation reasons, or require a permit to enter. Where this applies, information is given on how to obtain the relevant permit and any relevant charges.

MAPS
A guide to which maps cover the area, usually Ordnance Survey Landranger or Explorer series, with sheet numbers listed.

SITE DESIGNATION
Sites may carry formal conservation designations such as National Nature Reserve (NNR) or Site of Special Scientific Interest (SSSI). These highlight the ecological importance of the location and often influence how it is managed and accessed.

SPECIES
This section is not an exhaustive list of species that occur at a site. Rather, it lists the key species likely to be of most interest for the visitor, because the location is a particularly good one at which to see them. It does not guarantee that the species will be present when you visit, or in some cases that you will be able to see or hear if it is present.

Reference to seasons in the species lists and elsewhere in the text can be interpreted as:

All year – resident species; there may still be fluctuations in numbers; for example, numbers of breeding duck species may be supplemented by others visiting in winter.

Summer – birds that generally visit to breed only and then leave, generally arriving from and returning to wintering areas further south.

Winter – northern and, to a lesser extent, eastern migrants that come here to avoid colder conditions in the areas where they breed.

Migration/Passage – species that stop off in the area covered in the book in spring and autumn en route between breeding and wintering grounds. Spring passage can refer to any time from late February (e.g. Stonechats) to mid-June (e.g. Dunlin and Ringed Plover nesting in the Arctic). Autumn can also start in mid-June (e.g. departing Common Sandpiper) and last into November or even December (e.g. some gulls and wildfowl).

ACCESS

This section includes the main directions to get to the site for both car users and those arriving by public transport. It is obviously desirable for recreational birding to be undertaken by greener methods whenever possible, but in some cases this is not practical. Where car access is the only realistic option, that is stated, and car park locations will be indicated in all cases.

Public transport options are given whenever feasible. These will generally refer to the most obvious options in terms of convenience and frequency, but these may not be the cheapest. Whilst generally the main options will be rail and bus, the tram networks in Greater Manchester and on the Fylde Coast are included where appropriate.

Generally, it is assumed that on-site birders will be moving around on foot. In some cases where there is access for mobility scooters this is mentioned, and there is also a list on page 264 with details of sites best serving birders with restricted mobility.

YOUR VISIT

This section varies with the nature of specific sites. It can include times of year to visit but will also cover the best time of day to be present to optimise the birding experience. Any other information relevant to getting the best out of your visit will also be included here.

GENERAL REMARKS ON THE TEXT

Following the standard convention in natural history books, the English names of all species mentioned in the text (including other animals and any plants) are capitalised (e.g. Greylag Goose). Where reference is to a wider grouping than species, this is not capitalised (for example 'geese and swans', 'grey geese' and 'wintering geese'). In the small number of cases where a subspecies name is used colloquially as the species name in Britain, this is stated as the species name in this book (e.g. Pied Wagtail, which is the subspecies of the continental White Wagtail).

A NOTE ON BIRD RECORDING IN NORTHWEST ENGLAND

There are now many ways to record your bird observations to help build knowledge of population and range changes. These include relatively new national and international electronic platforms such as the BTO's Birdtrack system and eBird, hosted by Cornell in the USA. Much of the information underpinning this book comes from the local recording bodies though, and this is a brief note on them.

Bird recording in Cheshire is facilitated by the Cheshire and Wirral Ornithological Society (CAWOS). It was formed in 1988 as the successor to the Cheshire Ornithological Association. It produces the annual *Cheshire and Wirral Bird Report*. Further details on its activities and membership can be found at cawos.org.

Despite its name, the Lancashire and Cheshire Fauna Society now covers Lancashire and North Merseyside. The society was formed as long ago as 1914 and was originally based out of the area's universities. As well as the annual *Lancashire Bird Report* it produces periodic publications on other animal groups. The society's website is at lacfs.org.uk.

The Greater Manchester Bird Recording Group oversees bird recording in the Greater Manchester area. It was founded in 2002. The GMBRG is free to join and has produced the annual report *Birds in Greater Manchester*. Further information is available on the website at manchesterbirding.com.

All the above websites include details of how to submit your birding records. There are also several local bird clubs and societies that publish their own reports, particularly in Lancashire, and some nature reserves produce their own reports. Further details are provided in the 'online resources and contacts' section.

GLOSSARY OF SPECIES NAMES

In line with the *Where to Watch* series this book typically uses the names for species that are commonly used among British birders and the general public. In many cases these are shortened from the more widely used common name where usually only one species in the family occurs in Britain (for example, Kingfisher for Common Kingfisher, Wren for Eurasian Wren).

This glossary is used for two main reasons. Firstly, users of the series from overseas may benefit from the cross-reference to the International Ornithologists' Union (IOC)'s standardised international English names. Secondly, less experienced local birders may find it helpful when finding species referred to in Europe-wide and even some British-focused field guides, which may include other species from the same family, including rarities, and use the fuller names.

Avocet Pied Avocet *Recurvirostra avosetta*
Bittern Eurasian Bittern *Botaurus stellaris*
Blackbird European Blackbird *Turdus merula*
Buzzard Common Buzzard *Buteo buteo*
Cattle Egret Western Cattle Egret *Bubulculus ibis*
Coot Eurasian Coot *Fulica atra*
Cormorant Great Cormorant *Phalocrocorax carbo*
Crossbill Common Crossbill *Loxia curvirostra*
Cuckoo Common Cuckoo *Cuculus canorus*
Curlew Eurasian Curlew *Numenius arquata*
Eider Common Eider *Somateria mollissima*
Fulmar Northern Fulmar *Fulmarus glacialis*
Goldeneye Common Goldeneye *Bucephala clangula*
Golden Plover European Golden Plover *Pluvialis dominica*
Goldfinch European Goldfinch *Carduelis carduelis*
Goshawk Northern Goshawk *Accipiter gentilis*
Guillemot Common Guillemot *Uria aalge*
Herring Gull European Herring Gull *Larus argentatus*
Hobby Eurasian Hobby *Falco subbuteo*
Honey Buzzard European Honey Buzzard *Pernis aviporus*
Jay Eurasian Jay *Garrulus glandarius*
Kestrel Common Kestrel *Falco tinnunculus*
Knot Red Knot *Calidris canutus*
Lapwing Northern Lapwing *Vanellus vanellus*
Linnet Common Linnet *Linaria cannabina*
Marsh Harrier Western Marsh Harrier *Circus aeruginosus*
Moorhen Common Moorhen *Gallinula chloropus*
Night Heron Black-crowned Night Heron *Nycticorax nycticorax*
Nightjar European Nightjar *Caprimulgus europaeus*
Oystercatcher Eurasian Oystercatcher *Haematopus ostralegus*
Pochard Common Pochard *Aythya farina*
Quail Common Quail *Coturnix coturnix*
Raven Northern Raven *Corvus corax*

Redpoll Lesser Redpoll *Acanthis flammea cabaret* *
Redshank Common Redshank *Tringa totanus*
Redstart Common Redstart *Phoenicurus phoenicurus*
Reed Bunting Western Reed Bunting *Emberiza schoeniclus*
Scaup Greater Scaup Aythya marila
Shag European Shag *Phalacrocorax aristotelis*
Shelduck Eurasian Shelduck *Tadorna tadorna*
Shoveler Northern Shoveler *Anas clypeata*
Siskin Eurasian Siskin *Spinus spinus*
Skylark Eurasian Skylark *Alauda arvensis*
Sparrowhawk Eurasian Sparrowhawk *Accipiter nisus*
Spoonbill Eurasian Spoonbill *Platalea leucorodia*
Starling Common Starling *Sturnus vulgaris*
Stonechat European Stonechat *Saxicola rubicola*
Swallow Barn Swallow *Hirundo rustica*
Teal Eurasian Teal *Anas crecca*
Turtle Dove European Turtle Dove *Streptopelia turtur*
Wheatear Northern Wheatear *Oenanthe oenanthe*
Whitethroat Common Whitethroat *Silvia communis*
Wigeon Eurasian Wigeon *Anas penelope*
Wren Eurasian Wren *Troglodytes troglodytes*

* Lesser Redpoll is one of several previously separate species recently 'lumped' into one species known simply as Redpoll. In the text other subspecies less regular in the area will be referred to specifically (i.e. Common/Mealy Redpoll *A.f. flammea*, and Coues's Arctic Redpoll *A.f. exilipes*).

GLOSSARY OF TERMS AND ACRONYMS

Aythya – diving duck species including, most commonly in north-west England, Tufted Duck and Pochard.

BBRC – British Birds Rarities Committee, a committee of expert birdwatchers who assess all records of nationally rare species in Britain.

BirdTrack – an online system run by the BTO that allows birdwatchers to submit and store their sightings, helping to monitor changes in bird distribution and populations.

BTO – British Trust for Ornithology, a charity that undertakes national bird surveys and scientific research with citizen science contributions from birdwatchers.

CAWOS – Cheshire and Wirral Ornithological Society.

CP – country park.

CWT – Cheshire Wildlife Trust.

Darvic or Darvic ring – a coloured, coded, plastic ring on the leg (or less commonly neck) of a wild bird that helps track the movement of the bird.

eBird – a global recording platform run by the Cornell Lab of Ornithology, which enables birdwatchers to log sightings worldwide and contribute to international research and conservation.

GC – golf course.

Fall – the grounding of unusually high numbers of migrant birds as a result of particular weather conditions and wind direction.

Hirundine – a member of the bird family Hirundinidae, including in Britain Swallow, House Martin and Sand Martin.

In-off – a migrating bird seen to arrive off the sea and head inland.

Irruption – the movement of large numbers of a species of bird to an area where they do not normally occur, particularly due to food shortages in the usual range.

Larid – a gull of the family Laridae.

LNR – Local Nature Reserve.

LWT – Lancashire Wildlife Trust.

NNR – National Nature Reserve.

Nocturnal migration (or 'noc-mig') – bird migration in the hours of darkness, and particularly the identification of birds doing this by hearing or recording their call notes.

OS – Ordnance Survey, the national mapping agency for Great Britain.

Passerine – the large group of bird families that have feet adapted for perching, including all songbirds.

Patch – an area covered regularly by a birdwatcher and more systematically than other sites visited, often close to where they live ('local patch').

Ramsar Site – a wetland site designated to be of international importance under the Ramsar Convention.

RNLI – Royal National Lifeboat Institution charity, and the lifeboat stations it operates.

RSPB – Royal Society for the Protection of Birds.

SAC – Special Areas of Conservation as designated under several different government conservation regulations.

Sandwinning – commercial extraction of sand from beaches, as practised until recently on the Ribble Estuary at Lytham St Annes and Southport.

Seaduck – marine duck species including Common Scoter, Velvet Scoter, Long-tailed Duck, Scaup and Eider.

Seawatching – birdwatching consisting of staying at a fixed point on a headland or other coastal vantage point and recording passing seabird movement.

SPA – Special Protection Area, designated to protect one or more rare, threatened or vulnerable bird species.

SSSI – Site of Special Scientific Interest protected area.

Tern raft – an artificial floating island, usually on a nature reserve, installed to encourage tern species to nest on site.

Trektellen – an international website devoted to the recording of migration, including both 'noc-mig' and 'vis-mig'.

Vis-mig – the visible migration of birds.

WeBS or Wetland Bird Survey – national monthly counts of waterbird species on wetlands used to monitor population levels of wildfowl and other waterfowl.

WECG – Woolston Eyes Conservation Group.

Winter thrushes – Redwings and Fieldfares, normally winter visitors to the area.

WWT – Wildfowl and Wetlands Trust.

KEY TO THE MAPS

- Ⓟ Car park
- Ⓗ Hide
- ★ Viewpoint
- Ⓥ Visitor Centre
- ㎩ Public house
- 🗼 Lighthouse
- ♿ Public toilet
- CG Coastguard
- GC Golf course
- ✝ Church
- Towns
- Conifers

- Deciduous
- Marsh
- Scrub
- Reedbeds
- Lakes
- Sea
- Railway
- Main road
- Minor road
- Track
- Footpath
- Embankment

CHESHIRE

Cheshire and the Wirral Peninsula offer a wide range of habitats that make the region one of great interest to birdwatchers. To the west, the Dee and Mersey estuaries are internationally important for wintering and passage waders and wildfowl. A short but ecologically rich stretch of Irish Sea coastline along the Wirral includes saltmarsh, sand dunes and intertidal mudflats, providing year-round opportunities to observe both breeding and migratory species, and in the right conditions can be exceptional for petrels and other seabirds that are not dependent on deep water.

Inland, the landscape transitions into a broad, fertile plain of arable land and pasture, with hedgerows, wet ditches and scattered mature woodlands. In the east of the region, low moorland and gritstone ridges form part of the Pennine fringe, with upland grasslands and heather moor. Cheshire's glacial heritage is visible in its meres and mosses – open lakes fringed by reedbeds, fen and carr woodland, as well as rare raised bogs, which offer habitat for wetland species. The legacy of salt extraction has created saline flashes and lagoons that are now key stopover points for migrating waders and wildfowl.

Located on the west coast of Britain, the county is somewhat shielded from transatlantic vagrants by Ireland, and its position relative to the east coast means that drift migrants from the continent are less frequent than east or south coast counties. As a result, national rarities, with the possible exception of waders, are harder won here and somehow feel worth 'more', but the region rewards patient exploration across a richly varied landscape.

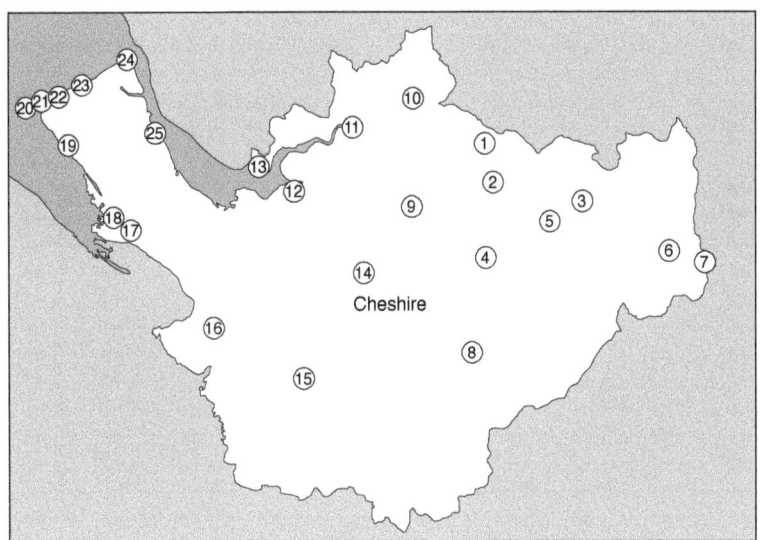

1 ROSTHERNE MERE

> **A.W. Boyd Bird Observatory Rostherne**
> Rostherne Lane, Rostherne WA16 6RS
> SJ743843
> joked.diamonds.blotchy
>
> Websites: gov.uk/government/publications/cheshires-national-nature-reserves/cheshires-national-nature-reserves, *cawos.org/rosmere.html*
> Opening times: Permit holders may visit during daylight hours; visitors without permits are generally admitted on Sunday mornings from 9am when wardens are present.
> Parking: Car park is closed at 6pm; non-members are required to apply for a £1 day permit.
> Permits: CAWOS members £12 single, £17 joint/family, £5 children 11–16 years; non-CAWOS members please use the form at cawos.org/contact.php, selecting Rostherne in the dropdown menu. The charge for groups is £1 for a visit to the observatory and £2 for a walk on the reserve. Unfortunately, single visitors who are not permit holders cannot be accommodated.
> OS Landranger Map 109, OS Explorer Map 268
> NNR, SSSI

Rostherne is the largest of the Cheshire meres and also the deepest, with the original basin having been deepened by salt subsidence. It covers 48ha and has a maximum depth of around 30m. Its size and depth mean it rarely freezes. The banks slope steeply for the most part, and although much of the mere is fringed by narrow reedbeds, there is little submerged vegetation, and few wildfowl breed. There is limited habitat for passing waders. Mixed woodlands run down to the mere edge for more than half of its circumference. Elsewhere it is bordered by pasture. The mere, woods and pastures together make up the Rostherne Mere National Nature Reserve, which comprises around 80ha of mere, woodland and willow beds, together with about 70ha of farmland around the mere. Birds can be watched from the comfort of the A.W. Boyd Memorial Observatory, which is run by the Cheshire and Wirral Ornithological Society.

SPECIES

Throughout the year, Great Crested Grebe, Cormorant, Canada Goose, Tufted Duck, Mandarin Duck, Sparrowhawk, Kestrel, Common Buzzard, Stock Dove, Great Spotted Woodpecker, Nuthatch, Jay and Cetti's Warbler can all be seen.

In winter, from December to February, the mere holds Cormorant, Wigeon, Teal, Pochard, Tufted Duck, Goldeneye, Goosander and other wildfowl. Gulls and corvids roost. Siskin and Redpoll often feed in the trees below the observatory.

As spring arrives between March and May, the gull roost is vacated and wildfowl disperse. Occasional Curlew, Common Sandpiper may appear, along with the possibility of other waders. Swift, hirundines and occasional terns may fly over the mere.

During early summer, in June and July, non-breeding waterbirds may include

Cheshire

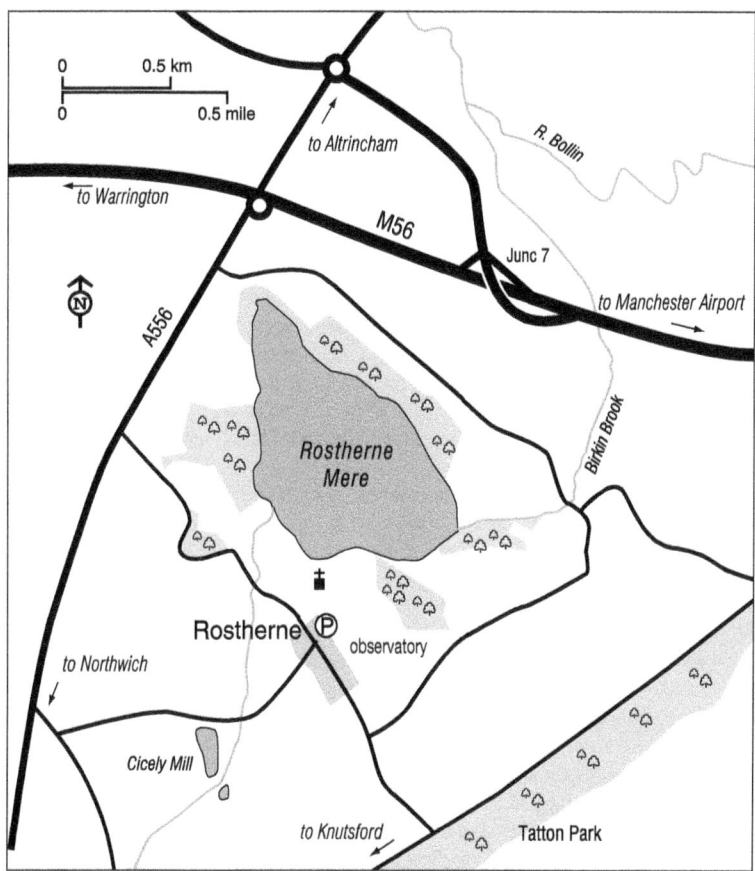

Wigeon, Shoveler and Cormorant. A moulting flock of Greylag and Canada Geese may be seen, and Pochard and Tufted Duck arrive in July. Reed Warblers are numerous in the reedbeds.

Later in the year, from August to November, Great Crested Grebe, Shoveler, Tufted Duck are numerous. Hobby, Peregrine Falcon visit. There are occasional terns and waders. Large numbers of Starlings roost in the reedbeds.

ACCESS

The nearest train station is Ashley, about 4km from the reserve. Rostherne is on regional route 70 (Cheshire Cycleway) of the Sustrans National Cycle Network. Bus services run along the A556, stopping at Bucklow Hill (1.6km from the NNR). A public footpath crosses a portion of the western boundary of the NNR. By car, drive through Rostherne Village and park in the Natural England Car Park (SJ744833); the entrance is opposite Egerton Hall. The car park closes at 6:00pm. From the car park walk back to the road, turn right and walk through the village, bearing right at the Rowans towards St Mary's Church. The observatory can be accessed along the permissive path adjacent to the graveyard. For non-permit holders, there is a permissive path around Wood Bongs; the wood is accessed from the top path. The top path and churchyard are also good places to view the mere. To use the observatory, you will need to buy a permit.

YOUR VISIT

For general viewing of wildfowl visit any time of day. During the winter months, afternoon and dusk visits are recommended for roosting birds. Severe weather concentrates wildfowl on the mere because Rostherne remains open long after other meres and flashes have frozen. This is a rare weather event these days. Westerly or north-westerly gales may blow in a seabird. The elevated position of the observatory makes it ideally situated for watching diurnal cold-weather movements. With the onset of severe frosts, Lapwings may be seen flying west, sometimes overtaken by faster, more pointed-winged Golden Plovers. Skylarks and thrushes also move south or west. Towards dusk in winter, the number of birds on and around the mere increases rapidly. Woodpigeons line the treetops in enormous numbers when there is a good acorn crop. Corvids – mainly Jackdaws, Rooks and to a lesser extent Carrion Crows – also congregate in considerable numbers.

Rostherne Mere is primarily of importance for its wintering wildfowl populations. In autumn and winter, hundreds of ducks may be seen, with occasional three-figure counts of Mallard, Wigeon and Teal, most prominent on the edges of the mere. Shoveler is most common during autumn, and a few remain through the winter, and small numbers of Gadwall may be seen all year round. Pintail is a scarce winter visitor. Mandarin Ducks are increasing, with up to 85 counted, and the species is now breeding on the reserve. Pochard is a winter visitor in declining numbers, with counts over 100 very unusual, and associated with icy conditions on other waters. Tufted Duck is regular all year round and has bred in recent years. It is frequently the most numerous diving duck with occasional post-breeding counts of over 100. One or two Shelduck usually visit the reserve each year. Common Scoter is an increasing passage migrant. Goldeneye are regularly seen during winter in small numbers and Goosander can also be reliably seen, especially during cold evenings. There have also been records of Smew, Green-winged Teal, American Wigeon, Ferruginous Duck, and a Hooded Merganser in 2021.

Several hundred feral geese may be present, and the proliferation of wildfowl collections has led to an increase in sightings of escaped birds; it is now not unusual to find one or two Barnacle Geese among the Canadas. Snow, Lesser White-fronted, and Bar-headed Geese have all turned up from time to time. Pink-footed Geese may also appear within the feral flock but are much more likely to be seen on overhead passage. Whooper Swans are recorded annually, usually resting for a short while before moving on. Egyptian Geese have been recorded more regularly, breeding nearby in Tatton Park. Up to 24 have been seen feeding on local pastures. Other exotic wildfowl include Black Swan, South African and Ruddy Shelduck, and both White-faced and Black-bellied Whistling Ducks. Water Rail can be heard 'sharming' in the reedbeds in winter and most probably breed, while Coot and Moorhen are plentiful, although the former has decreased in recent years.

Rostherne Mere is one of the 21 lakes in England and Wales which can be classed as 'immemorial' – never without Great Crested Grebes as long as can be remembered. Numbers peak in late summer with up to 100 seen recently. Several pairs breed each year. Little Grebe is also a breeding resident and small numbers are present throughout the year, while Black-necked Grebe is fairly regular, most reliably recorded in late summer, highlighting the proximity of the nearby breeding population at Woolston Eyes. Slavonian and Red-necked Grebes have been recorded but are extremely rare at Rostherne.

A particular feature of Rostherne is the breeding colony of Cormorants, including many continental *sinensis*, in trees along the edge of Harpers Bank Wood. Up to 190 nests have been recorded, with large numbers of birds leaving the mere by day to catch fish in the rivers, meres and flashes of eastern Cheshire and parts of Greater Manchester to bring back to the hungry chicks. After the chicks fledge, they form fishing parties of up to 50 birds, honing their hunting skills on the mere. Bitterns are recorded during most winters but are notoriously hard to see. Kingfishers are recorded frequently, and Grey Wagtails can often be seen on Rostherne Brook. Great White Egret and Little Egret are being seen more frequently along the edge of the mere, and Cattle Egret was recorded for the first time in 2024.

An impressive gull roost can form, especially in the winter, with up to 3,000 Black-headed Gulls, though such numbers are exceptional and 1,000 is a more typical count. Common Gulls are usually present, but three-figure counts are now infrequent. A few hundred larger gulls may also roost, and Mediterranean, Yellow-legged Gull and Caspian records are increasing. Parties of Common, Arctic and occasional Black Terns may visit on passage, especially after easterly winds, when they can often be seen feeding on midges. Gales may blow seabirds inland, including Great Skua, Arctic Skua, Kittiwake and Fulmar, as well as one astonishing record of a Barolo Shearwater, an exceptionally rare bird in British seas, never mind inland. It was present from 29 June–3 July 1977. Passage waders may occasionally be seen on the shoreline of the mere, with Common and Green Sandpipers the most frequent. Woodcock may be encountered along the woodland edges in winter.

From late summer to early winter, thousands of Starlings wheel over the mere at dusk before roosting in the reedbeds. The gathering roost is often harassed in spectacular fashion by a Peregrine Falcon or, more likely, a Sparrowhawk. Swirling flocks of hirundines and Swifts feed over the mere during spring, summer and early autumn. Hobby has been seen with increasing frequency and is now breeding nearby. Food passes between adults and juveniles can often be seen from the observatory. Woodland birds move through the alder trees at the mere edge, giving views of Siskin, Redpoll, Goldfinch and woodpeckers. Lesser Spotted Woodpecker used to be heard in spring, but in the last few years has become very scarce although it is still usually recorded each year. Nuthatches, Long-tailed and Coal Tits, and Jays take food from a bird table just below the observatory. One of the first signs of spring is Stock Doves displaying in front of the observatory. Bullfinch, Treecreeper and Goldcrest breed in small numbers. Willow Tit is still recorded in most years but the population is small and requires a renewed monitoring programme.

A well-studied population of Reed Warblers breeds in the reedbeds. Nests were formerly parasitised by Cuckoos. Though this is now a much rarer event, it was last recorded as recently as 2022. The Cetti's Warbler population grows from strength to strength with birds being heard all around the edge of the mere. It is one of the best-studied populations in Cheshire. A few pairs of Sedge Warbler and Common Whitethroat also breed, and Grasshopper Warbler and Lesser Whitethroat are recorded most years. Good numbers of Chiffchaffs and Blackcaps hold territories in the wooded areas but Willow Warbler is becoming a rarer sight.

The surrounding fields may hold large numbers of winter thrushes, a regular flock of Linnets, small numbers of Tree Sparrows and Grey Partridges and a few

Yellowhammers. Buzzards are a regular sight, and Golden Plover may join the regular Lapwing flocks. Barn Owls are frequently seen hunting across the unimproved pastures and breed in boxes provided for them. Little Owl is usually recorded most years at the edge of the reserve.

A new wetland habitat has been created in Dolls Meadow to the left of the observatory. Recent sightings have included Great White Egret, Little Egret, Stonechat, Lesser Whitethroat, Yellow-browed Warbler and Bearded Tit.

2 TATTON PARK AND KNUTSFORD MOOR

North entrance
Rostherne Drive, Rostherne WA16 6SG
SJ748827
lamppost.offline.articulated

Knutsford entrance
Knutsford Drive, Knutsford WA16 6HS
SJ751791
cupcake.plodding.along

Website: *tattonpark.org.uk/home.aspx*
Opening times: From October to March, 11am–5pm (last admission 4pm; gates locked at 5pm) Tuesday to Sunday. From April to September daily 10am–7pm (last admission 6pm, gates locked at 7pm). Admission is free for pedestrians and cyclists.
Parking: available via the Knutsford entrance, plus at the Old Hall and near the Mansion/Stableyard area. There is a £6.50–£8 per vehicle admission charge.
Phone: 01625374400
Email: tatton@cheshireeast.gov.uk
OS Landranger Map 109/118, OS Explorer Map 268
SSSI, Ramsar

The first recorded reference to the estate was in 1185, when it was owned by the de Tatton family and the Deer Park was created by Royal Charter in 1290. In 1598, the estate was acquired by Sir Thomas Egerton. The Egertons made significant changes to Tatton Park during their ownership. They built the present mansion, which was designed by the architects Samuel Wyatt and Lewis William Wyatt in the neoclassical style. They also landscaped the gardens, which were designed by Humphry Repton. In 1958, the last Lord Egerton died without an heir. The estate was bequeathed to the National Trust and run on their behalf by Cheshire County Council. Tatton Park is a biodiverse estate with a variety of habitats, including two large meres, Tatton Mere and the smaller Melchett Mere, 202ha of woodland, numerous ponds and rough grassland areas. The main attractions for birdwatching are Tatton Mere, the smaller Melchett Mere and the woodlands surrounding them.

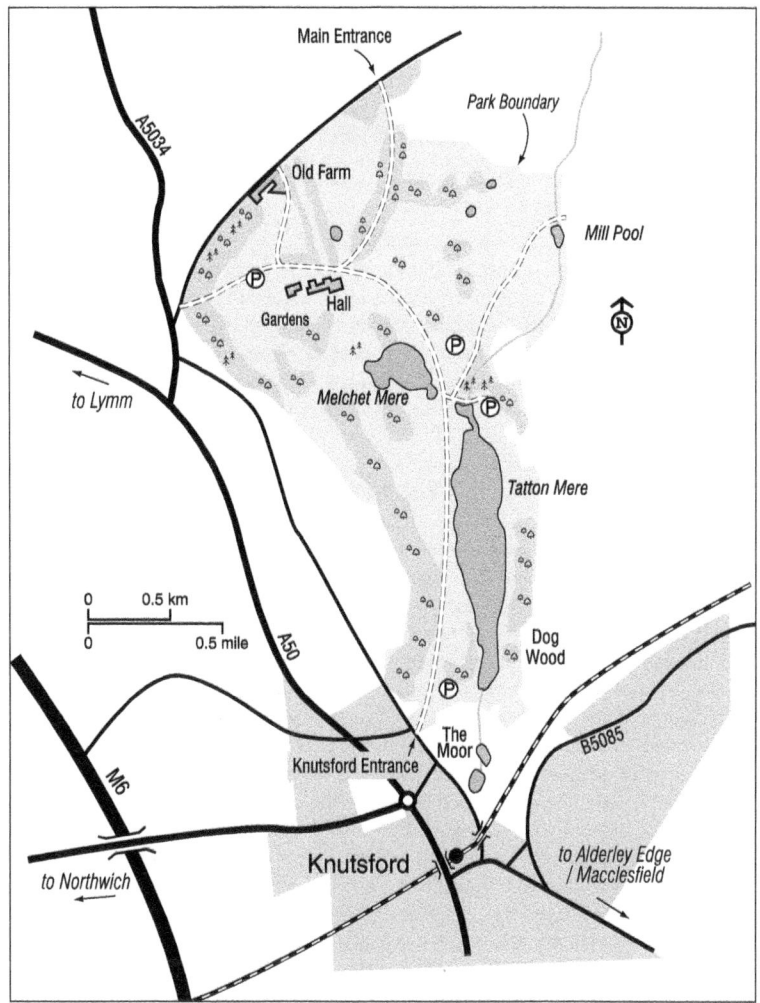

SPECIES

Little Grebe, Great Crested Grebe, Canada Goose, Tufted Duck, Buzzard, Woodcock. Stock Dove, Little Owl and woodland species are seen throughout the year.

In winter, from December to February, Teal, Pochard and Goldeneye gather on the meres. Water Rail and Common and Jack Snipes inhabit the marshy ground. Brambling feed under beeches, Redpoll in birches and Siskin in alders. Thrushes come to roost at the Moor.

As spring unfolds, Goldeneye display conspicuously and Shoveler frequent the reedbeds. Flocks of pipits, wagtails and Reed Bunting should be checked for Water Pipit and White Wagtail. Wheatear and Sand Martin appear from late March, with other hirundines in April and Swifts around the end of that month. Terns may visit from late April.

During early summer (June to July), broods of ducklings, Canada Goose and grebes can be seen, and Curlews may have young. Reed, Sedge and Garden Warblers nest on the Moor, while Willow Warbler, Whitethroat, Blackcap and Chiffchaff are found in the woodlands; and Tree Sparrow in the parkland.

Later in the year, from August to November, hirundines are sometimes plentiful. Terns and perhaps Little Gull are occasionally seen over the meres. Pied Wagtails and Swallows roost on the Moor. Wildfowl numbers increase sharply from October.

ACCESS

Knutsford train station is about 0.8km from the Knutsford entrance. The closest bus stop is on Queen Street. Admission is free for pedestrians and cyclists. There is a £6.50–£8 per vehicle admission charge. The park is open during the low season (October to March) from 11am until 5pm (last admission 4pm, gates locked at 5pm) Tuesday to Sunday. During the high season (April to September) the park is open daily from 10am to 7pm (last admission 6pm, gates locked at 7pm).

YOUR VISIT

Sunny weekends, bank holidays and special events days are best avoided. Morning visits are best as disturbance increases during the day and waterfowl may then leave the park and fly to Rostherne Mere. Evening visits in spring can be rewarding. Showery weather at passage times produces most migrants, particularly when the wind is from a southerly quarter in spring.

The larger area of woodlands comprises mixed oak, chestnut, beech and pine. Grazing by sheep, cattle, and both Red and Fallow Deer has prevented the development of a scrub layer in all but a few fenced woods. Tatton Mere was dammed by the monks of Mobberley Priory in medieval times and has little marginal vegetation. The mere extends to become Knutsford Moor to the south where the River Lilley enters. An extensive reedbed separates the mere from Knutsford Moor Pool, which lies outside the park. The smaller of the two meres, Melchett Mere, has boggy areas of rushes along its banks and acts as a refuge for wildfowl when boats and windsurfers force them from the main mere. At the north-east corner of the park is a mill pool backed by a large bed of sedges. The fast-flowing brook downstream from the mill attracts riparian species. To the north of the mill is the deer enclosure, a secluded area of grassland with scattered old trees standing on ancient field boundaries.

The meres and Dog Wood are dedicated SSSI and RAMSAR sites and nationally important habitats for wildlife. They provide nest sites and winter refuge for a large number and a good selection of wildfowl, as well as an abundance of wildflowers and plant life. Conservation work at Dog Wood has increased soil biodiversity and encouraged boggy areas, creating a sensitive wet woodland habitat where wildlife flourishes. The woodland areas are home to many different mammals, such as badgers and foxes. Nine species of bat have been recorded as well as a typical selection of woodland birds. Fallen branches are left in place to attract invertebrates and fungi. The Allen Bird Hide is on the edge of Melchett Mere.

Winter is arguably the best season for a visit to the meres and is likely to produce a selection of wildfowl including Teal, Pochard and Goldeneye, while Tufted Ducks are always present. There is a slim chance of a wintering Smew although it is no longer as regular as it once was. In contrast, Black-necked Grebes occur more frequently as they wander from other Cheshire breeding sites. There are movements between here and Rostherne Mere, so the likely species are similar, including Goosander, Common Scoter, Gadwall and Wigeon. Other exotic wildfowl seen here with some regularity are Mandarin, breeding Egyptian Goose and recently Black-bellied Whistling Duck.

The plentiful mature woodland makes Tatton outstanding for parkland birds. All three British woodpeckers have bred, though in keeping with other Cheshire sites, Lesser Spotted has declined to the point of possible local extinction. The ground-feeding Green Woodpecker favours the combination of permanent pasture and old trees. Nuthatch and Treecreeper are abundant. Jackdaws and Stock Doves nest in holes in trees. Dog Wood, along the eastern side of Tatton Mere, has in the past held breeding Marsh and Willow Tits but now they are at best occasional visitors.

Hirundines and Swifts appear over the meres in spring. At this time the surrounding woodlands are alive with songbirds joining the resident species. Goldcrests nest in the scattered pines. Blackcaps sing from the canopy, Whitethroats dance above the more scrubby areas and Garden Warblers occasionally breed. Willow Warblers sing from the birches and there is also a small chance of hearing a reeling Grasshopper Warbler on a spring evening. The reedbed at the Knutsford Moor end of the main mere hosts small numbers of Reed, Sedge and, recently, Cetti's Warblers. Knutsford Moor Pool holds a pair or two of Water Rails, which may be heard as the light fades. Curlews may be seen around the reserve, though they no longer breed, but Lapwings still tumble in wild display flights over rushy patches. Breeding waterfowl include Mallard and Tufted Ducks, which also nest on the mill pool where Grey Wagtails breed and Kingfishers are often seen fishing. Both Little and Great Crested Grebes breed on the meres whilst Canada Geese nest along the rushy fringes and beneath trees. In several years, escaped Barnacle Geese have paired with Canada Geese. Buzzard is seen very regularly and breeds in the parkland. Red Kites, though still rare, are seen more regularly and Ospreys are seen occasionally on passage. Barn Owl has bred on site recently, Tawny Owls do so regularly, though Little Owls are only recorded rarely.

Winter brings an increase in Great Crested Grebe numbers. Very occasionally these are joined by a rarer species. Slavonian and Red-necked Grebes, and Black-throated and Great Northern Divers have also occurred. As spring approaches, Goldeneyes regularly display here and numbers increase as passage to the north-east commences. Common Snipe creep around the edges of Melchett Mere and the occasional Jack Snipe is flushed from this area. Water Rail is often present. Rails are more easily seen around the Knutsford Moor reedbed, where Bearded Tit and Bittern have occurred and Woodcock may be encountered. In autumn the reedbed holds roosts of Pied Wagtail and Swallow. In winter, Redwings and Blackbirds roost in birch scrub and rhododendrons nearby. Also in winter, Siskins and Redpolls feed in the mereside alder and birch trees. Greenfinches and Chaffinches gather in the stubble fields around the park, and Brambling often appears either there or under beech trees. March brings the first passerine migrants to the mereside, when parties of Meadow Pipits, Pied Wagtails and Reed Buntings comb the short turf, and Wheatears sometimes flit along the shoreline. During showery weather in April the flocks are joined by White and Yellow Wagtails and perhaps a Water or Rock Pipit. Common Sandpipers rise from the water's edge and fly away low across the mere. The first Sand Martins are seen in mid- to late March, to be joined by Swallows and then House Martins during April and Swifts by the end of that month. Terns are recorded annually on either passage, with Black, Common and Arctic possible. There is a chance of a Little Gull, especially in the autumn. Migrant parties of hirundines appear over the meres in September and October, when their numbers may change markedly from hour to hour as flocks arrive in dribs and drabs from the north, feed for a while as they reassemble, then depart *en masse* to the south.

Tatton is not without its rarities. Records from the past have included Common Crane on a couple of occasions, Rough-legged Buzzard, Manx Shearwater, Grey Phalarope, Ferruginous Duck, a Red-breasted Flycatcher in November 1983, a Nutcracker in October 1968 and a Hooded Merganser in January 2021.

3 ALDERLEY WOODS

Alderley Edge
Macclesfield Road, Nether Alderley, Over Alderley SK10 4UG
SJ859772
broom.closet.ranged

Website: nationaltrust.org.uk/visit/cheshire-greater-manchester/alderley-edge-and-cheshire-countryside/woodland-valley-walk
Phone: 01625 584412
Email: alderleyedge@nationaltrust.org.uk
Opening times: Accessible at all times
Parking: Pay and display at the National Trust car park. For non-members the charge is £5 all day. Free parking for National Trust members (with valid sticker).
OS Landranger Map 118, OS Explorer Map 268
SSSI

The Edge at Alderley is an escarpment of red sandstone that contains veins of copper ore, mined from prehistoric times until early in the last century. Evidence of these mining activities is widespread in the form of square-sided cuttings and tunnels. On a fine day, the escarpment offers impressive views of the surrounding countryside across Cheshire towards the Peak District. The mature, planted woodland that covers much of the edge consists principally of oak, Scots pine and birch, with some chestnut and beech on the thinner soils overlying rocky outcrops, and larch. The ground flora is rather sparse, with wavy hair grass, bracken and brambles dominant. An extensive network of paths and bare, eroded areas around mine entrances give the woodland an open character. On the site of the old west mine, where a spoil tip stood into the 1960s, an area of open ground is developing birch scrub that is attractive to small finches and warblers.

SPECIES

The woods hold a good variety of common woodland birds throughout the year.

In winter, large flocks of mixed tits roam the woods alongside other woodland species. Green Woodpecker can be found. Redwings feed in holly, and Bullfinch, Redpoll and Siskin frequent the birches. Bramblings may join Chaffinches under beeches.

Summer visitors arrive during April and by early May most warblers will be

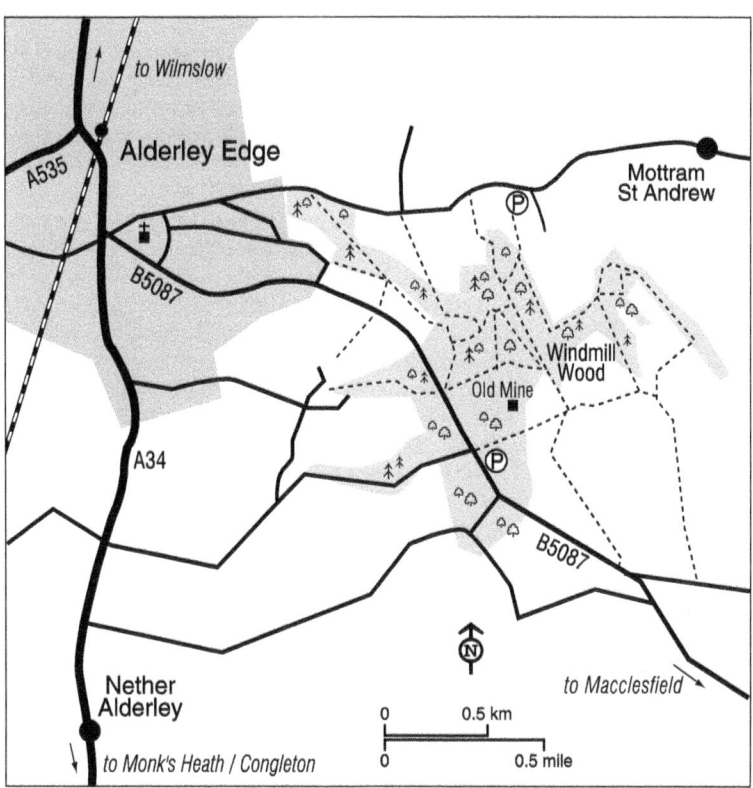

present. Roding Woodcocks are conspicuous in March and April. By May, young Tawny Owls call to be fed.

During June and July, residents and summer visitors are busy with nesting duties. Spotted Flycatcher are present in woods with Chiffchaff and Blackcaps, while Redpoll, Linnet and Whitethroat are on the heathland.

From August to November, migrants depart to be replaced by finches and thrushes for the winter.

ACCESS

The closest train station is Alderley Edge, about 4km away. By car, the woods straddle the B5087 Alderley Edge to Macclesfield road about 2km from Alderley. Several lay-bys border the road. The large car park just to the east of the Wizard Inn may be locked at dusk. Numerous paths criss-cross the woods and it should be no trouble for visitors to remain on paths to avoid contributing to the increasing erosion problem. Most of the woodland species can be located in the woods to the north of the road. To reach the West Mine area, enter the southern part of the woods by a track opposite an unmade car park some 200m west of the Wizard Inn. Follow this main track past the conifer plantations until you reach the gorse.

YOUR VISIT

The woods are owned mostly by the National Trust and attract many visitors. At times, particularly at the weekend and on sunny evenings in spring and summer, the area becomes a little crowded with visitors. As with all woodland birding, early morning visits are preferable. In spring birdsong is at its best early in the day, although evening visits generally give sufficient time to locate most species. In winter timing is less important.

During the winter months, the woods contain large flocks of tits, generally accompanied by a few Goldcrests, Nuthatches and Treecreepers. Coal and Great Tits in particular feed on beech mast; both species may outnumber Blue Tits in winter. In years with a good mast crop, Chaffinches and Bramblings gather under the beeches. Parties of Redpolls feed in the birches, a food source also used by small groups of Bullfinches. Siskins also feed here, but move into larches towards spring. Redwings visit to feed on holly berries, so long as these are available. Green Woodpeckers are seen each winter, often remaining into April. As spring approaches, their ringing 'yaffle' echoes through the woods with increasing frequency, but the birds usually disappear without nesting. Great Spotted Woodpecker nests commonly in the less accessible reaches of the woods.

Wood Warbler is a former breeder, but like Redstart and Pied Flycatcher, it is a scarce migrant now. Treecreeper and Nuthatch, though, are both plentiful breeders, the latter even occasionally plastering up fissures in the exposed rock faces to nest. From the middle of May, Spotted Flycatchers stammer their weak song while Chiffchaffs, which arrived in early April, are still singing vigorously. Coal Tits nest in mouse holes beneath the pines where the numerous Goldcrests hang their flimsy nest structures. The heath and scrubland area at the West Mine site supports breeding Whitethroats, Linnets and Redpolls. Adjacent pastures support a few pairs of Lapwing. Buzzards and Kestrels often hunt the area and Hobby can be seen in the summer. Although Alderley is not often noted as a Woodcock site, one or two birds rode over the woods in spring, occasionally passing over the main car park. Tawny Owls are present though hard to locate.

4 RUDHEATH SAND QUARRIES

Rudheath Sand Quarries
Sandy Lane, Allostock CW4 8HT
SJ742704
flexibly.womb.swipes

Websites: woodlandtrust.org.uk/visiting-woods/woods/rudheath/, cheshirewestandchester.gov.uk/residents/leisure-parks-and-events/parks-and-open-spaces/parks-and-open-spaces-northwich-and-winsford/shakerley-mere

Opening times: Shakerley Mere Car park closes at 8pm in summer and 5pm in winter
Parking: Shakerley Mere
OS Landranger Map 118, OS Explorer Map 267

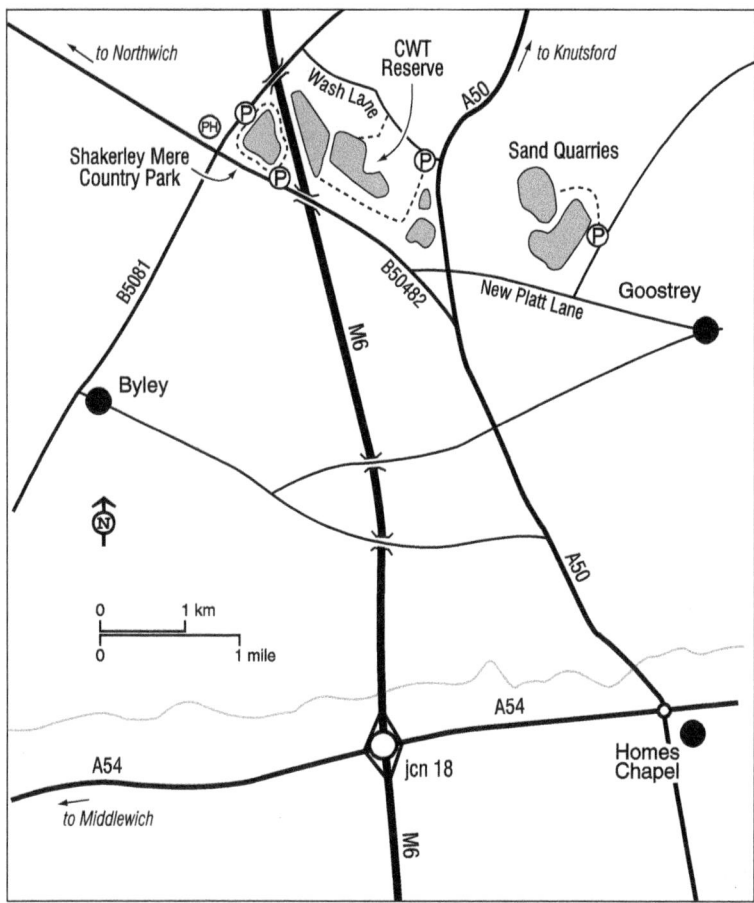

Rudheath, north of Holmes Chapel, was once an extensive heathland. Most of the land has been turned over to agriculture, but small areas of heath, birch woodland and pine plantations remain. Several flooded sand quarries add to the diversity of the area. Newplatt Wood Sand Quarry is a large, shallow, flooded sand quarry used for angling. Rudheath Woods are managed by Cheshire Wildlife Trust.

SPECIES
Great Crested Grebe, Canada Goose are present throughout the year, along with woodland birds including Green Woodpecker and Redpoll.

From December to February, waterfowl including Teal and Tufted Duck gather on the water, and Snipe are seen on exposed mud. Siskin frequent the alders, and flocks of tit and Redpoll in birch woods and mixed finch flocks more generally.

ACCESS
The area is approximately 4.6km from Goostrey or Holmes Chapel stations. The 219 bus route, operated by D&G bus from Sandbach, runs Goostrey. The closest stop is on the corner of Booth Bed Lane. Newplatt Wood Sand Quarry lies to the

Cheshire

north of New Platt Lane and east of the A50 Holmes Chapel to Knutsford Road just south of Allostock village. A minor road runs up the eastern side of the site. From here a well-worn path skirts the eastern edge of the lake, giving interrupted views across the water. For the woods, take Wash Lane west off the A50 immediately south of Allostock village. Park on the unmade track that leads south. This continues as a bridlepath along the eastern edge of the reserve, which is accessible through stiles and along paths.

YOUR VISIT

The woods include both dry and wet woodland with enclosed grassland and scrub. The mainly birch woodland provides a great place to explore with a path around the edge, and for the more adventurous there are small paths into the centre. Springs emerge around the reserve and provide wetland areas where you can find alder and willow. Shakeley Mere is managed for recreation by Cheshire West and Chester Council. There is a small amount of heathland, a designated site of biological importance, in the south-east corner, and a circular path runs around the perimeter of the lake, a distance of 1.4km.

Canada and Greylag Geese are resident, and small numbers of other wildfowl species are usually present, especially between autumn and spring. These include Mute Swan, Teal, Mallard, Gadwall, Shoveler, Pochard, Tufted Duck, Goldeneye and Goosander. Garganey have visited. Great Crested Grebe and Little Grebes, Moorhen and Coot breed. Areas of sandy mud are often exposed around the gently shelving shoreline. This mud attracts waders such as Oystercatcher and Snipe, and occasionally other species such as Common Sandpiper, Redshank, Greenshank and Dunlin. The plentiful fish attract Grey Heron, Cormorant and occasionally Kingfisher and migrant Osprey.

Rudheath Woods hold typical woodland species and regularly attract moderate numbers of Redpolls. Siskins and mixed flocks of tits occur, and occasionally large numbers of Chaffinches with a few Bramblings can be seen. Hobby is seen quite regularly in summer, Peregrines breed nearby and may be seen overhead, whilst Buzzards are regular. Goldcrest, Chiffchaff and Blackcap are regular. Green Sandpiper, Red Kite, Marsh Harrier, Black Tern, Sandwich Tern and 'Arctic' Redpoll have also been recorded.

5 CHELFORD SAND QUARRIES

Mere Farm Quarry
Alderley Road, Chelford SK11 9AP
SJ819746
crunched.grasp.dialects

Acre Nook Sand Quarry
Lapwing Lane, Lower Withington SK11 9AD
SJ820727
chair.cuter.flickers

Lapwing Hall Pool
Lapwing Lane, Lower Withington SK11 9AD
SJ819731
dinner.depth.spud

Parkland, via Lapwing Lane
Lower Withington SK11 9AD
SJ814728
legroom.stems.identity

Website: cawos.org/chelford.html
Parking: Available at each location
OS Landranger Map 118, OS Explorer Map 267

The Chelford Sand Quarries consist of quarries from the Dingle Bank complex at Lower Withington and the Mere Farm complex to the East of Chelford. The former complex has played a significant role in industrial sand extraction in the UK for over 80 years and was regarded as the most important industrial sand site in the country. Quarrying operations there, primarily managed by members of the Sandbach-based Sibelco group, extracted over one million tonnes of silica sand

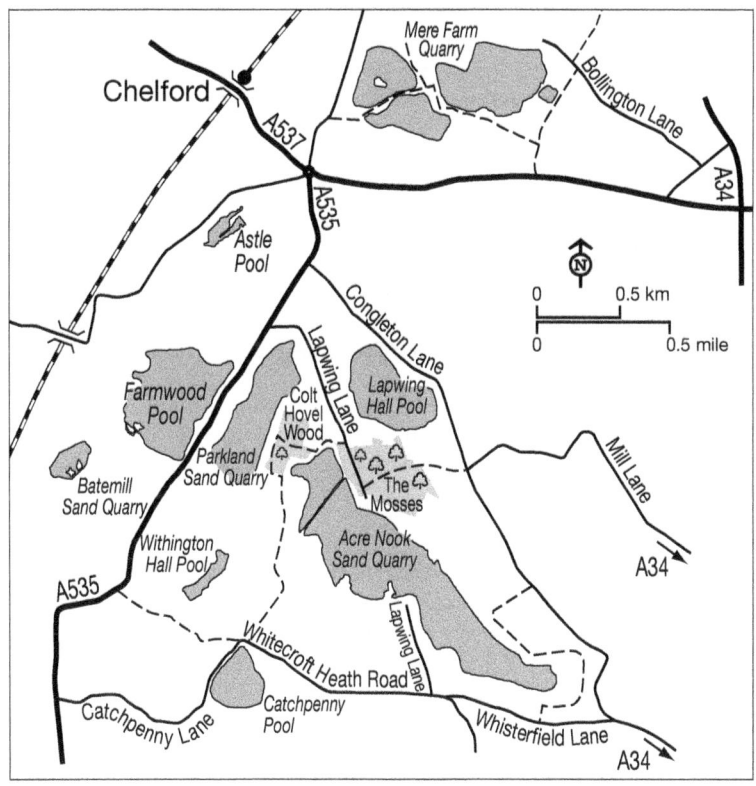

annually, predominantly for glass manufacture. The Mere Farm Quarry complex was managed separately for sand for industrial uses. This account deals with the four most recently quarried sites.

ACCESS

The closest station is in Chelford and all sites are in easy reach by bicycle from there. Buses between Macclesfield and Knutsford stop in Chelford. Mere Farm is viewable from the footpath off Alderley Road.

Acre Nook can be viewed from the bridleway accessed from Lapwing Lane, Lower Withington SK11 9AD, SJ820727, ballots.studs.splint; an access point for the south-east section of Acre Nook is the footpath off Whisterfield Lane, SJ832711, repeated.lecturing.tins. There are two access points for Lapwing Hall Pool: off Lapwing Lane SJ819731, dinner.depth.spud, and off the bridleway through The Mosses, SJ825726, sage.croutons.carefully. The pool is viewable from a permissive path between these two points. Parkland is reached via the bridleway at the North end of Acre Nook and viewed from legroom.stems.identity.

SPECIES

Throughout the year, a great variety can be seen: Great Crested and Little Grebes, Cormorant, Grey Heron, Egyptian Goose, Mandarin, Tufted Duck, Gadwall, Teal, Lapwing, Buzzard, Raven, Sparrowhawk, Tawny Owl, woodland birds, Skylark, Tree Sparrow, Bullfinch and Reed Bunting.

From December to February, dabbling and diving ducks are numerous, especially Wigeon, Pochard, Gadwall, Goldeneye and Goosander, with occasional scarce waterfowl, including Smew, Long-tailed Duck and Scaup. Green Sandpiper, Snipe. Stonechat, winter thrushes and Goldfinch are occasionally seen.

From March to May, Whooper Swan and Black-necked Grebe may be seen. Oystercatcher and Lapwing start nesting, and passage waders, especially Ringed Plover, Whimbrel and Common Sandpiper, move through. Hirundines, Yellow Wagtail and breeding warblers arrive, including Garden Warbler and occasional Grasshopper Warbler. Passage terns and Little Gull move through and a Cuckoo calls but rarely stays these days. White Wagtail and Wheatear pass through.

During June and July, breeding wildfowl, waders, woodland birds and summer migrants are active. Common Scoter may appear on overland passage, while Hobby hunt overhead. The first returning Green Sandpipers and Curlews are noted, and gull roosts start to build.

In late summer and autumn, wildfowl numbers increase, with Shoveler reaching their peak. Whooper Swan and Pink-footed Goose pass through, while Kingfisher and passage waders can be seen. Black-necked Grebe, Snipe numbers increase. A few Golden Plover join flocking Lapwing. Skylark and Meadow Pipit on passage, and winter thrushes arrive.

YOUR VISIT

The former Dingle Bank Quarry area has undergone significant changes as quarrying has ceased and these changes are affecting the make-up of the bird population. The final active quarry, Acre Nook, has now closed, and restoration work is complete. The cessation of quarrying has left three substantial lakes within the former Dingle Bank complex, reshaping the landscape into diverse habitats. Acre Nook Lake, the largest, covers approximately 72ha. Parkland Lake, covering 17ha, is adjacent to the A535 and once the water has reached its expected level it will be controlled by a sluice; viewing is from a bridleway beside its east bank. The third water body, Lapwing Hall Pool, located at the northern boundary near Congleton Lane, is also expected to cover 17ha and has been designated primarily for nature conservation. The restoration vision for the sites included in the original planning documents aimed to create a varied landscape that supports biodiversity, incorporating semi-natural grasslands, woodlands and wetland margins around the lakes: the habitat created is still developing. However, the future uses of all the waters will depend upon the wishes of the site owner(s) and any planning application being granted. As the water levels rise what were once sandy shores have been inundated, resulting in grassy banks to the water's edge in places, whereas elsewhere planted or self-sown scrub and trees are now partly under water. A network of footpaths provides birdwatchers with opportunities to view the sand quarries, all of which are in private ownership. The Mere Farm Quarry pools is the only site where the water has reached its expected level and extent. Permission for watersports on one pool here was granted in 2020 but has yet to be taken up.

The Chelford Sand Quarries are primarily of interest for their waterbirds, but the area contains several areas of woodland, including The Mosses, which is mainly birch and is surrounded mainly by pasture land. Mere Farm, Lapwing Hall, Parkland and Acre Nook described here can be observed from the network of footpaths that criss-cross the area. They are the most recent of the quarries to be excavated and restored. Three other pools at sites quarried and restored in much earlier years are all on private land and inaccessible for birding.

In winter, the primary interest is wildfowl. Wigeon, followed by Mallard, are the most numerous of all the ducks, the pools holding in excess of 500 birds at times, but other dabblers are well represented, including Teal, many of which only reveal themselves in icy weather. Of the diving ducks, Tufted are easily the most common, although peaking post-breeding in some years, followed in descending order by Pochard, which can occasionally top 100 around the waters, Goldeneye and Goosander appear in low double figures. In recent years a Tundra Bean, Pink-footed in varying numbers, White-fronted, Brent and a Red-breasted Goose have been present among the large numbers of Greylag and Canada Geese present in winter. Tufted Ducks breed annually, usually on multiple waters. Teal and Pochard have been proven to breed just twice and once respectively. Gadwall may be working themselves up to it but post-quarrying habitat changes may be negative. Common Scoter appear almost annually on passage, Garganey rarely. Little and Great Crested Grebes breed widely and Black-necked Grebe is becoming a regular visitor, mainly on passage but has been recorded in all months.

The pools regularly attract scarcer species, including Slavonian Grebe, Great Northern Diver, Smew, Long-tailed Duck, Scaup, and in 2022 a female Ring-necked Duck which spent a month on Acre Nook Quarry and was joined by a male after a couple of weeks.

There may be substantial gull roosts on the pools from late summer, with Black-headed and Lesser Black-backed predominating. There are four-figure counts of the former and several hundred of the latter, while Yellow-legged Gull is annual. Terns are seen on passage, with Common much the most frequent being almost annual; Arctic and Black Terns are noted much less often, and a Caspian Tern was present at Acre Nook for six days in 2013.

Great White Egret is recorded with increasing frequency. Glossy Ibis and Bittern have occurred. Buzzard, Sparrowhawk, Kestrel, and Tawny and Barn Owls are all present, while Raven and Hobby breed nearby. Peregrines are seen most often from August to January but Osprey is only a rare visitor.

Lapwing and Oystercatcher breed but following the cessation of quarrying suitable Little Ringed Plover habitat has all but disappeared. Up to 1,000 Lapwing build up in post-breeding flocks and these may be joined by an occasional Golden Plover. Curlew are most often seen from July to late autumn when small overnight roosts form. Many of the wader species visit during the passage periods and may not remain for long. Those seen annually or in most years include Dunlin, whose numbers rarely exceed single figures; they occur most often in spring but have been seen in each month. Ringed Plover and Black-tailed Godwits favour the summer months, Greenshank are most likely from July to September, Whimbrel appear in both passage periods and most often among the Curlew roosts, and Ruff which favour August to October. Green Sandpiper has been recorded in every month but is most likely to be seen in March and April and particularly from July to October (less often in winter). Common Sandpiper is a regular visitor on passage. In contrast, there are very few records of Little Stint and Wood and Curlew Sandpipers. However, Grey Phalarope and, more remarkably, Purple Sandpiper have occurred.

The immediate surrounds of the sand quarry pools have breeding Reed and Sedge Warblers and Reed Buntings, while in the winter months Water Rails may call from cover, unseen, and Snipe venture out into the open to probe for their food. On the slopes warbler species are moving into the developing scrub and woody areas while in the clearer areas of the slopes birds feeding in winter may include Meadow Pipits, flocks of Goldfinches and occasional Stonechats.

The mixed arable/pasture land between the quarry pools, with its scrub and hedges, has a good range of species including Tree Sparrow, Yellow Wagtail, Skylark, Garden Warbler, Lesser Whitethroat and Grasshopper Warbler. The occasional small field left to weedy seeds can provide sizeable mixed finch flocks including Brambling and Linnet. The mostly small areas of woodland away from the quarry boundaries, often including conifers planted as shelter belts, may appear relatively unexciting ornithologically but have included an occasional Redstart pausing on passage or a longer-staying Spotted Flycatcher on territory, while winter often brings mixed Siskin and Redpoll flocks to brighten a dull winter's day.

6 MACCLESFIELD FOREST, LANGLEY RESERVOIRS AND TEGG'S NOSE COUNTRY PARK

Langley reservoirs
Holehouse Lane, Sutton, Langley SK11 0NB
SJ945717
steady.rewrites.bets

Ridgegate Reservoir
Standing Stone Road, Sutton, Langley SK11 0NE
SJ953715
walnuts.inhales.factoring

Trentabank Reservoir
Standing Stone Road, Sutton, Langley SK11 0NS
SJ963711
captive.cubs.crabmeat

Tegg's Nose Country Park
Saddler's Way, Macclesfield SK11 0AP
SJ950732
running.width.tennis

Websites: *macclesfield-forest.co.uk, teggsnose.co.uk*
Phone: 01625374833
Opening times: Accessible at all times
Parking: Pay and display parking available at the main car park
Landranger Map 108, OS Explorer Map 268

Macclesfield Forest, located just 5km south-east of Macclesfield town, is a diverse and historically rich area situated on the slopes of the Pennines. It is a working forest owned by United Utilities, featuring conifer plantations predominantly of pine, spruce and larch, alongside some sycamore and beech trees, particularly along the roadsides. Originally part of the expansive Royal Forest of Macclesfield, the area was once under the ownership of the Earl of Chester and stretched from the Pennines to the Staffordshire Moorlands and the High Peak area. The forest is home to four reservoirs, two of which, Ridgegate and Trentabank, supply drinking water to Macclesfield. Ridgegate, constructed in the late nineteenth century, is a long, narrow reservoir created by damming the valley. Trentabank, established in 1929, is known for its steep, rocky shores and limited waterfowl activity. Farther down the valley near the village of Langley, two smaller reservoirs add to the area's charm. These reservoirs have fluctuating water levels that limit the development of marginal vegetation. Notable landmarks include Shining Tor, Cheshire's highest point at 559m, and Shutlingsloe, a 509m peak affectionately nicknamed the 'Cheshire Matterhorn' due to its distinctive shape. Shutlingsloe offers stunning views and is steeped in history, with its name tracing back to the Celtic tribe 'Shutta', and 'Lowe', meaning hill or mound. The forest also encompasses Tegg's Nose Country Park, which features mixed broadleaved woodlands of oak, ash and beech, along with alder carr at lower elevations. Grazing cattle coexist with a thriving scrub layer, including brambles, while bracken and heather dominate higher altitudes. Close by is St Stephen's Church, or 'Forest Chapel', rebuilt in 1834 and known for its annual August rush-bearing ceremony. Macclesfield Forest is a haven for ramblers and nature enthusiasts. The area's blend of natural beauty, historical significance and diverse ecosystems makes it a unique destination.

SPECIES

Tufted Duck, Buzzard, Sparrowhawk, Kestrel, Tawny Owl, Goldcrest, Raven and Crossbill can be seen throughout the year.

From December to February, Goldeneye, Pochard, Teal, Wigeon and Goosander are present on the reservoirs. Goshawk is possible, along with common raptors.

From March to May, Goldeneye display, Canada Geese take up territory and Grey Herons renovate their nests. Woodcocks start to rode and Crossbills may be nesting. Passing migrants include Brambling. Finch flocks in early spring include Siskin, Redpoll and Goldfinch. In April hirundines and Yellow Wagtail pass through,

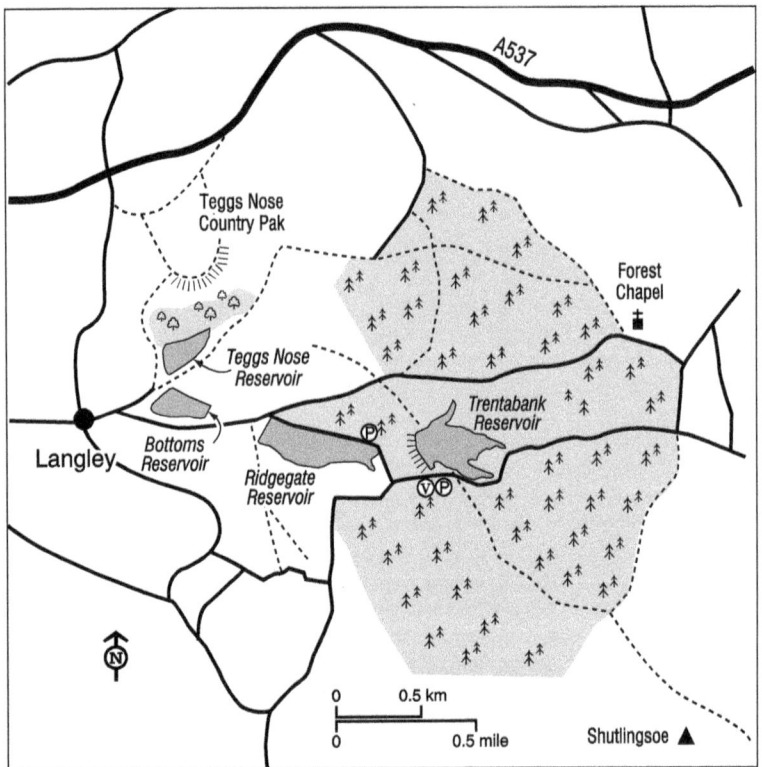

and summer visitors arrive, including Common Sandpiper, Pied Flycatcher and Tree Pipit.

During June and July, Grey Herons fledge, and Great Crested Grebes, Tufted Ducks and Common Sandpipers may have young. Redstart, Tree Pipit and Pied Flycatcher can be found at Tegg's Nose, and Wood Warbler is also possible.

Later in the year, summer visitors depart. Goldeneye may return from August but are more likely in October. Diurnal passage includes Meadow and Tree Pipits and White Wagtails in August and September, with Meadow Pipits continuing to pass through in October. Pied Wagtails are more numerous in October and a few Grey Wagtails are likely; Siskin from mid-September, and Chaffinch and Brambling in October. Large-scale thrush movements in October and November include Redwing, Fieldfare, Blackbird and Song Thrush. Goosander numbers may build up at Trentabank.

ACCESS

Macclesfield station is 4.5km from the top car park at Tegg's Nose. The distance may not be an issue, but the 738m climb is not for the faint-hearted. It is 7.7km to the viewpoint at Trentabank and a slightly less intimidating 472m climb. The 14A bus service, run by High Peak, runs from Macclesfield to Langley, though the service is infrequent. By car, the forest lies east of Langley village south-east of Macclesfield. Immediately through Langley village, take a left turn. Holehouse Lane leads to a car park at the foot of Tegg's Nose. Follow the path up the hill and explore the scrubby areas to the right. A larger car park at the summit of Tegg's Nose is signposted off the A537 Macclesfield to Buxton road and provides an easier walk to look out over the valley from the promontory near the quarry exhibition. This is a good vantage point for Ravens and on days when there is visible migration. Otherwise continue up the main road. Bottoms Reservoir appears on the left. Carry on up the road for Ridgegate and Trentabank Reservoirs. A lay-by alongside Trentabank allows a good view of the heronry and is one of the best viewpoints for the forest as a whole. For watching autumn migration continue up the road past Trentabank to the hilltop, then walk right along the footpath to the ridge and wait. If there is a passage, birds will pass low over this ridge. The Forest Chapel road is generally most productive for Crossbill, although they may be seen from any of the tracks through the forest.

YOUR VISIT

Evening visits are advisable for watching roosting and crepuscular species. Clear mornings are best for raptors. On cloudy days, mist may obscure the valley, although on cold winter days when fog blankets the lowlands the sun may well be shining on the hills. Diurnal passage over ridges above the forest is heaviest on autumn days with mist or hazy clouds or when it is overcast, although very large movements have occurred with a clear sky and a light wind from the east. Passage stops abruptly in heavy or persistent rain or high winds. On bright, sunny days migrants fly too high to be seen and are detected only by their calls. Tegg's Nose and Trentabank can be crowded at weekends, especially in summer, but few people wander far from the roads into the forest.

Despite an overall impression of quiet, the forest does have some specialities and unusual birds turn up with surprising frequency. During the winter months flocks of Goldcrest and Coal and Long-tailed Tits roam the plantations, their thin calls breaking the misty silence. With perseverance, Crossbill is quite likely to be seen in flight across the valley, and sometimes in roadside trees. Though present all year, the opening of Larch cones attracts small finches that arrive to feed on their seeds. Siskin is most numerous, with flocks of up to 150, and there are also smaller numbers of Redpoll and Goldfinch. Bramblings may appear with Chaffinches under the beeches at Tegg's Nose, which also has an excellent range of breeding woodland species. Wood Warblers occasionally trill beneath the oaks and may still breed. Redstart, Pied Flycatcher, Garden Warbler, Blackcap and Tree Pipit are all vocal contributors to the soundscape of the park. Green and Great Spotted Woodpecker are seen, though Lesser Spotted appears to have been lost as a breeder, in common with most of Cheshire.

The west–east oriented valley forms a corridor for migrant birds crossing the Pennines. In early spring parties of Brambling and Chaffinch, the former detected by their nasal calls, fly at tree-top height up the valley. Jackdaws move east at greater altitude; their winter numbers greatly exceed summer totals, and this regular passage gives some indication of the origin of the winter flocks. Scanning the valley sides can be productive and the area remains the most reliable site in Cheshire for Goshawk. Disturbance by forestry activities has led to a reduction in records, but the best chance to see this magnificent bird is a bright day in late February or early March, when pre-breeding displays may take place. Ravens, Sparrowhawks and Buzzards are regular overhead whilst Peregrines breed nearby. Hen Harrier, White-tailed Eagle, Rough-legged Buzzard and Red Kite have also been recorded, the latter with increasing frequency.

Crepuscular and nocturnal species include roding Woodcock, best seen from the road by Trentabank. Their curious display flight begins in the last few days of February in mild springs but is better witnessed in May or June when, with the hours of darkness restricted, the performance begins while the light is still good. At a distance the only call to be heard is a sharp squeak, but as the birds pass overhead with deep, slow wingbeats listen for a series of three or four deep grunts. Tawny Owl is common and makes use of old crow nests due to the lack of cavities in trees that provide the usual nest-site. Long-eared Owl probably breeds in the area and hunts the moors above the forest but is very difficult to locate. By day, Kestrels hover above the ridges. Hobby and Osprey are seen fairly regularly, and there has been a record of Honey Buzzard. There are April records of Firecrest and Great Grey Shrike. The east Cheshire hills are sadly no longer the haunt of Black Grouse, but Nightjar has been recorded recently, and Curlew can still be heard bubbling on the moors.

Autumn passage brings the occasional rarity to the forest. Wryneck and Firecrest have been noted – but the area is known best for heavy movements of commoner species, including thrushes, larks, pipits, wagtails, finches and Woodpigeon passing overhead between September and November. Migrant Brambling are often recorded here when they appear to be absent elsewhere in the county. Easterly winds in October and November bring the largest movements with thousands of Redwings and Fieldfares, hundreds of Blackbirds and dozens of Song Thrushes on the best days. Starling and Jackdaw are important passage species. Tired migrant thrushes and finches will often settle in the scrub and woodland where they make easy pickings for the resident Sparrowhawks.

A few common waterbirds winter on the reservoirs, including Little Grebe, Coot, Canada Goose, Mallard, Pochard and Tufted Duck. Up to a dozen Goldeneye are usually present, and in late winter there is no better place than Ridgegate for watching their display. Goosanders are regular, especially at Trentabank. The ducks arrive in late afternoon from rivers, fish-ponds and lakes over a wide surrounding area. Thirty or more are sometimes present, although the roost may settle at Lamaload Reservoir, a short distance to the north if that water is not disturbed. Dippers are seen occasionally and more unusual species seen on the reservoirs in winter have included Smew and Slavonian Grebe.

A few waterfowl breed on the reservoirs. Great Crested and Little Grebes may nest on Bottoms Reservoir, and Tufted Duck may rear ducklings on Ridgegate. Canada Geese sit conspicuously on nests beneath trees near the water. Common Sandpipers, which feed along the shoreline, evade anglers by nesting inside the plantations. As the young hatch, the adults perch on roadside walls, bobbing and calling anxiously. There is a small heronry in trees beside Trentabank. Renovation of old nests may start in February, with new nests still being built in April. Grey Herons fly out to fish in brooks, pools and rivers in the surrounding hills. Returning birds are greeted with much bill-clattering by their mates. During May and June, the growing young birds gabble and squawk noisily. The heronry is much reduced compared to last century though Cormorants are now regular breeders.

Cheshire

7 THE DANE VALLEY

Cat and Fiddle
Buxton New Road, Macclesfield SK11 0AR
SK001718
same.snooping.darting

Shining Tor
Buxton New Road, Macclesfield SK11 0AR
SJ994737
relay.auctioned.rate

Danebower
Buxton Road, Macclesfield SK11 0BQ
SK009699
nylon.envelope.animate

Danebridge
The Falls, Wincle SK11 0QE
SJ964652
firming.pizzas.spilling

Cutthorn Hill and Three Shires Head
Wildboarclough, Macclesfield SK11 0BQ
SK002681
clearly.winters.unzipped

Opening times: Accessible at all times
Parking: There's free parking for customers at the Cat and Fiddle. For Cutthorn Hill and Three Shires Head, the main option is Clough House car park (free).
OS Landranger Map 118/119, OS Explorer Map 268

The River Dane is a clean trout stream forming the boundary between Cheshire and Staffordshire. Its waters are rich in invertebrates that provide food for a good number of riparian birds. Danebower is a gritstone hill topped with peat moorland. Rocky slopes and mixed woodlands, including those at Shell Brook, support a diverse range of breeding species.

SPECIES
All year, Red Grouse, Little Owl, Dipper, Grey Wagtail and Kingfisher can be seen.
In spring and summer, Ring Ouzel and Wheatear appear on scree slopes, Common Redstart and Pied and Spotted Flycatchers, Tree Pipit and warblers sing in woods. Mandarin Duck arrive and breed.

ACCESS
The area is 8–11km steeply uphill from bus stops on Buxton Old Road or stations in Macclesfield and Congleton. For the upper Dane Valley, the public footpath north from Danebower to the Cat and Fiddle (SK009700 to SK000719) offers a good chance of moorland species in early spring before the rambling season

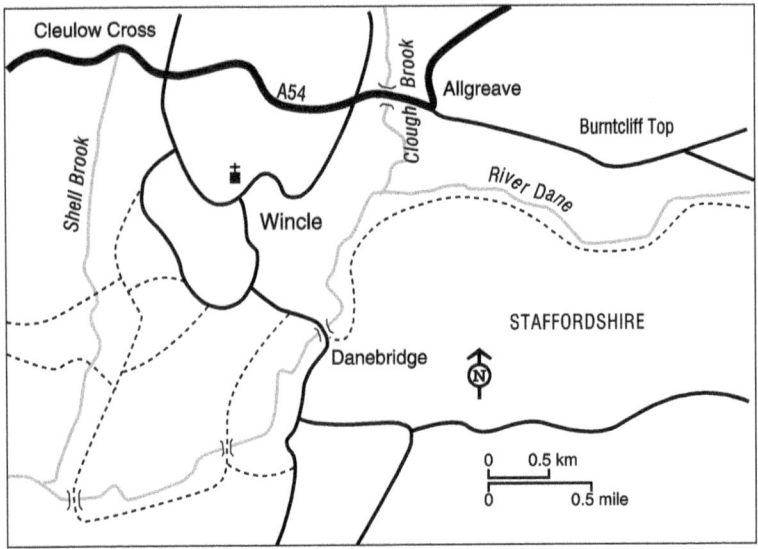

intensifies. Look for Cheshire's only wild thyme on the roadside verges at the northern end. The lane below Cutthorn Hill and its continuation as a footpath up the brook to Three Shires Head is a good place for moorland-edge species, as is the rocky slope alongside the minor road running east from the Rose and Crown at Allgreave (SJ973669). Access to the valley farther west is easy from Danebridge (SJ964651). A footpath follows the river downstream for 1.6km before crossing into Staffordshire (SJ955641). The Shell Brook woods are accessible by footpaths in places; for example, north out of Danebridge.

YOUR VISIT

The valley is visually attractive at all seasons but is far richer in birds during the early summer months, especially April to June. Many birdwatchers visit the Danebower chimney in March to look for early migrants. The moors are quiet in winter, but persistent visits should eventually be rewarded. Red Grouse can still be seen occasionally around the Cat and Fiddle, though Twite, Curlew and Golden Plover are no longer counted in the range of breeding species. Similarly, the Ring Ouzels that used to feed in the pastures around the quarry chimney have become scarce, transient visitors. A few Wheatears still nest along rocky slopes. Birds soaring above the Allgreave–Gradbach ridge may include Buzzard, Raven and Peregrine. Barn and Tawny Owls breed in the valley. In recent years, Red Kites have been seen, and the trout pool at Danebridge has been known to attract a passing Osprey. Dipper, Grey Wagtail and Kingfisher are seen commonly along the river, especially below Danebridge, and Mandarin Duck breeds. Common Sandpiper is an occasional passage migrant. Goosander sometimes visit in winter in pursuit of fish. Spotted and Pied Flycatchers, Redstart and Tree Pipit breed in woodlands and scattered stands of sycamore and other trees. The Shell Brook woods hold a wide range of woodland species. In winter, when many birds have left the moor, there is a remote chance of encountering a wandering predator such as a harrier or Short-eared Owl.

8 THE SANDBACH FLASHES

Foden's Flash
Oakwood Lane, Stud Green, Moston CW11 3PR
SJ730613
cubed.bonus.earliest

Watch Lane Flash
Watch Lane, Moston CW11 3PD
SJ727606
shippers.lateral.rainy

Elton Hall Flash
Clay Lane, Moston CW11 3QY
SJ724595
stealthier.knocking.locals

Website: *sandbachflashes.co.uk*
Email: info@sandbachflashes.org.uk
Opening times: Accessible at all times
Parking: Parking is best on the verge along Clay Lane at SJ724505.
OS Landranger Map 118, OS Explorer Map 268
SSSI

Sandbach Flashes are a series of shallow pools formed during the last century as a result of salt mining. The flashes lie in a triangle between Sandbach, Middlewich and Crewe. The main flashes are in an area designated as an SSSI. Although there are 14 separate flashes, most of the birding interest is concentrated on just two of those, Elton Hall Flash and Pump House Flash. Most of the flashes were formed in the 1920s due to subsidence caused by wild brine pumping for the salt industry. They are all on private land but adequate viewing is possible from roads and public footpaths. 249 species of birds have been recorded in the area, making it one of Cheshire's premier birding areas.

SPECIES

Great Crested and Little Grebes are present year-round, along with Mute Swan, Lapwing, Water Rail, Barn and Tawny Owls, Skylark, Cetti's Warbler, Bullfinch and Reed Bunting.

In winter, Black-tailed Godwit, Green Sandpiper, Yellow-legged Gull, Long-eared Owl, Sparrowhawk, Peregrine, Redpoll and Siskin are regular visitors.

As spring arrives, Garganey, Little Ringed Plover, Oystercatcher, Black-tailed Godwit and Common Sandpiper appear, and wader passage can include Dunlin, Sanderling, Ringed Plover, Redshank, Ruff, Whimbrel and Greenshank; Wood Sandpiper and rarer species are possible. Mediterranean Gull, Common Tern, White and Yellow Wagtails, Lesser Whitethroat, Grasshopper Warbler, Wheatear and Reed Bunting are also present.

Little Ringed Plover, Green Sandpiper, Hobby, Skylark, Sedge and Reed Warblers and Lesser Whitethroat can be seen during June and July.

From August to November, Great White Egret may be seen, and wader passage resumes with Dunlin, Common Sandpiper, Green Sandpiper, Redshank, Greenshank,

Cheshire

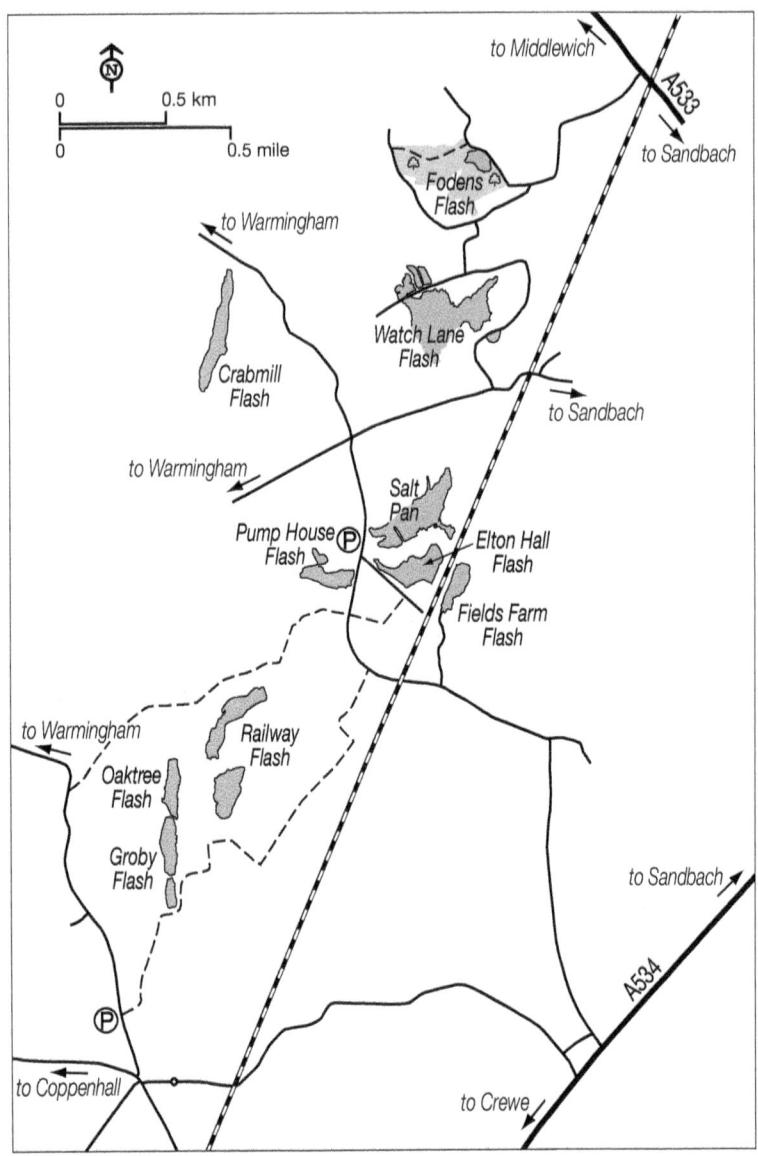

Black-tailed Godwit, Snipe, Jack Snipe, Whimbrel, Curlew, Little Ringed Plover and occasional rarities. Returning ducks occasionally include Pochard, Garganey and Pintail. Visible migration brings Meadow Pipit (September), Woodpigeon (October), Redpoll, Siskin and winter thrushes.

ACCESS

Most sites are easily reached from Sandbach station and all flashes are within 8km of the mainline station at Crewe. By car the best access is from the A534 between junction 17 of the M6 motorway and Crewe. Turn north-west from this road into Clay Lane at SJ735572. Continue along Clay Lane for 3km and then park on the verge on the left at SJ724505.

YOUR VISIT

Any time of the year can be great at the Flashes. The area makes a wonderful local patch but like any patch regular visits are necessary to connect with the star species. Many of the locals are keen patch year-listers and visit daily; their reward is a year-list of perhaps 140 species, including the odd rarity. A telescope and tripod are essential.

Once you have parked your car or walked or cycled to the birders' parking verge at SJ724505 carefully cross the road and locate the small wooden gate leading 30m to the viewpoint. You are now overlooking the old saltpan, these days referred to as Elton Hall Flash, the premier site at the Flashes. Time spent here will eventually reward you. In winter you will see large flocks of Greylag and Canada Geese. Wigeon will likely be in flocks totalling many hundreds, and witnessing a close flock grazing in the fields here on a crisp winter's morning will be a highlight. Teal have declined a little but you should have no trouble seeing more than 100. During dry periods sandbars will appear to the right of the dead tree and these will attract loafing gulls and waders. Watching a mixed flock of waders pitch in on the sandbars in spring is a rare event but one you will not forget; the Dunlin will be coming into their summer dress, accompanied perhaps by a Sanderling or Ringed Plover. Scanning for raptors can be very fruitful, with Hobby, Osprey, Red Kite and Marsh Harrier all recorded more or less annually. The feeding station around the viewpoint operates during the non-breeding season and attracts a nice selection of commoner species including Greenfinch, Reed Bunting and the odd Redpoll, Siskin or Brambling. The roll call of rarities here is a long one and includes Crane, White Stork, Glossy Ibis, Night Heron, Cattle Egret, Sabine's and Franklin's Gulls, Caspian, Gull-billed and Whiskered Terns, White-rumped and (two) Stilt Sandpipers, Black Kite, Alpine Swift, Golden Oriole and Wryneck.

Next, walk back across Clay Lane and stand by the iron gate. This spot overlooks an old hawthorn hedge and a small subsiding pool that has been nicknamed Rosemary's Flash by local birders. This pool does dry out in summer at the moment but with continuing subsidence it may become an excellent site. The old hedge has held Pied Flycatcher and Redstart and even a Great Grey Shrike on one occasion. Rosemary's Flash holds increasing numbers of ducks in winter with a Green-winged Teal the rarest so far.

Continue by walking back along Clay Lane and after 30m turn left down the side lane. This lane is nice and quiet with no traffic. At times it floods and you will need wellington boots but during heavy winter rains it often becomes impassible. If passable, it is well worth wading through the muddy water. The back pool can be viewed from several spots during the next 300m, giving good views of many ducks and the odd wader. The willows flanking the lane are a favourite site for Chiffchaff, Cetti's Warbler and the odd Water Rail in winter. Walk as far as the railway embankment, which is always worth checking in autumn for Spotted Flycatcher and Garden Warbler. Rarities recorded over the years down here have included Leach's Storm-petrel, Gannet, two Wilson's Phalaropes, Lesser Yellowlegs, Spotted Sandpiper, American Wigeon, Lesser Scaup, Hoopoe and Yellow-browed Warbler.

Walk back to Clay Lane and turn left. After a further 70m you will see Pump House Flash on your right. It is at its best during winter floods when the adjacent field also becomes submerged. This is a good site for gulls and waders and always worth checking. The hedgerow at the back of the field can be good for Lesser Whitethroat in spring and Redstart occasionally in autumn. Scarce and rare birds recorded here have included Kentish Plover, Marsh Sandpiper, Black-winged Stilt, Crane, Cattle Egret, Glossy Ibis, Night Heron and Guillemot.

On the opposite side of the road you will see some small subsidence pools that locals call Hancock's Flashes. This whole area is still subsiding and bodes well for the future of bird recording at the Flashes. In winter it is proving popular with ducks and a Curlew flock. The rarest bird seen on this flash was an Upland Plover in December 1983.

The Railway Flashes and Maw Green Tip area can be accessed two different ways. From Pump House and Hancock's Flashes you can carry on walking down Clay Lane for another 500m. Before reaching the railway bridge and the farm tracks off to the left and right, look for half a dozen steps leading up the bank and through the hedge on the right at SJ724589 (removal.squaring.finger). This footpath skirts Railway Farm and runs parallel with the railway line for about 0.8km. At SJ719581 (belly.petty.kings) look for a plank across a ditch by stiles and then a proper wooden bridge over the Fowle Brook. The footpath then turns right and follows the river and you will soon spot a large circular pool on the right. This is Railway Flash Number 2, which is best viewed by climbing up the slope on the left to get a higher vantage point. Explore the area fully. There are two more flashes nearby – Oaktree and Groby – but they are usually heavily disturbed by fishermen. This whole area provides good birding. Railway Flash Number 2 often holds good numbers of ducks and has a muddy edge on the east side. The slopes of the tip have some planted saplings and naturally regenerating scrub. Large areas of Blackthorn and Hawthorn scrub are present around Groby Flash and along the Fowle Brook. With bird recording going back to 1935, this area has a long bird list. Railway Flash Number 1 has had Spotted Sandpiper, and a Corncrake was heard in 1937, while Number 2 has recorded Glossy Ibis, White-winged Black Tern, Spoonbill, Red-necked Phalarope, Grey Phalarope, Lesser Scaup, Great Grey Shrike and Little Bunting. Garganey have bred here. The tip itself has recorded Crane, White Stork, two Richards Pipits, Red-backed Shrike, Wryneck, Hawfinch, Black Redstart and Lapland Bunting. This whole area is good for owls with five species present some years. Jack Snipe is regular in the winter on the tip or around Railway Flash. In spring the area comes alive with birdsong from Skylarks and a variety of warblers. Wheatears, Stonechats and Whinchats pass through in April and again in autumn. Regular observation of visible migration in autumn has resulted in notable counts of Meadow Pipits, Woodpigeons and winter thrushes. Return the way you came. The other way to access this area is by car to Groby Road about 8km from Elton Hall Flash. Park on the side of Groby Road at SJ712574 or on the nearby estate. Access the footpath through the kissing gate on the left (the east side of Groby Road) and follow the footpath around the old tip for 1.6km.

Other sites worth visiting if you have time or like exploring are Watch Lane Flash. Park at SJ731607 in a lay-by and then walk down the causeway to SJ728607. This flash is heavily disturbed by fishermen so it is best to arrive at dawn and have a quick scan of the water. Common Reed has become well established and is spreading around the flash; it is liked by Sedge, Reed and Cetti's Warblers. The causeway continues as a bridle path for 0.8km, passing Watch Lane Farm on the right. The traditional orchard here has produced Hawfinch and Turtle Dove. Watch Lane Flash used to be one of the premier sites at the flashes and its long bird list includes two

Great Northern Divers, Red-throated Diver, Eider, Red-necked and Slavonian Grebes, Purple Heron, Lesser Yellowlegs, Spotted Crake, Wryneck, Great Grey Shrike and Red-rumped Swallow.

Foden's Flash, or The Moat, is an interesting small pool with a bulrush bed backed by a carr of willow and alder. Park by the canal at SJ732611 and walk down Oakwood Lane to the flash at SJ730513. Short circular walks on quiet lanes, field footpaths and the canal towpath are the best way to bird this area. Goosander sometimes frequent the pool in winter, and the woodlands here used to have breeding Willow Tit, Lesser Spotted Woodpecker and Turtle Dove. The habitat is no different today so one day they may recolonise. Woodcock winter here in numbers and are best seen at dusk when they fly out from their day roost; the best spot to witness this is from a vantage point along the canal. This site has produced Bittern, Corncrake (1938), wintering Long-eared Owls, Firecrest and a singing Nightingale on one occasion.

Crabmill and Warmingham Flashes are both under-watched. Crabmill Flash can be accessed by a footpath from Crabmill Lane at SJ716609 (wellington boots needed). Head south-west down this footpath and just before reaching the river turn sharp left, then walk parallel with the river. The footpath then passes between the flash and the river, allowing access to another area that is still subsiding: Warmingham Flash, which is narrow and about 1.5km long and can be viewed from Green, Dragon's and Tetton Lanes. Birders should be on the lookout for Little Owls and Tree Sparrows. Please consult a good map to explore rights of way.

9 MARBURY COUNTRY PARK AND WITTON LIME BEDS

Marbury Country Park
Marbury Lane, Anderton, Chester CW9 6AS
SJ652763
sung.scale.sake

Website: *northwichwoodlands.org.uk*
Annual bird reports: *foam.merseyforest.uk*
Opening times: 9am-8pm in summer (1 April–30 September), 9–5pm in winter (1 October–31 March).
Parking: 24-hour parking is at lines.dance.broke, although this is still pay and display during normal car park opening hours.
Budworth Mere: Accessed via Marbury Country Park career.small.stage

Witton Lime Beds

Neumann's Flash Hide: launch.songs.pads.
Ashton's Flash: viewed from the bund between the two flashes: prices.yard.modes
Parking (both sites): 24-hour parking at clip.snacks.saying.
OS Landranger Map 118, OS Explorer Map 267
SSSI

Cheshire

The cluster of sites lying to the north of Northwich offer a considerable diversity of habitats, encompassing woodland, fresh water, reedbeds and industrial sludge tanks. Access has been improved greatly in recent years as part of the Mersey Forest initiative. Northwich lies at the heart of the Cheshire salt and chemical industry. Consequently, a great deal of land has been left in a rough state, with plentiful scrub and coarse grassland. Flashes and subsidence pools formed as a consequence of underground brine pumping have been used as sludge beds for lime-rich waste. They may contain pools of shallow water following rain and are then attractive to waders and wildfowl. Budworth Mere and Pickmere may also have formed as a consequence of the dissolution of rock salt, but by natural means and much longer ago.

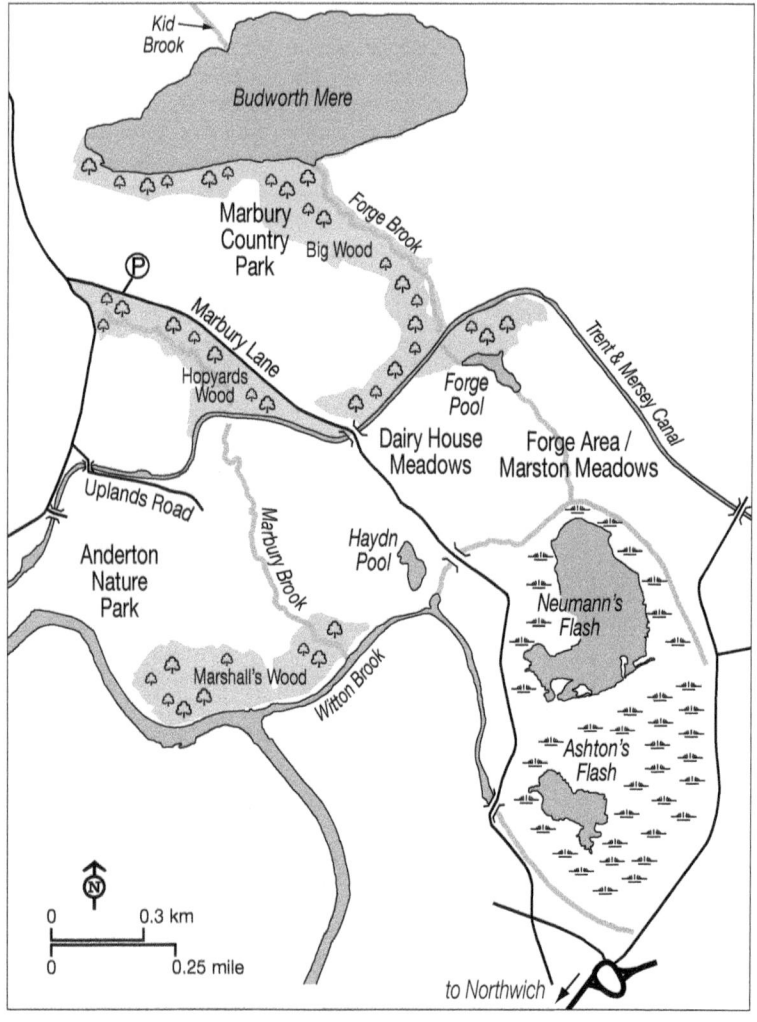

SPECIES

Common wildfowl and grebes frequent the water bodies all year unless driven off by frosts. Kingfishers are seen regularly from Budworth Mere hide and Water Rails from the hide at Neumann's Flash. There are woodland species in the parkland.

From December to February, at least one Bittern usually resides in the Coward Reedbed, sometimes showing well from the hide. Curlew roost at the flashes and often feed at Budworth Mere by day. Wildfowl on Budworth Mere include Goosander and Goldeneye, whilst Wigeon, Shoveler, Gadwall and Teal can be found at the flashes. Spectacular Starling murmurations at the mere and flashes, with attendant raptors. Jack Snipe and Woodcock numbers peak.

In spring, wildfowl and Curlew flocks disperse. Light wader passage often includes small flocks of Whimbrel. Avocet, Oystercatcher and Little Ringed Plovers display. Woodland resounds to the drumming of Great Spotted and yaffling of Green Woodpeckers. Scrubland comes to life as warblers arrive and take up territory. At Budworth Mere, hundreds of Sand Martins arrive in early spring, with other hirundines and Swift following in April. Passage Whinchats, Wheatears and Yellow Wagtails can be found in suitable habitats. Cuckoos arrive by late April. Favourable conditions can attract Garganey, Common Scoter, Kittiwake, Black and Arctic Terns and Little Gull at the main water bodies. Budworth Mere's Grey Herons are at their busiest, although the heronry has much declined in recent years. Marsh Harriers are regular spring visitors.

In late spring to summer, breeding wildfowl and waders may have young. Grasshopper, Reed, Sedge and Garden Warblers and Lesser Whitethroats sing well into June. Cuckoos call to mid-June and Hobbies can be found at dragonfly hatchings. Lapwing flocks start to gather by the end of June. The first passage waders appear in July and Curlews return. Little and Great White Egrets regularly visit the mere and the flashes. Common Tern, and perhaps other tern species, visit occasionally.

Wader passage is at its most varied between August and November but is dependent on water levels. Juvenile warblers roam the scrub, sometimes with mixed flocks of tits, and in September take berries from elder and other bushes. Shoveler, Gadwall and Teal numbers peak and there are huge overnight roosts of geese, Mallard and Black-headed Gulls. Terns pass through in appropriate conditions until September – a few waders such as Dunlin, Redshank and Green Sandpiper until October, when winter wildfowl return. In invasion years, Hawfinch may appear in the parkland in late autumn.

ACCESS

It is 3.2km from Northwich station to Marbury Country Park and half that to the Witton Flashes. Access to Witton Flashes is from a roundabout on the A559 to the east of Northwich town centre; take Leicester Street west and turn right to the waste disposal site and continue over the metal bridge to the Witton Mill car park (restricted opening hours). The flashes can also be visited from the east side with parking at the lay-by along New Warrington Road (no timing restrictions). The flashes are visible behind an embankment that runs between the two parking areas. Paths encircle both Neumann's and Ashton's Flashes with various viewpoints, screens and hides. Do not venture onto either of the flashes. All gateways are used by tenant farmers and rangers and should not be blocked, nor should the roadside passing places. A footpath has been constructed alongside Witton

Brook and reedbed. This continues through scrub towards Anderton Nature Park.

Marbury Country Park is signposted off the minor road between Anderton and Comberbach (SJ648325) and has a pay and display car park with restricted opening times. Parking is also available along the road before the car park and is open 24 hours, although payments apply for timings as per the main car park. There is open access to the southern shores of the mere through the park. To reach Pick Mere, which holds similar species but is more disturbed than Budworth Mere, either follow minor roads from the park, turning right in Comberbach and continuing through Great Budworth to Pickmere village, or take the B5391 east from the A559 Lostock to Warrington road. From Pickmere village follow signs for Pickmere Lake. Park at the fairground end and walk the very muddy footpath along the south bank of the mere, examining fields for migrants.

YOUR VISIT

The huge Witton Flashes have been filled with alkali waste and now resemble sludge beds. Neumann's Flash remains flooded all year, with muddy shallows developing during dry summers. Ashton's Flash is shallower and can dry out completely from July to October. A bund separates the more vegetated Ashton's Flash from the relatively bare Neumann's Flash to the north. The flashes support a range of lime-loving flowers that are otherwise hardly known in Cheshire: Fragrant Orchid, Hoary Mustard, Ploughman's Spikenard and Yellow-wort are all abundant. The area in general is well endowed with hawthorn and other scrub. Immediately to the north-west of Witton Flashes, a narrow tributary flows underneath the road at Butterfinch Bridge to join Witton Brook. To the west of the road it widens out, bordered on one side by *Phragmites* reeds. To the other side a stand of Sea Club-rush, Sea Aster, Lesser Sea Spurrey and several other species forms one of the best inland saltmarshes in the county. To the north of Neumann's Flash, Dairy House Meadows is maintained as an area of scrubland and flower meadows. Marbury Country Park lies within the grounds of the now-demolished Marbury Hall. It consists for the most part of dog-walking grassland. The scrub that provided much ornithological interest has now largely been cleared away, but there is a mature planted woodland of mixed broadleaved trees, with some conifers, that encircles the park. The mere, called Budworth Mere to distinguish it from Marbury Mere near Whitchurch, is fringed by a narrow belt of *Phragmites* reeds for much of its perimeter. In places, these reeds thicken out to form substantial beds, notably at the western end where a small woodland reserve has been set aside as a memorial to T.A. Coward, the great Cheshire naturalist. Other parts of the northern shoreline are subject to grazing and trampling by cattle. A sandy spit by the mouth of the Kid Brook on the north bank attracts bathing gulls and waders. The mere is subject to disturbance from a sailing club based at the eastern end and by members of the public visiting the park. Pickmere is a few kilometres to the east and holds a similar range of species to Budworth Mere but is more prone to disturbance.

In winter, small flocks of ducks of various species frequent the flashes, although visiting flocks of Wigeon can number hundreds. Budworth Mere seldom supports very large numbers of wildfowl, but most of the common species occur regularly in winter; Goosander have become particularly common and a few Goldeneye are usually present. The large numbers of Canada and Greylag Geese that feed by the mere during the day should be checked for vagrants and often include a few stray

Pink-footed Geese. Garganey are recorded on the flashes in most years during spring and autumn. Shelduck also breed on the flashes and a Blue-winged Teal was present on Neumann's Flash in October 2021.

Grey Herons are seen daily and 20 or more pairs still nest at Bedworth Mere, although this is much reduced from the heronry's heyday. Bittern may be seen in winter and usually through to March in the reedbeds at Budworth Mere, and Water Rail is common around the flashes and with patience can show well from the hides. Little and Great White Egret are frequently recorded across the sites (mostly late summer and autumn), and Spoonbills and Cattle Egret more occasionally. Night Heron has occurred on Budworth Mere and Glossy Ibis has been seen a few times on the flashes, the best Cheshire site for the species.

A large number of Lapwings occur, with counts of 1,000–3,000 present in most years. Three-figure counts of Curlew are not uncommon at their flashes roosts or feeding by day in the fields north of Budworth Mere. As long as the weather remains mild, dozens of Snipe may feed on the flashes, but with the onset of frosts they disperse. Oystercatchers and Little Ringed Plovers breed and Avocets have made attempts to do so. Passage waders are frequently recorded in the right conditions, with a good variety including Black-tailed Godwit, Whimbrel, Ruff, Greenshank, Green and Wood Sandpipers, and occasionally Little Stint and Curlew Sandpiper. Coastal species such as Sanderling, Knot and Turnstone occur in some years. The list of rarities and scarcities has included Stilt Sandpiper, Temminck's Stint, Red-necked Phalarope, Long-billed Dowitcher and Black-winged Stilt, the last of which attempted breeding but failed to get any young to the fledging stage.

Black Terns appear over the mere when conditions are right in spring and autumn, and Arctic Tern flocks can occasionally be impressive on spring passage. Little Gull is also seen annually. Caspian Tern has visited on no fewer than four occasions, and other records include Cheshire's first and (at the time of writing) only American Herring Gull, Franklin's Gull, and White-winged and Whiskered Terns.

Barn and Tawny Owls are recorded frequently and breed locally. Visiting raptors include occasional records of Red Kite and Osprey, while Marsh Harrier, Hobby and Peregrine breed nearby and are seen more frequently, in addition to the more expected Kestrel and Sparrowhawk.

Scrubland species nesting in and around the flashes include Common Whitethroat and Garden Warbler, with a few pairs of Grasshopper Warbler in bramble thickets and Reed Warbler beside Witton Brook. Budworth Mere reedbeds hold breeding Reed Warblers, which visit from late April. During winter Stonechats may be present at Ashton's Flash (a Siberian Stonechat wintered in 2019–20), and Fieldfare, Redwing and Skylark flocks may be encountered in appropriate habitats. Kingfishers and Grey Wagtails also nest in the vicinity; the former may be seen dashing low over the water within a few metres of the mereside footpath, their approach heralded by a ringing call. Spring passage over and by the mere brings Sand Martins in early March in better numbers than at any other inland site in the county; three-figure counts are regular and counts of 1,000 were made in both 2023 and 2024. The woodlands fringing the park hold a good variety of woodland birds. Lesser Spotted Woodpecker is becoming vanishingly rare and breeding was last noted in 2019. The park is now one of the most reliable Cheshire sites for Hawfinch in invasion years.

From October to February there can be spectacular murmurations of Starlings at either the Coward Reedbed at Budworth Mere or at Witton Flashes, and often at both. These can number tens of thousands, with the wheeling flocks often attended

Cheshire

by Sparrowhawks and Peregrines, whilst even a wintering Bittern has been seen to catch some roosting in the reedbed.

10 WOOLSTON EYES

Woolston Eyes
Martinscroft, Woolston, Warrington WA1 4QH
SJ653887
incur.examine.late

Website: woolstoneyes.com
Opening times: All times, but strictly by permit, though access may be arranged when rarities present.
- Parking: Thelwall Lane alongside the north bank of the Manchester Ship Canal at Latchford Locks (WA4 1PD).
Permits: £20 for a single individual permit and £30 for a family permit
OS Landranger Map 109, OS Explorer Map 275/276
SSSI

Woolston Eyes, alongside the Manchester Ship Canal, is a remarkable nature reserve and Site of Special Scientific Interest (SSSI). Managed by the Woolston Eyes Conservation Group in agreement with the Manchester Ship Canal Company (MSCCo), this historic site serves as both a haven for wildlife and a functional area for canal dredging under a Waste Management Licence issued by the Environment Agency. Its name derives from the Anglo-Saxon word ees, meaning land near a

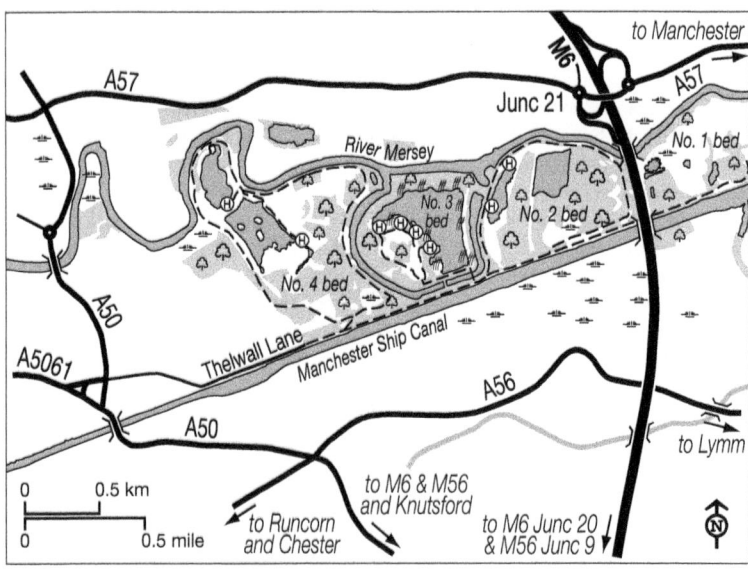

loop in a river, aptly describing its geography. Designated as an SSSI in 1986, Woolston Eyes was recognised for its importance to wintering wildfowl such as Teal, Shoveler and Pochard. Subsequent revisions acknowledged nationally significant breeding populations of Black-necked Grebe, Pochard and Gadwall. Woolston Eyes also supports a thriving population of amphibians, including Great Crested Newt.

SPECIES

Teal, Shoveler, Pochard, Tufted Duck, Water Rail, Snipe, Black-headed Gull, Kingfisher and Willow Tit can be seen throughout the year.

From December to February, wildfowl include Pintail and diving ducks. Pink-footed Goose are sometimes overhead. Short-eared Owl, Peregrine Falcon and Merlin occasionally hunt over the reserve. Magpie roost in scrub where a few Chiffchaffs often winter.

Black-necked and Little Grebes return between March and May. Breeding ducks take up territory, while winter wildfowl may linger and show signs of nesting. Reed Buntings appear from March and warblers arrive in April and May.

From June, wildfowl broods include Tufted Duck and Gadwall. The gull colony bustles with activity. Hundreds of Swifts may wheel overhead. Warbler song tails off in July, and Starling roosts in thousands.

Shoveler numbers increase in September and October, and Snipe return. Warblers depart during August and September when birds from outside the area pass through. Swallow roosts in early autumn are replaced by incoming thrushes in October. Goldcrest and Reed Bunting pass through in October. By November, large numbers of Teal have usually returned.

ACCESS

By train, Warrington Central and Bank Quay stations are 4km and 4.5km west of the reserve with the bus interchange directly opposite the Central station. Buses do run regularly to within 0.4km of Latchford Locks on the western side and close to Weir Lane on the north-east side from the interchange opposite Warrington Central Station. Anyone intending to visit the reserve using public transport is advised to check up-to-date information from Warrington Transport before setting out. The site can be accessed on foot by the vehicle route from Thelwall Lane. From the east it can be reached via a public footpath that runs from the end of Weir Lane across Woolston Weir up the hill and down to the Ship Canal track with the entrance to No.3 bed on your right. Please do not use Weir Lane to park vehicles as it restricts access for residents. A public path skirts No.2 Bed but viewing other than from hides is likely to disturb the birds and should not be attempted. Car access is via Thelwall Lane to the west of the reserve alongside the north bank of the Manchester Ship Canal at Latchford Locks. Use Sat-Nav Post Code WA4 1PD. Your reserve key will open the track barrier to reach the parking areas for No.4 bed and No.3 bed.

Permits allow visits to No.3 and No.4 beds 365 days per year. Permit prices are £20 for a single individual permit and £30 for a family permit (entitles any members of the family to visit) from *permits.woolstoneyes.com*.

YOUR VISIT

The area is divided into four distinct 'beds', each with unique ecological features. These beds are surrounded by steep embankments and bordered by the

Manchester Ship Canal and the River Mersey, both of which have benefited from improved water quality over time. This improvement has contributed to the reserve's rich biodiversity, especially its bird populations.

No.1 Bed: located east of the Thelwall Viaduct. This area comprises rough grassland and willow scrub interspersed with reedy pools formed by sand extraction. It supports dragonflies, damselflies, butterflies and a variety of birds, including Little Ringed Plovers, Reed Buntings and Swallows. Unfortunately, access to this bed is currently restricted.

No.2 Bed: periodically used for pumping dredgings, this bed transforms seasonally. When wet, it attracts wintering waterfowl such as Teal. The bed also hosts breeding warblers, Reed Buntings and Willow Tits, but access is limited due to the presence of invasive Giant Hogweed and restrictions by the landowner.

No.3 Bed: this bed is an island formed by the redirection of the River Mersey and is a focus of conservation efforts. The eastern and northern areas are shallowly flooded, with extensive reedbeds, while the rest features dense vegetation, including wildflower meadows and winter feed crops. These enhancements attract pollinators, finches and passage waders. Kingfishers and other aquatic species are frequently observed.

No.4 Bed: situated to the west of the reserve, this bed consists of willow scrub, nettles and rank vegetation. Woolston Eyes Conservation Group has leased part of this bed to develop wetlands with ponds, islands and scrapes, and a larger wetland project covering 80ha is under way.

Only the two westernmost beds – Nos.3 and 4 – are accessible to permit holders. This is because Nos.1 and 2 beds are still in use to accommodate dredging from the Manchester Ship Canal.

More than 240 species have been recorded at the Eyes, including all five European grebes, the three woodpeckers and four species of owl. The principal interest lies in the huge populations of breeding warblers and in the waterfowl that favour the shallow pools. Lying next to the Mersey, the flooded tanks attract large numbers of dabbling ducks that move inland from the estuary. The Mersey Valley Pochard flock also spends time here in winter. Up to 1,000 or more Teal are present at this season with hundreds of Mallard, Shoveler, Gadwall, Tufted Duck and Pintail. Other ducks occur less frequently or in smaller numbers and have included scarce Ferruginous, Ring-necked and Long-tailed Ducks. Green-winged Teal, Lesser Scaup, Red-crested Pochard, Common Scoter and Smew have been noted. In late autumn and winter, large numbers of Pink-footed Geese may fly over and have been known to alight briefly.

The extensive shallow marshes form excellent nesting habitat for wildfowl. Teal, Shoveler, Mallard, Pochard, Tufted Duck and Gadwall are all present through the summer, the last in increasing numbers; post-breeding flocks of over 1,000 are known. Garganey have lingered in some years with breeding proven. Little and Great Crested Grebes nest, and the site is rightly famous for a nationally important number of breeding Black-necked Grebes.

More than 100 singing Sedge Warblers and over 40 Reed Buntings are present in summer with smaller numbers of Reed and Cetti's Warblers. Water Rail may be heard squealing at any time of the year and may sometimes nest. Bearded Tit has bred and Marsh Harrier has become established. Emergent tussocks provide nest-sites for a large colony of Black-headed Gull, whose presence may attract a migrant Little or Mediterranean Gull in spring, the latter in sufficient numbers and regularity for display to be noted.

Hundreds of Swifts often feed overhead. Rank grass and brambles on the embankments hold Grasshopper Warblers and Whitethroats. Willow Warblers sing from taller scrub. Blackcaps and Willow Tits nest along the wooded stretches of riverbank, where Kingfishers are often seen and Sand Martins nest in the banks of the Ship Canal. The scrub and *Typha* on No.3 Bed hold huge roosts of Starlings, comprising thousands of birds. In most years a Swallow roost may contain hundreds of birds, occasionally even more. A Hobby often attends the roost and has provided some spectacular performances. From October to early winter, hundreds of Redwings and other thrushes may shelter by night. Mist-netting by day has revealed a significant passage of Reed Bunting and various other passerines through the Eyes, and an impressive number of warblers and Goldcrests passes through. The huge amount of cover means that rarities are usually first found by mist-netting. They have included multiple Penduline Tits, Blyth's Reed Warbler, Bluethroat (White-spotted), Little Bunting and White-crowned Sparrow. Chimney and Alpine Swifts have also been seen.

In winter, roosts dwindle as food supplies diminish, although Magpies seldom seem to go hungry and continue to roost in the scrub. One or two Chiffchaffs remain in the willow scrub during most winters, sometimes joined by a Siberian bird. Merlin is seen annually in very small numbers and on occasion a Short-eared Owl hunts over grassy areas. A Peregrine Falcon may visit from time to time.

The Eyes can be excellent for waders. Large numbers of Snipe and smaller numbers of Jack Snipe can be present as can hundreds of Lapwings. Green Sandpipers are regular, and Wood Sandpipers are seen most years. Rarer birds include Buff-breasted Sandpiper, Wilson's Phalarope, Cheshire's first White-tailed Lapwing and Glossy Ibis.

11 MOORE NATURE RESERVE

Moore Nature Reserve
Lapwing Lane, Penketh, Warrington WA4 6XE
SJ578855
bucket.formed.perky

Website: facebook.com/moorenr
Opening times: Accessible at all times
Parking: Lapwing Lane
OS Landranger Map 108, OS Explorer Map 265/175

Situated between the Manchester Ship Canal and the River Mersey, Moore Nature Reserve and has been managed as a nature reserve since 1991. Previously used for farmland and sand quarrying, the site now boasts five large lakes surrounded by extensive woodland, meadows and wetlands, offering a diverse range of habitats. The reserve also includes alder woodland, rough grassland, and pools fringed by reeds, alongside areas of arable land, coarse grassland and sparsely vegetated sandy ground. The mudflats of the River Mersey, the Manchester Ship Canal and a flooded sand quarry to the south provide additional habitat. Cared for by wardens from the Waste Recycling Group Limited, Moore Nature Reserve is open to the public year-round, with a network of paths and nine bird hides and viewpoints to make wildlife more accessible.

SPECIES

All year round, common waterbirds include Little Grebe, with resident woodland species such as Willow Tit and woodpeckers.

Winter wildfowl include Gadwall, Teal, Wigeon, Pochard and Tufted Duck. Flocks of finch, bunting and sparrow gather in weedy areas. There is also the possibility of Peregrine, Buzzard and Short-eared Owl. Lapwing, Golden Plover and loafing gulls can be seen on the Mersey.

Between March and May, Sand Martins arrive along the ship canal and Mersey. Passage Wheatears are on sandy areas. Breeding warblers arrive, and passage birds visit the woodlands and scrub.

During June and July, Reed, Sedge and Grasshopper Warblers are breeding, along with Whitethroat and Willow Warbler. Little Grebe and other waterbirds can be seen with their young.

Wildfowl return between August and November, including Shoveler. Warblers mingle with tit flocks in scrub. Lapwing flocks build up on the Mersey, where there may be additional wader species.

ACCESS

Buses from Runcorn and Warrington call at Moore. The closest stations are Warrington and Runcorn East, both around 6km from the site. By car, from Moore village take Moore Lane in a north-westerly direction, signposted to various haulage firms. At the end of this road cross the swing bridge over the ship canal into Lapwing Lane, taking the rough track that continues straight ahead, rather than following the surfaced road to the right. There is a car park immediately on your right. From here follow Lapwing Lane North until you reach further paths that

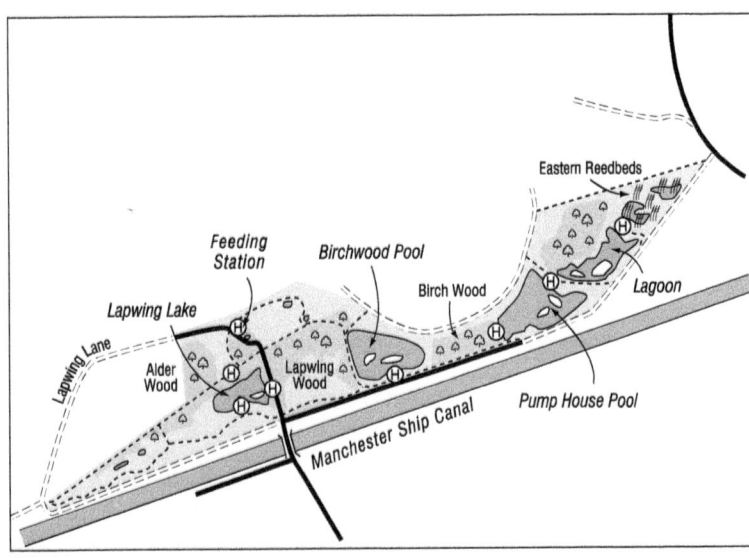

lead west along the bed of a dry canal – passing a screen from which Lapwing Lake can be viewed – or east towards the reedbed pool 2km distant. Alternatively, from the A56 at Lower Walton follow the main road north across the ship canal, taking the second turning on the left into Taylor Street then first right into Eastland Road. This continues as a track under the railway bridge where it is a short walk west to the reedbed pool (SJ5986). From Lapwing Lane footpaths lead west past the arable fields of reclaimed Norton Marsh and along a rough track parallel to the ship canal. After 1.5–2km view the Mersey mudbanks.

YOUR VISIT

The huge network of footpaths attracts many dog-walkers, who tend to flush wary birds off the western lake. Morning visits are advisable. However, this is an extensive area and quiet places can always be found to watch birds in scrub, woodland or reedbed and on the mudflats at Randell's Sluice, though the last is dependent on the state of the tide.

The alder woodlands have closed high canopies and attract a limited range of species including Willow Tit, for which this is one of the few reliable sites left in Cheshire, Treecreeper and all three woodpeckers, though Lesser Spotted did not breed in 2024, following a long decline; one bird held territory, though. In winter, large flocks of Siskins are generally present in the woodland. Reed Buntings feed on grass seeds in rough, open areas. In summer, Whitethroats are plentiful where scrub and rank grassland meet, while Grasshopper Warblers favour stands of bramble and willow herb. The reedbeds around the pools shelter Sedge and Reed Warblers in summer. Several pairs of Little Grebe nest, and though Black-necked Grebe no longer breed, they do still occur. Gadwall is seen with some regularity, but waterfowl are more varied in autumn and winter when small parties of Teal, Wigeon and Shoveler visit. Goosander, Pochard and Tufted Duck are more likely to be seen on the deeper western pool. The sand spit south of the ship canal attracts Tufted Duck. Goldeneye and infrequent Scaup, Long-tailed Duck and Smew have also been recorded.

To the west of the reserve, a bend of the River Mersey approaches to within 20m of the ship canal by Randell's Sluice. The river is estuarine here and low tide reveals areas of exposed mud. Flocks of up to 2,000 Lapwings with up to several hundred Golden Plovers are present through the winter. Other waders occur according to season and depending on tides elsewhere in the Mersey estuary. Up to 200 Dunlin and parties of Curlew visit. Redshank, Ruff and Black-tailed Godwit also turn up. Less common species occur occasionally. Predators such as Peregrine and Short-eared Owl sometimes wander by, attracted by the wader flocks along the river or small mammals in the rough grassland. Cormorants and Great Crested Grebes now fish in both the river and the ship canal – evidence of improving water quality from the industrial parts of the catchment area; Kingfisher may also be seen. Diving ducks occasionally visit from the lagoons of Fiddlers Ferry power station to the north of the river. Gulls visit the river and the pools to bathe. Glaucous, Iceland, Caspian and Ring-billed Gulls are seen from time to time, and there is a single record of Franklin's Gull. The closure of the tip has reduced the number of gull records. Arable fields attract Stock Dove, the occasional Tree Sparrow, Linnet and Yellowhammer. Fieldfare and Redwing feed on berries in the tall hedges.

Oystercatcher, Lapwing and Little Ringed Plover sometimes nest. Grey Heron breeds and Little Egret may be on the cusp of joining them. Shelduck can be seen prospecting for nest-sites in spring, while Gadwall, Mallard, Tufted Duck, and Little and Great Crested Grebes also breed regularly. Scarce visitors have included Yellow-browed Warbler, Bearded Tit and Great Grey Shrike.

12 FRODSHAM MARSH AND THE WEAVER BEND

Frodsham Marsh
Marsh Lane, Netherton, Overton, Frodsham WA6 7DZ
SJ511779
trace.kinds.rank

Website: birdingplaces.eu/en/birdingplaces/united-kingdom/frodsham-marsh
Opening times: Accessible at all times
Parking: It's best to park outside the marsh area along Marsh Lane or nearby roads, then continue on foot or by bicycle to the Weaver Bend and sludge tanks.
OS Landranger Map 117, OS Explorer Map 275

The Mersey Estuary supports internationally important populations of wildfowl and waders, but for the most part it remains inaccessible to the casual birdwatcher. A notable exception is Frodsham Marsh, a flat expanse of grazing marsh located between the M56 motorway and the Manchester Ship Canal. This area also includes the 'estuary' of the River Weaver, which flows into the canal, providing a rich and varied birdwatching habitat.

SPECIES
Lapwing, Redshank, Dunlin, Marsh Harrier, Little Grebe, Canada Goose and Shelduck are present throughout the year.

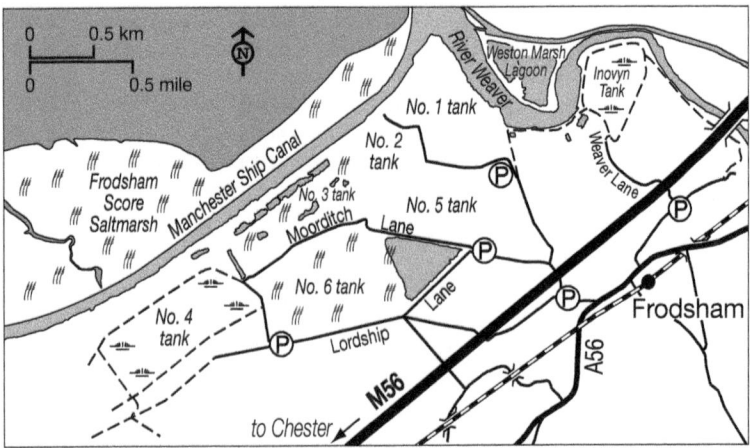

In winter, Wigeon, Teal, Pintail, Pochard, Tufted Duck, Goldeneye, wild geese and swans are present, along with scarcer waterfowl in frosty weather. Peregrine, Merlin and Short-eared Owl hunt the marshes with perhaps a Hen Harrier. Flocks of Lapwing and Golden Plover are often joined by Ruff, Black-tailed Godwit and other waders, with Snipe, Curlew and Water Pipit on damper ground. Stubble fields hold large numbers of pigeons, finches and, after snow, Skylarks. Fieldfares and Redwings can be seen in fields and hedgerows, Water Rail in reedbeds and Water Pipit in flooded fields.

Winter flocks disperse and raptors move back to breeding haunts in March and April, when Wheatears, pipits, wagtails and Reed Buntings appear with other incoming migrants. Golden Plovers remain into May in breeding plumage. Breeding Ringed and Little Ringed Plovers arrive. Passage waders include more Ringed Plover, Sanderling, Common Sandpiper, Ruff, Greenshank and others. Terns move through quickly. Garganey are likely from March onward.

In summer, Meadow Pipits and Grasshopper Warblers are around the tanks, while Sedge and Reed Warblers occupy the reeds. Marsh Harriers are breeding at this time. Return wader passage begins in July, when Curlew Sandpiper, Spotted Redshank, Greenshank, Ruff, Dunlin, Redshank and Curlew increase. There is a Grey Heron day roost.

Wader passage is best from August to early October with Dunlin, Ruff, Redshank, Black-tailed Godwit, occasional Curlew Sandpiper, Little Stint and Turnstone, Sanderling or a rarer species. Estuarine waders such as Ringed Plover and Knot are on tanks at high tides. Garganey are regular among Teal and other ducks in August and September. Black and other terns pass through, perhaps Little Gull. Hobby is likely at Swallow roosts in August and September. Warblers and wagtails move out. Water Rail increases in October, when there is a small passage of Rock Pipits, and a larger movement of Skylarks and Meadow Pipits. Winter thrushes and finches arrive in October and November.

ACCESS

Frodsham Station is 1.6km from the Weaver Bend and 3.2km from the far end of No.6 tank. The site lends itself to exploration on foot or by bicycle. There is a frequent bus service from Warrington, Chester and Runcorn. By car, leave the M56 at Junction 12 and follow signs into Frodsham. After passing beneath a footbridge, the road descends towards a set of traffic lights. To reach the Weaver Bend, turn right shortly before the lights into Ship Street and follow this lane until a bridge over the motorway appears on the left (SJ520785). Turn left over the bridge and continue along the unmade road until a gate bars the way. A footpath continues from here up to the bend. Once over the motorway, the roads on the marsh are private. For the main marsh and sludge tanks, continue through the lights in Frodsham and pass along the main shopping street before turning right down Marsh Lane. This lane passes over the motorway where, now a private road, it forks (SJ512779). Take the right-hand fork to reach the Weaver Bend. This track runs alongside then climbs the bank of No.5 tank. From here the path to the Bend is obvious. The left-hand fork leads along the southern side of the embankment of No.5 Bed; another fork climbs the bank to the right and skirts No.6 tank. This is a single-track road with passing places. Take care not to cause obstructions, or better still park outside the area and walk.

YOUR VISIT

Much of Frodsham Marsh is occupied by large sludge tanks, into which dredged material from the Manchester Ship Canal is pumped to drain. Over time, these tanks revert to pasture, but while still wet, they offer outstanding habitats for wildfowl and waders. The most productive area at present is No.6 tank, which is particularly suited to ducks in its open, flooded eastern section, while the western area attracts harriers. The newly developed No.3 mitigation area, located behind No.6, has already begun to draw large flocks of Lapwing and Golden Plover. The banks at the junction of tanks Nos.3, 5 and 6 provide an ideal vantage point for birdwatchers, with the secluded 'Phalarope' pool hidden in the south-west corner of No.3 and offering further opportunities for observation of No.2 and No.5 tanks. No.4 and No.5 tanks are grassed and house an onshore wind farm.

Established banks within the marsh develop a cover of scrub and rough grassland. The low-lying fields that make up the rest of the marsh are bordered by tall, berry-bearing thorn hedges and are drained by deep ditches containing stands of reeds. Some fields are cultivated for cereal crops, and the resulting stubble provides essential winter feed for many bird species.

The Weaver Bend is a 1.2km meander of the River Weaver, extending from the motorway bridge to the ship canal, where it flows into the estuary via the Weaver Sluice. Once a prime location for spotting rare American waders, bird records from this area have declined significantly over the past decade. Water levels now remain consistently high, reducing the extent of the muddy edges and banks preferred by waders. However, the Weaver Bend still attracts wildfowl in significant numbers and variety.

Frodsham Score lies to the north of the Manchester Ship Canal and consists of a closely cropped turf area bordering the tidal mudflats of the Mersey. Despite severe pollution from nearby chemical works, the estuary's mudflats support considerable numbers of Shelduck and waders. The saltmarsh at Frodsham Score is in a constant state of flux, with riverbank edges eroded by tides, frost, water, wind and rain. Large clumps of marsh periodically collapse into the river, where they are broken down into silt and transported by the tides. This natural cycle results in the creation of sandbars or the rebuilding of riverbanks elsewhere. The process embodies the organic nature of the Mersey marshes, a continuous cycle of erosion and regeneration. Views of Frodsham Score are limited to a 2km stretch west of the Canal Pools, where the Score's banks slope down to merge with the open, flat expanse of the Mersey marshes. This vantage point provides birdwatchers with the opportunity to observe the natural ebb and flow of the marshland environment and the diverse array of bird species it supports.

At the time of writing, there are plans to turn over much of the dry, flat areas, including all of No.2 tank, to a solar farm, which would be mitigated by the creation of a tailor-made wetland reserve on No.3 bed.

In winter, a wide range of wildfowl visit. Tufted Duck and small numbers of Pochard feed near the junction between the Weaver and the ship canal. Scaup are seen occasionally, and Goldeneye are regular. Scarcer species may occur, especially in severe weather when the flow of the river keeps the Weaver Bend open, long after Cheshire's meres have frozen. These have included Smew, Long-tailed Duck and Ring-necked Duck as well as good numbers of Great Crested and Little Grebes and Cormorants. Scarce waterbirds do turn up in other seasons: a Lesser Scaup was present on No.6 tank for a month in late spring 2014; Red-necked Grebe, and with increasing frequency, Black-necked Grebe are seen on No.6 tank. Dabbling ducks frequent the saltmarsh but do also visit No.6 in significant numbers. Pintail, Gadwall, Wigeon and Teal are regular. Mallard and possibly Shoveler are thought to breed, and in recent years Garganey, too, have been seen in suitable habitat in the right season. Frodsham is one of the more reliable places in Cheshire to see this charismatic little duck. Green-winged Teal is recorded about every other year and American Wigeon has been seen twice. Shelduck are numerous on the Score and a few pairs breed.

A flock of Canada Geese has taken up residence and sometimes contains feral Barnacle Geese. A few thousand Pink-footed Geese are becoming increasingly regular in winter, both overhead and grazing on the marsh, and occasionally very small numbers of White-fronted or Brent Geese occur. Mute Swan has bred, and small numbers are frequent on the marsh. Bewick's Swan is now less than annual, but Whooper Swan numbers are increasing, with up to 45 seen on the marsh.

Frodsham is rightfully most famous for its waders, with a remarkably high percentage of the species on the British list being recorded here. It is not uncommon to see 23 species in a day. Before it grassed over, No.5 tank was arguably one of the best places in the country for rare waders, with the Weaver Bend running it a close second. In recent years, No.6 tank has been the best location. There are good birds to be seen at almost any time and plenty of habitat to cover beyond the Weaver Bend and No.6 tank. The highest tides are likely to push waders off the Mersey and at such times the sludge beds should be checked carefully. Consult tide tables for predicted high tide times.

The Mersey holds one of the largest concentrations of Dunlin in the country, with five-figure counts frequently reported from Frodsham Score. Grey Plover, Curlew and Oystercatcher are also recorded in high numbers, and high tides will result in thousands of birds flying into the sludge tanks to roost. Depending on the season, a wide variety of other species may join them. Traditionally coastal waders such as Turnstone, Bar-tailed Godwit, Sanderling and Knot are most regular in autumn and winter. Curlew Sandpiper and Little Stint are recorded in most years, especially on autumn passage. Common Sandpiper is a regular passage visitor, with double-figure counts not uncommon. Green Sandpipers, too, are seen regularly; they tend to stay farther up the Weaver towards the motorway bridge. Greenshank and Spotted Redshank are frequently recorded and Wood Sandpiper is more or less annual. More than 1,000 and sometimes in excess of 4,000 Black-tailed Godwits can be seen, especially on No.6 tank in early autumn, and Ruff are often present with them. Large numbers of Snipe can be seen in the wet areas, with more than 100 present in winter along with smaller numbers of Jack Snipe. Avocet, Ringed Plover, Little Ringed Plover and Oystercatcher all breed locally.

Up to 1,000 Golden Plover roost on the sludge tanks and feed on the fields or Score along with similar numbers of Lapwing. The Golden Plover may still be present in late April, by which time they are showing the extensive black on the underparts that characterise the northern subspecies *altifrons*. The wind farm mitigation area on No.3 tank has not so far led to an increase in plover numbers. Frodsham Marsh has attracted American Golden Plover and Dotterel in the past.

Frodsham's list of vagrant waders is remarkable: Baird's Sandpiper, Black-winged Stilt, Broad-billed Sandpiper, Buff-breasted Sandpiper, Collared Pratincole, Great Snipe, Lesser Yellowlegs, Long-billed Dowitcher, Marsh Sandpiper, Semipalmated Sandpiper, Sharp-tailed Sandpiper, Stilt Sandpiper, Terek Sandpiper and multiple records of all three phalaropes. Pectoral Sandpipers are seen every couple of years, a little more frequently than Temminck's Stints.

Terns pass through on spring and autumn passage. Common Terns are regular, Arctic Terns are scarcer and Black Tern may be reliably seen following easterly winds. Little Gulls are also recorded regularly and like the terns may linger over the Weaver or flooded sludge-beds, picking food items from the surface. There have been several White-winged Black Terns, two Gull-billed Terns and a Whiskered Tern, and whilst gull-watching might not be the first thing on a visitor's mind, Ring-billed and Franklin's Gulls have been recorded.

The open views and concentration of prey species make Frodsham excellent for raptors. Several times a day the alarm calls of Lapwings and the excited whistling of Golden Plovers signal the approach of a predator. Peregrine is a regular visitor, as is Merlin in winter. Short-eared Owl is a passage migrant, with most birds now favouring the Stanlow area further west, following the completion of the wind farm. Long-eared Owl is now very rare in the area. Kestrel, Sparrowhawk, Buzzard and Marsh Harrier breed on the marsh, whilst Barn Owl has bred at Marsh Farm. Hen Harrier is seen most years, Osprey is approximately annual, and both Hobby and Red Kite are seen with increasing regularity. Montagu's Harrier, Honey Buzzard and White-tailed Eagle have also been recorded, the last spending two weeks on the Score in July 2023. One of the more remarkable features of Frodsham is the number of Ravens: the highest count to date is more than 300.

The flooded, reedy ditches and tanks attract such southern marshland species as Spoonbill. As well as Little Egrets, double-figure counts of Cattle Egrets and multiple Great White Egrets are increasingly regular. Grey Herons breed on No.6 tank, and it may not be long before one of the egrets does too. Bittern is a rare winter visitor, but Water Rails are fairly common and Spotted Crake has been recorded occasionally. Cetti's Warblers are abundant, and in spring they are joined by Sedge and Reed Warblers. In winter, Water Pipits are reliably recorded in wet meadows, most frequently from Lordship Lane.

In winter, the stubble fields attract flocks of Woodpigeon, Stock Dove and finches, including Chaffinch, Brambling, Greenfinch, up to 300 Linnet and occasionally Twite. In snowy weather hundreds of Skylarks from farther inland gather on the stubbles. Cold weather may drive in thousands of Fieldfares and other thrushes, which strip any remaining berries off the tall thorn hedges before being forced to feed on the ground. Released Grey and Red-legged Partridges may be seen. The hedges and tank margins attract warblers and chats in migration periods, and all common migrants may occur. Grasshopper Warblers may hold territory in bramble patches and Whitethroats are everywhere. The open spaces and viewable perching spots make Frodsham the best non-coastal site in the county to see Whinchat, Wheatear, Ring Ouzel and Common Redstart on passage, while Stonechats breed. There have been three Black Redstarts since 2018, four Great Grey Shrikes since 2003 and both Red-backed and Woodchat Shrikes. Cuckoos are regular. On overcast late autumn days with an easterly breeze, southward passage of pipits and Skylarks may be heavy, and small parties alight to feed beside the tanks. Reed Buntings may also be encountered at such times. Rarities have included three Aquatic Warblers, Savi's Warbler, Rose-coloured Starling and Barred Warbler.

13 HALE HEAD

Hale Lighthouse
Lighthouse Road, Hale, Halton L24 4AZ
SJ471809
goals.splashes.handicaps

Within Way
Manor Farm, Hale, Halton L24 4BZ,
SJ471821
insurers.nurtures.sprayed

Carr Lane Pools Town Lane
Hale, Halton L24 4AQ
SJ471828
plugs.skill.tailing

Pickering's Pasture
Mersey View Road, Widnes, Halton WA8 8LL
SJ488835
elaborate.types.things

Website: thefriendsofpickeringspasture.org.uk
Opening times: Manor Farm in Hale is open daily from 10am to 5pm, closing for Christmas and New Year; Carr Lane Pools has footpaths accessible year-round, though vehicle access is restricted; and Pickering's Pasture is open 24/7 as a LNR.
Parking: Limited parking at the southern end near Hale Lighthouse (SJ473814); Pickering's Pasture LNR has a car park via Mersey View Road (SJ484838).
OS Landranger Map 108, OS Explorer Map 275
LNR, SSSI, SPA

Hale Head is a low sandstone bluff that juts out into the Mersey estuary from the north bank opposite Frodsham Marsh and offers an alternative viewing point for estuarine species. Approach via a minor road south from Hale Village. There is limited parking at the southern end. From here a track leads down to a disused lighthouse. Hale Head, now in Cheshire, was the southernmost point in the historic county of Lancashire. A lighthouse was established here in 1838; the original octagonal structure was superseded by a taller cylindrical tower in 1906.

SPECIES

From December to February, wildfowl and shore waders are present, especially on the Mersey where there is also a gull roost. Great White and Cattle Egrets may be seen, while finches and buntings feed on the arable fields. Water Pipit occurs some years on Hale Marsh.

Between March and May, Garganey, Wheatear, Whinchat and other passage chats and warblers pass through, along with wagtails and occasional Ring Ouzel. Little Ringed Plover return, and passage waders may be seen. Herons and Cormorants commute to nesting sites at the duck decoy.

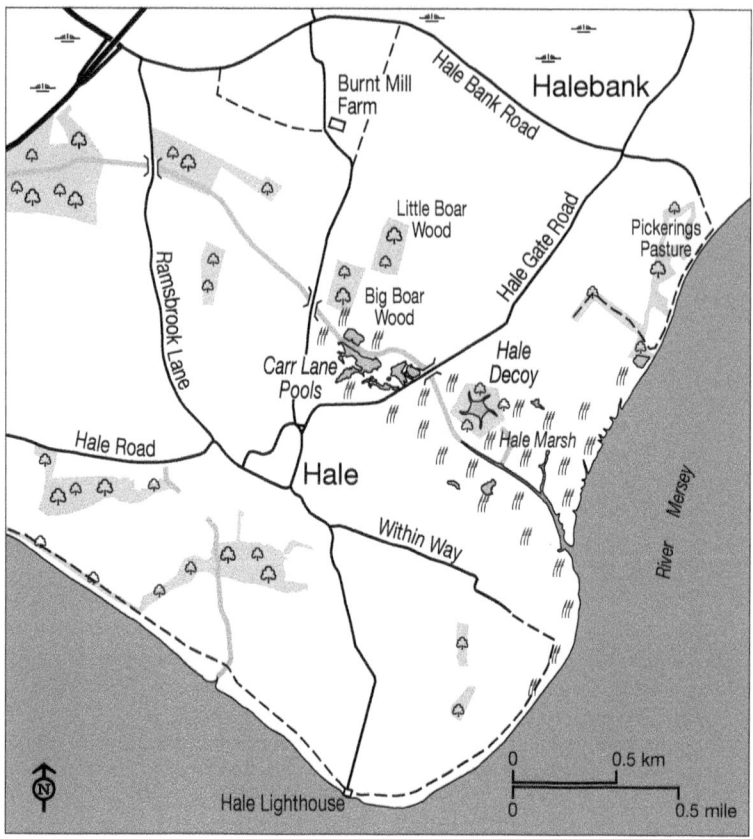

During June and July, returning waders are seen. Herons and Cormorants commute to the Duck Decoy. This time of year offers the best chance of spotting Spoonbill or Quail.

From August to November, Little Stint, Curlew Sandpiper and other waders pass through. Visible migration occurs from September, especially in south-easterly or southerly winds, involving thrushes, Woodpigeon, Buntings, including Lapland, and finches, including Hawfinch.

ACCESS

The 38, 45, 82A and C45 buses pass through Hale village and Hale Bank. The closest train station is Halewood, about 6km from Hale Lighthouse. Hale Point is approached by a minor road that leads south from Hale village. There is limited parking at the southern end (SJ473814). From here a track leads to the lighthouse. Footpaths follow the coast east and west. Pickering's Pasture LNR has a car park, accessed via Mersey View Road. The Mersey View public house on Hale Gate Road is a useful landmark to locate the correct road (SJ484838).

YOUR VISIT

Hale Marsh and Pickering's Pasture LNR are located to the east of Hale village between Hale and Widnes. Halegate Marsh, which covers 345ha, is flooded on spring tides, which reach up to the road on such occasions. There is no access onto the marsh itself, which can be viewed from Withins Way from Hale village, from the road between Hale and Halebank and from the hide at Pickering's Pasture LNR. From the bottom of Within Way, an outcrop of rocks at the edge of the saltmarsh attracts various waders and ducks and is always worth checking out. Paths leading to Hale Lighthouse and along Hale shore can be muddy and uneven. Nearby, Pickering's Pasture LNR, a former landfill site, has been reclaimed and an excellent wildlife area created in its place. The reserve has wide views across the River Mersey and offers great opportunities for birdwatching. The wildflower meadows make a colourful sight throughout spring and summer and boast many species, including orchids. The reserve has large areas of woodland and shrubbery. Oak, elm, Alder, Wild Cherry and Sycamore are common species, with Guelder Rose, Alder Buckthorn, hawthorn, Blackthorn, Hazel and Dog Rose providing lookouts and hiding places for smaller birds. The rangers and friends of the pasture have put up nestboxes to encourage breeding on site. A small feeding station is situated by the scrape hide.

A feature of the marsh is the duck decoy, one of a small number still in existence in the UK. The decoy was an ancient method of catching ducks for food, consisting of a pond where ducks swam underneath a series of hoops and nets, attracted by a small dog (piper) running along the bank. It is thought that the ducks regarded the dog as a predator such as a fox, and swam towards it as a method of drawing attention away from nesting or resting areas. Access is restricted. The Friends of Hale Decoy Group organise occasional public visits, though places are limited and need to be booked in advance. The site now hosts a significant heronry with more than 80 pairs of Little Egret in 2024. Surrounding the decoy is a saltmarsh, which provides feeding and roosting sites for waders and wildfowl. There are two hides on the decoy, one looking over the inner pond of the decoy, the other onto the marsh.

Carr Lane Pools is the name given to a set of shallow flooded freshwater scrapes enclosed by Carr Lane, Town Lane and a narrow tributary of the Mersey, Ramsbrook. Please note that the pools are on private land and can only be viewed from the road. There are two main areas to view the pools. The first is on Carr Lane itself and provides good views of the pools especially in the afternoon and evening when the sun is behind you. The second viewing area is from a small gated area on Town Lane (park in Curlender Way and cross the road). This provides good views of the main scrape and is usually the most productive for viewing passage waders.

The area around Burnt Mill Farm and Carr Lane is mostly arable, with a mix of crops and set-aside depending on the time of year. The local farmers have preserved a lot of the natural hedgerows and regularly put seed down during the winter periods. An open horse paddock to the north of Burnt Mill Farm is grazed throughout the year and provides a unique habitat. Further south along Carr Lane (towards Hale village) there are two woods (Big Boar's Wood and Little Boar's Wood), which are made up of a mix of deciduous trees and hold most of the expected woodland species.

Hale Head is a favoured crossing point of the wide Mersey Estuary and can be excellent for overhead passage in autumn. Dawn visits are essential to witness visible migration, and mornings are best for passage migrants.

In winter there are good numbers of Teal and Wigeon, with smaller numbers of Shoveler and Gadwall on Carr Lane Pools. A significant flock of Canada Geese on Hale Marsh occasionally attracts scarcer species, and with the help of a telescope, Whooper Swan and much more occasionally Bewick's Swan can be seen on the other bank of the Mersey (Frodsham Score). In spring, Garganey are regular and can turn up from mid-March on Carr Lane Pools. Rare and scarce wildfowl have included Green-winged Teal, Bean and Brent Geese on the pool and marsh, and assorted sea duck on the Mersey.

Winter is the best time to see Great White Egrets feeding on the banks of the Mersey or the marshes, or roosting at the duck decoy, but egrets can be seen all year. The decoy holds a large Cormorant colony, more than 80 pairs of Little Egret in 2024 and smaller numbers of Grey Heron. There have been double-figure counts of Cattle Egret in recent years. Scarcer species have included Spoonbill in most springs in recent years, Glossy Ibis and Crane on occasion, and Bittern on three occasions since 2000.

Wader numbers and variety can be spectacular in any season. Common and Jack Snipe can be seen on Carr Lane Pools and Hale Shore in winter. Although waders are present at low tide, about two hours before high tide produces the best opportunity to see waders at closer quarters. Golden Plover regularly roost on the flats, whilst Little Stint occasionally winter. From late April the rocky shore under Hale Lighthouse can provide refuge for large numbers of passage waders, often including Whimbrel, Sanderling, Little Stint, Curlew Sandpiper and impressive numbers of Dunlin and Ringed Plover, while Carr Lane Pools hold Little Ringed Plover, Black-tailed Godwit, Ruff and Dunlin. You have a good chance of picking out Little Stint, Curlew Sandpiper, Spotted Redshank, Greenshank, Bar-tailed Godwit, Knot, Wood Sandpiper or something scarcer. Timing visits around high tide on the Mersey can result in larger numbers of roosting waders. Summer can be quiet, but from early August, return wader passage starts to pick up with often even larger numbers of arctic waders passing through. Carr Lane Pools are worth checking for waders, though water levels are sometimes too high. Scarce passage waders are recorded on Carr Lane Pools with some regularity, including a remarkable hat-trick of Temmick's Stint, Pectoral Sandpiper and Wood Sandpiper together in 2023. Broad-billed and White-rumped Sandpipers have been recorded on Hale Shore, and though still being assessed by the BBRC, a Hudsonian Godwit was present in 2024.

Hale Shore and the coastline towards the Runcorn-Widnes Bridge offer the best opportunity to see scarce gull species. Little Gull is seen on passage and both Sabine's and Bonaparte's Gulls have been recorded. Yellow-legged Gull can be seen reliably and Caspian Gulls are present here more frequently than any other site in the county. Wintering Glaucous, Iceland and Kumlein's Gulls have also been recorded.

Winter is the best time to see raptors – either on Hale Marsh or more distantly on the opposite shore. Short-eared Owl, Merlin, Kestrel, Sparrowhawk, Marsh Harrier, Buzzard and Peregrine are the most regular, but Hen Harrier may be seen and Barn Owl is present. From Pickering's Pasture up to three Peregrines can be seen roosting on Runcorn Bridge. Hobby is recorded with some regularity in summer, and Ospreys are seen annually on passage. Rarer raptors include Honey Buzzard and White-tailed Eagle.

The hedges around Hale Head provide good cover for passerines, and the arable fields and paddocks are always worth a look. Stonechat can be seen all year round, whilst migrant Wheatear, Whinchat and Redstart are seen on passage. Ring

Ouzels are more likely in early spring but can be seen in autumn too. Good numbers of White and Yellow Wagtails and occasional 'Channel' and Blue-headed Wagtails can be seen, for example around Carr Lane Pools or on Hale Marsh. In winter both areas have been reliable sites for Water Pipit in the recent past, though records have been more intermittent since 2021.

The fields around Burnt Mill Farm can be productive for winter bunting flocks, which will often contain Corn Bunting, a county scarcity, as well as Yellowhammer and Reed Bunting. Corn Bunting has been lost as a breeding species, though it may occur on passage. A few Grey Partridge still breed. Corncrake and Quail have also been recorded on the Hale Head. Big Boar's and Little Boar's Woods will often hold good-sized finch flocks with Brambling, Siskin and Redpoll regularly recorded, alongside the more expected woodland species. The woodland along Carr Lane and at Pickering's Pasture can be productive, with both Spotted and Pied Flycatchers and Redstart possible alongside the more expected migrants. Yellow-browed Warblers are seen in some autumns.

Visible migration at Hale is exceptional, although good weather conditions and an early start are a necessity – otherwise birds will be too high to see or hear. Peak conditions occur from mid-September through to early November when there is a light south to south-easterly and bright conditions. Then, up to 20,000 birds can be seen moving south. Scarcer species can often include Hawfinch and Lapland Bunting, with large numbers of other finches, thrushes and Woodpigeons moving through.

14 DELAMERE FOREST

Delamere Switchback
Ashton Road, Delamere WA6 6NG
SJ550718
thudding.knee.titles

Blakemere Station Road
Delamere WA6 6NX
SJ556711
grounding.machine.firms

Hogshead Wood
Hogshead Lane, Delamere CW8 2ET
motorist.gurgled.searches
SJ548703

Nunsmere
Tarporley Road, Sandiway CW8 2ET
SJ590690
brief.huddling.residual

Abbots Moss
Whitegate Way, Whitegate CW8 2EA
SJ594687
chatting.beam.unusable

Cheshire

Petty Pool and Pettytpool Woods
Dalefords Lane, Marton, Sandiway CW8 2BN
SJ612694
disbanded.ballparks.widely

Newchurch Common
Whitegate Way, Littler, Whitegate CW7 2BF
SJ605688
backyards.barks.encoder

Opening times: All these sites are generally open year-round, with most being public footpaths or nature reserves accessible at any time.

Parking: Generally unrestricted along Forestry Commission ride

OS Landranger Map 117/118, OS Explorer Map 267

SAC, Ramsar

Geologically, Delamere Forest is largely composed of sand and gravel deposited during the last glacial period. It developed into a mix of sandy dry heaths, with sphagnum-filled boggy hollows and stands of pine trees. Though agriculture and conifer plantations have subsequently removed much of this botanical diversity, Abbots Moss is a superb example of a floating 'schwingmoor' bog and is home to several delicate and specialised plants and wildlife, including 150 species of spider and scarce Odonata such as Downy Emerald and White-faced Darter. Efforts are being made to restore other areas of bog within the area. Delamere's natural meres and lakes were formed by glaciers retreating after the last ice age, when large blocks of ice became detached from the glacier and melted, forming depressions in the ground which filled with water. These are augmented by flooded sand quarries with peripheral beaches and reedbeds.

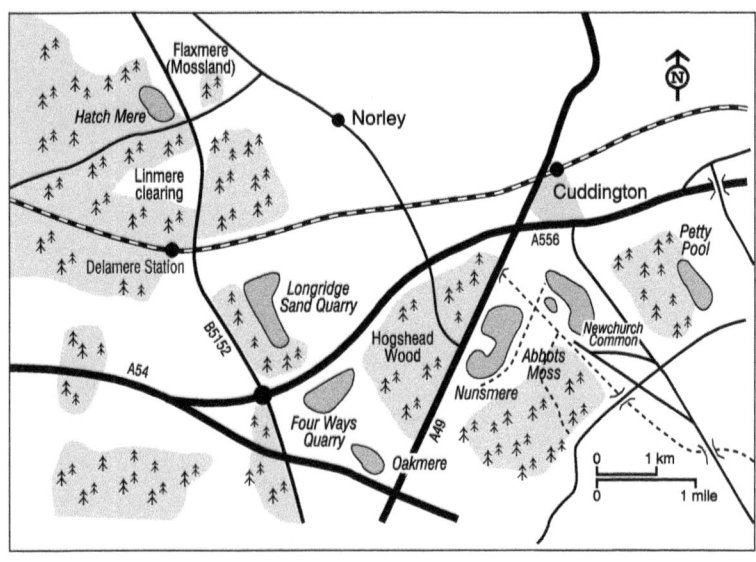

SPECIES

Throughout the year, Sparrowhawk, Buzzard and Kestrel can be seen passing by day, while Tawny Owl is present at night. Woodland species include Green and Great Spotted Woodpeckers, Nuthatch, Treecreeper, Long-tailed and Coal Tits and Goldcrest. Redpoll and often Bullfinch feed in birches, while Crossbill and occasionally Siskin are found in pines. Stock Dove and Corn Bunting occupy arable areas, and Little Grebe and Tufted Duck are seen in quarries and pools.

From December to February, Teal, Pochard and Goldeneye gather on the water, while Common and Jack Snipe can be found by the quarries and pools. Bramblings may feed under beeches.

In spring, from March to May, Brambling and plentiful Siskin occupy the mature conifers, and crossbills may have young. Summer migrants, including warblers, move into young plantations.

During June and July, Grebes and Tufted Ducks may have young. Mediterranean Gull may appear in Blakemere among the Black-headed Gull colony. Young plantations hold Whitethroat and possibly Grasshopper Warbler, as well as Linnet, Redpoll, Reed Bunting and Yellowhammer. Local post-breeding Crossbills may be joined by migrants; Tree Pipit, Redstart, Pied Flycatcher may have young.

From August to November, summer migrants depart and plantations become quiet; waders may drop in by quarries, where gulls may gather.

ACCESS

Delamere station is a good central place to explore from by bike. The 82 bus between Chester and Northwich has numerous stops along the A556. By car, the forest area straddles the A54 Chester to Winsford and A49 Warrington to Whitchurch roads. Access to the plantations is generally unrestricted along Forestry Commission rides. See below for specific site details

YOUR VISIT

Though the mature plantations support relatively few species, Crossbills are seen occasionally and may breed. Siskin breeds in very small numbers, but is far more numerous in winter and especially in spring. Bramblings can be numerous, with up to 300 joining Chaffinches on Newchurch Common in 'big' years. Goldcrest and Coal Tit can be found at all seasons, while Blackcap, Willow Warbler and Chiffchaff are widespread in summer. Pied and Spotted Flycatchers also breed in small numbers. Otherwise, typical woodland species may be found, though Wood Warbler is only a scarce passage migrant, and both Lesser Spotted Woodpecker and Willow Tit have also been lost as breeders.

New plantations and heathland areas attract migrant Tree Pipit and Redstart, which breed sparsely. Willow Warbler is abundant, along with Whitethroat, and there are smaller numbers of Garden Warbler. Yellowhammer and Cuckoo can still be heard singing, especially on Newchurch Common and both Shemmy and Abbots Mosses. Following a long absence, Nightjar has bred in recent years in the remaining heathland areas around Primrose Hill. Hobbies breed locally in some years, so it's worth keeping an eye open in the wetter areas, where they may be seen hunting for dragonflies. The lakes and sand quarries may attract passage waders or terns, while Little and Great Crested Grebes breed and small numbers of wildfowl may be seen.

The part of the forest closest to Delamere station attracts large numbers of day-trippers, especially on sunny days, but more remote areas always remain relatively free from disturbance. Morning or evening visits in spring are best for birdsong. Dusk visits in summer might reveal some of the nocturnal species. Evening visits to the sand quarries in winter may produce an unusual gull attending a pre-roost gathering.

DELAMERE SWITCHBACK
Take the B5152 Frodsham road north from the A556. Pass over the railway bridge and continue for about 1.5km before turning left onto a minor road just before Hatchmere (SJ552713). From this road, tracks lead north and south into the plantations. Explore this area for woodland birds: Nuthatch, Treecreeper and Green Woodpecker all year; Siskin in winter and spring, Crossbill best from November to March; Redstart and Pied Flycatcher.

BLAKEMERE
A peaty area of wet heath has now developed into a large lake that holds a sizeable colony of Black-headed Gulls and in some years a pair or two of Mediterranean Gulls.

HOGSHEAD WOOD
This area of young plantations to the west of the A49, opposite Nunsmere, tends to be less disturbed than other parts of the forest and supports good numbers of scrubland species. Bulldozed areas are particularly attractive for butterflies. Part of the wood has been taken for a new quarry, which remains dry at the time of writing.

NUNSMERE
This disused quarry is adjacent to the A49, 0.8km south of the A556 crossroads. The banks are overgrown with birch scrub. Garden Warbler, Bullfinch and Redpoll nest; Green Woodpecker feeds in more open areas. Breeding waterbirds include grebes and Tufted Duck despite the occasional water-skiing. Winter wildfowl are few in number but have included Long-tailed Duck.

ABBOTS MOSS
A forestry track runs east from the A49 immediately to the south of Nunsmere. Walk down the track, turn left at the end and then bear right where a striped pole bars the road. This ride leads through conifer plantations and between two floating bogs ringed with Alder and birch trees. Green Woodpecker and Tree Pipit nest, as well as Redstart in some years. Crossbill, Siskin and Sparrowhawk are possible at any time of year. Hobbies have been seen.

PETTY POOL AND PETTYPOOL WOOD
Travel east along the A556 from the A49 crossroads and turn right shortly after at another set of lights. After 2km, at the end of a line of large houses, a footpath leads into the plantations. Parking is not possible immediately by this entrance, so park elsewhere and walk back. Rides lead through the plantation to the pool. Siskin is plentiful in spring, and a few Wood Warblers appear on passage. Rhododendrons by the pool hold roosting finches, including Brambling, in winter.

NEWCHURCH COMMON
Continue south along the minor road past Pettypool Wood and take the first on the right, Sandy Lane. Turn right again at the end. Parking is limited but the road widens at SJ609686 (brotherly.manicured.cracking). A bridle path runs between the two quarries, which are now largely hidden by trees and scrub containing nesting Linnet and Yellowhammer. Goldeneye, Tufted Duck and Pochard favour these pools. There are a few Teal in winter, when Jack Snipe may occur in the reedmace. Large numbers of gulls may also be present in winter. Scaup and Common Scoter have occurred. Stubble fields in the vicinity hold finch flocks in winter. Green Woodpeckers feed in areas of short grass. Follow Whitegate Way, which passes the southern end of the smaller pool, west from Newchurch Common to view the pool, which has breeding Little Grebe, Tufted Duck and scarce dragonflies. From here it is possible to follow paths through to Abbots Moss and Nunsmere.

15 BEESTON CASTLE AND PECKFORTON HILLS

Beeston Castle car park
Chapel Lane, Beeston CW6 9TR
SJ540590
claw.eyelid.upholds

Peckforton Hills
Waste Hill Road, Higher Burwardsley CH3 9PD
SJ522564
educates.courage.erupt

Bulkeley Hill Viewpoint
Stone House Lane, Bulkeley SY14 8BQ
SJ524550
longer.propelled.sandpaper

Bickerton Hill
Allmans Lane, Duckington, Brown Knowl SY14 8LH
SJ497528
unafraid.usages.verve

Opening times: The car park is open year-round.
Parking: Beeston Castle has car parks at the foot of the outer wall; Burwardsley Woods has limited parking on a no-through lane; Bickerton Hill has a National Trust car park by a roadside pond and additional signposted access points along Goldford Lane.
OS Landranger Map 117, OS Explorer Map 268
SSSI

Cheshire

The Peckforton Hills, a striking sandstone ridge, offer a rich tapestry of historical, geological and ecological significance. Stretching from Frodsham to Malpas as part of the Mid-Cheshire Ridge, these hills include notable high points like Bulkeley Hill (224m) and Peckforton Point (203m). The ridge's Triassic sandstone strata, laid down around 245–250 million years ago, bear witness to ancient geological processes and glacial influences, including the Irish Sea Ice Sheet.

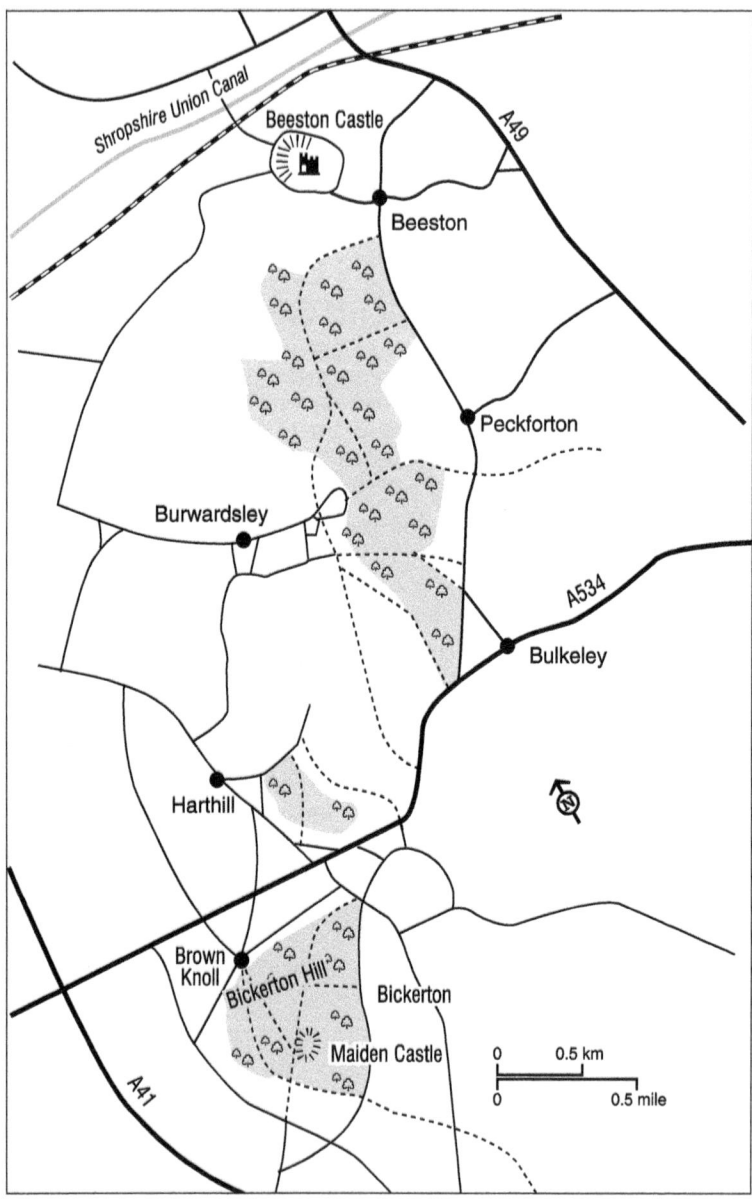

Beeston Castle, perched on a nearby hill and once a royal stronghold, incorporates remnants of an Iron Age hillfort and echoes centuries of tumultuous history. In contrast, Victorian-era Peckforton Castle stands as the last fortified home built in the UK. The hills, crisscrossed by the Sandstone Trail, are a haven for biodiversity, with areas such as Peckforton Woods designated as Sites of Special Scientific Interest and recognised for their ancient woodlands and grasslands.

SPECIES

Throughout the year, Buzzard, Peregrine, Raven, woodpeckers, Marsh Tit and general woodland birds are present. Redpoll and Bullfinch, among other scrubland birds, can be seen on the heaths.

From December to February, territorial Marsh Tits may join mobile flocks of resident woodland birds.

In spring, Siskin and Redpoll are in larches, and Brambling may frequent pines. Raptors display, and spring migrants include flycatchers and warblers. There is a chance of migrant Ring Ouzel in April and Nightjar in late May.

Breeding species are less conspicuous by July. Stock Dove and other birds feed on bilberries at Bickerton.

Woodland is quiet from August to November. Migrant warblers join tit flocks.

ACCESS

The nearest bus stops are in Tattenhall and Tarporley about 6km away. The nearest train stations are at Nantwich (16km) and Chester (19km). Beeston Castle (SJ537593) can be seen from afar, lying just to the west of the A49 south of Tarporley. Minor roads are well signposted and there are car parks at the foot of the outer wall. English Heritage manages the castle; there is an admission charge. The keep is very popular with day-trippers but provides an excellent platform from which to view raptors. Minor roads skirt the Peckforton Hills on either side and there is a well-marked network of public footpaths. Burwardsley woods are best approached from the west. Follow signs for the Candle Workshops into the village but turn left past the Pheasant Inn, opposite a telephone box (SJ523565). This is a no-through road and parking is difficult. The lane deteriorates into a track that soon passes an area of old oak woods on the right. Explore the footpaths through the woods. Bickerton Hill is best approached from Bickerton village, just south of the A534. Take Goldford Lane, which runs beside the churchyard. There are signposted access points at various points along the lane. The entrance to a National Trust car park is beside a roadside pond. Footpaths and bridlepaths cross the heath and woods. An area of oak woodland at the south-west corner of the hill may be rewarding.

YOUR VISIT

The tops of the hills contain some heathland areas, but most of the summit of Peckforton Hill has been planted with conifers. There are considerable tracts of oak woodland on the steeper slopes. Birch and bracken are spreading extensively across the remaining open areas. Cattle have been allowed to browse on Bickerton Hill to stem the loss of heathland to scrub. Bulkeley Hill has some old chestnut and oak trees, but is largely covered by birch woodland. Climbing Corydalis is a characteristic plant of the woodlands and Common Cow-wheat is locally plentiful.

The older tracts of oak are particularly rich in woodland bird species, with a notable concentration of Pied Flycatchers, especially in the Burwardsley woods. Marsh Tits and a few Common Redstarts are also present, though Wood Warbler is probably only a passage migrant now. All three species of woodpecker formerly bred, but Lesser Spotted Woodpecker has not been recorded since 2015. Pied Flycatcher and Wood Warbler may also be seen within the grounds of Beeston Castle. Warblers often join mixed flocks of tits in autumn, but viewing can be difficult in the dense birch scrub. Siskins have nested in the conifers along the ridge and Crossbills visit.

Bickerton Hill, in particular, supports a few pairs of Tree Pipits, and Ring Ouzels are seen with some regularity on passage. Nightjars continue to be reported on passage in May or June, and much suitable habitat remains for this species. Cuckoos are still regular. Garden Warblers and Spotted Flycatchers breed in many of the wooded areas. The isolated nature of the hills attracts migrant raptors. Peregrine and Raven both nest; in fact, Peregrines are known to have nested here in medieval times. Goshawks display in the area occasionally and Hobbies breed. Buzzards can usually be seen somewhere along the ridge. Tawny and Barn Owls are recorded regularly, and a few Little Owls are still seen.

Trippers invade many accessible sites in Cheshire at weekends, not least Beeston Castle, but the view across the Cheshire Plain towards the Dee and Mersey estuaries makes a visit worthwhile, especially with additional incentive of the chance to see a head-level Raven or Peregrine. It is usually possible to find relatively quiet footpaths on the Peckforton Hills. Spring visits are optimal for woodland species.

16 THE RIVER DEE

Overleigh Cemetery
Cemetery Lodge, Handbridge, Chester CH4 7HW
SJ403652
mutual.banks.focal

Chester Meadows
Grosvenor Park Terrace, Newtown, Chester CH4 7BB
SJ414661
nails.mason.improving

River access, Eccleston
Paddock Road, Eccleston CH4 9HP
SJ414621
rate.crawled.reception

River access, Churton
Knowle Lane, Churton CH3 6LG
SJ416564
blaze.wanting.immune

Opening times: Accessible at all times
Parking: Park in the village or on Ferry Lane.
OS Landranger Map 117, OS Explorer Map 266
SAC

The meandering River Dee forms the Wales–England border from the south-west corner of Cheshire at Shocklach north to Aldford, where the boundary swings west around Eaton Park. Willows overhang the river and exposed sandy banks on the inside of bends provide good areas for rare and scarce dragonflies and damselflies. This section of the river is prone to flooding, and flooded arable land and pasture can attract wildfowl from the estuary.

SPECIES
Throughout the year, Grey Heron, Canada and Greylag Geese, Mandarin Duck, Kingfisher, Nuthatch, Buzzard and Raven can be seen. In winter, from December to February, the Cormorant roost reaches its peak. Feral geese may attract wild relatives. There are Mandarin Duck and Kingfisher on the river. Floods may attract Pintail, Wigeon and other wildfowl. In spring, from March to May, woodpeckers drum near the river in early morning. By June and July, Mandarins may have young, and geese are now moulting. From August to November, Cormorant roost reassembles, and geese become more mobile.

ACCESS
Buses from Chester stop at Aldford and Churton. The nearest railway station is Chester, approximately 5km from Eccleston village and without any steep inclines. By car take the A413(T) south from the A5(T) roundabout SJ390625. Turn left after 0.8km onto a minor road signposted to Eccleston. Park in the village or on Ferry Lane, from which footpaths follow the riverbank in either direction. Eaton Park is strictly private, but interrupted views of the park lake can be obtained by following the path south for a little over about 2km. Viewing becomes more difficult when the trees are in full leaf. Geese may be visible at times from the Aldford to Huntington road (B51310) near the Crook of Dee (SJ425615). Alternatively, from Churton village take the lane beside the White Horse public house and walk down the farm track to the river. A footpath follows the meanders of the riverbank to north or south.

YOUR VISIT
The riverbank footpath at Eccleston attracts its share of human visitors. This is a very pleasant walk at all seasons. The riverside path from Churton is best walked in frosty weather, especially when inland meres and flashes are frozen. Farmland is likely to flood at any time following heavy, prolonged rain, or when a quick thaw melts snow off the Welsh hills higher up the Dee catchment.

Small-leaved Lime trees, rare in Cheshire, grow on the riverbanks between the Iron Bridge and Eccleston. The village of Eccleston lies on the west bank of the river Dee. It is characterised by lichen-rich sandstone walls and a mixture of mature evergreen and broadleaved trees. Historically, Hawfinch frequented the area, and in invasion years the Overleigh cemetery in the southern suburbs of Chester can still attract this stunning finch.

The mature timber of Eaton Park houses many hole-nesting birds: Mandarin, Stock Dove and Jackdaw at the larger end of the scale and Nuthatch, Lesser Spotted Woodpecker and Marsh Tit at the smaller end, though the last two are declining so fast they are likely to be extinct within the next few years. From autumn until early spring, the lakeside trees at Eaton hold one of the county's main inland Cormorant roosts. Several dozen birds are regularly present. A smaller roost is established in riverside trees near Churton. Many Cormorants fly out to the Dee

Cheshire

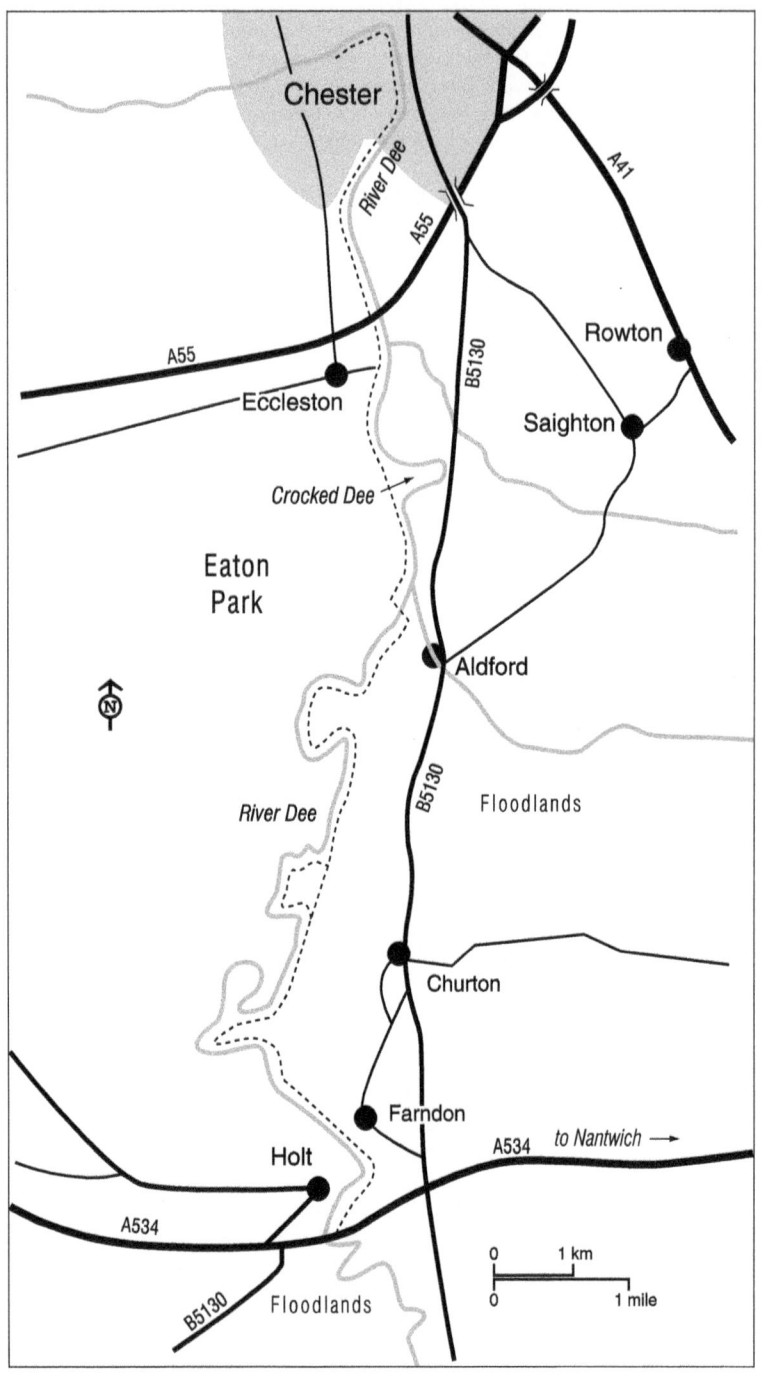

estuary to feed and can be seen commuting over Chester. Winter waterfowl include occasional small numbers of Goldeneye, Pochard, Tufted Duck and Goosander. Mandarin may be seen in winter on tree-lined stretches of river, particularly northwards from Farndon: counts are usually in single figures but up to 40 have been seen at Eaton Hall, with a maximum of 137 in December 2011. Large numbers of Greylag and Canada Geese are regular, with occasional Pink-footed Geese.

When the river floods following heavy rain or after a rapid thaw of snow on the Welsh hills, flocks of up to several hundred Pintail, Teal, Shoveler and other ducks fly in from the estuaries to flooded farmland upstream of Farndon or along Aldford Brook. Snipe can also be recorded in large numbers. As the days lengthen, warblers and other migrants appear in riverside trees, and Common Sandpipers skim low across the river. In the early morning, Nuthatches pipe from the treetops and woodpeckers drum on hollow branches. There is a large heronry in the secluded woods of Eaton Park. The birds radiate to feed in marl pits, by rivers or beside the saltmarshes of the Dee estuary. Mandarins may be seen with ducklings along quieter reaches of the river. Kingfishers breed locally and may be seen at any season.

Careful observation along the banks of the Dee, a large river with a north–south orientation, might produce surprises at migration time. Common and Green Sandpipers follow the riverbanks, and with the onset of autumn parties of Siskin and Redpoll appear in the Alders. Ravens are seen with increasing frequency and have nested on Chester Town Hall. Peregrines also visit regularly and are sometimes seen roosting on electricity pylons in winter.

17 BURTON MERE WETLANDS

RSPB Burton Mere
Puddington Lane, Burton, Neston CH64 5SF
SJ314735
desktop.unfounded.lipstick

Website: rspb.org.uk/days-out/reserves/
dee-estuary-burton-mere-wetlands
Opening times: 9am–dusk, though check the website for opening times of the visitor centre. Entrance charges apply for non-RSPB members, with adults charged £7.00 and children (5–17 years) £3.50. RSPB members and children under 5 enjoy free entry. Fully accessible.
Parking: RSPB car park
OS Landranger Map 117, OS Explorer Map 266
SPA, Ramsar, SSSI

Burton Mere Wetlands (BMW), managed by the Royal Society for the Protection of Birds (RSPB), is a vibrant nature reserve located on the Dee Estuary, straddling the border between England and Wales. The reserve encompasses a diverse mosaic of freshwater wetlands, mixed farmland and semi-ancient woodlands, providing a haven for a wide variety of wildlife throughout the year.

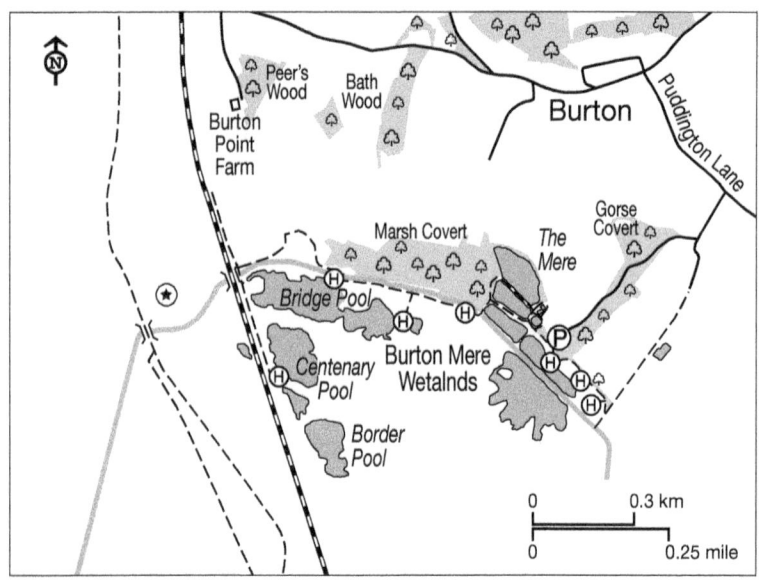

SPECIES

Shelduck, Shoveler, Gadwall, Teal and Tufted Duck are present throughout the year, alongside Water Rail, Redshank, Lapwing, Black-headed Gull, Black-tailed Godwit, Marsh Harrier, Peregrine, Kestrel, Water Rail, Kingfisher, Cetti's Warbler, Reed Bunting, Skylark, Grey Heron and all three egret species.

In the winter months, Water Pipit are on wet grassland areas where they may be hard to see. Pink-footed Geese and large numbers of ducks are common, with wild swans appearing occasionally. Ruff and Spotted Redshank are visitors, and Merlin, Hen Harrier and Short-eared Owl hunt the marshes. Avocets arrive in February.

Some wildfowl depart in March and April but Garganey are likely. Black-tailed Godwit numbers increase and passage stints and sandpipers are present in April or May. Marsh Harriers display. Easterly winds in May can bring marsh terns in good numbers.

During early summer, small numbers of Pintail, Wigeon and Teal breed. Small numbers of non-breeding waders and 100–200 Avocets are present. Reed Warblers feed young, and the heronry is a hive of activity. Return wader passage also gets going.

Wader passage in August may bring Spotted Redshank and Greenshank in large numbers. Rarer waders are most likely in September and October when wintering raptors also start to return. Spotted Crake is almost annual in August or September.

ACCESS

The nearest train stations are at Neston 5.5km and Hooton 7km. The 487 bus route goes from Liverpool to the Botanical Gardens at Ness, which is 5.5km from the reserve. The Burton Marsh Greenway is a scenic cycle route to Burton from Chester and Wales and links with the Wirral Way in Neston. The reserve is sign-posted off Puddington Lane, Burton, accessed via the A540 and 10 minutes from the M56.

YOUR VISIT

The origins of Burton Mere Wetlands trace back to the late nineteenth century when the area, initially tidal mudflats, was reclaimed during the construction of the Wrexham to Bidston railway line. The reclaimed land was subsequently used for grazing, duck shooting and, later, arable farming. In 1986, the RSPB purchased part of this land, establishing the Inner Marsh Farm reserve, which opened in June 1992 with three freshwater pools and a hide for birdwatching. Over time, the reserve expanded with additional land acquisitions, including Burton Marsh Farm in 2006 and Burton Mere Fisheries in 2008. These expansions facilitated the creation of new wetland habitats and visitor facilities. In September 2011, the reserve was officially reopened as Burton Mere Wetlands, marking a significant milestone in its development. The reserve is renowned for its rich biodiversity, particularly its avian inhabitants. Seasonally, visitors can observe species such as Avocets, egrets, harriers, Redshanks, Swallows and Swifts. The wet grasslands, lagoons and scrapes provide ideal habitats for wading birds and wildfowl. In spring, reedbeds come alive with warblers, while Grey Herons, egrets and Kingfishers are frequently seen hunting for food. The ancient woodlands are a spectacle in themselves, with bluebells carpeting the forest floor in spring and various fungi emerging in autumn.

The visitor centre provides panoramic views across the wetlands and houses informative displays about the reserve's wildlife and conservation efforts. There are extensive nature trails which meander through diverse habitats, including wetlands and woodlands, offering excellent wildlife-viewing opportunities. Several hides and viewing screens are strategically placed for birdwatching, allowing visitors to observe wildlife without causing disturbance.

The reserve is designed to be accessible, with flat routes from the parking area to the entrance, level access to the reception, toilets and information points, and wheelchairs available for hire. On-site parking is available, including designated spaces for visitors with disabilities, located within about 50m of the main entrance. A new café within the reserve offers a range of hot and cold food and drinks for visitors to enjoy. Throughout the year, the reserve hosts various events and activities aimed at educating and engaging visitors about the wildlife and ongoing conservation work. Burton Mere Wetlands stands as a testament to successful habitat restoration and conservation, offering visitors a chance to immerse themselves in the natural beauty and wildlife of the Dee Estuary.

The Dee was historically famous for its wild geese. These were lost to excessive disturbance from shooting. The progressive protections from successive purchases in the Burton area and of the establishment of nearby Gayton Sands RSPB reserve mean that numbers of wildfowl, including geese, wintering on the Dee Estuary have risen markedly. More than 10,000 Pink-footed Geese now winter in the area and these attract smaller numbers of scarcer species, especially European White-fronted Geese and more rarely Greenland White-fronted Geese and Tundra Bean Geese. The provenance of rarer geese is hard to prove, and a Lesser White-fronted Goose in 2023 was thought not to be wild. However, lone and unapproachable single Snow Geese which arrive with Pink-feet relocating from Norfolk, rather than joining the large flock of feral Greylag and Canada Geese, have better credentials. A large herd of Mute Swan, often feeding out on the Dee Marshes, is joined by wild swans in winter. Bewick's Swan numbers have been declining over the last 10 years and now the species can be very hard to see, with low single-figure counts only since 2020. The default wild swan is now Whooper and around 20 are present.

Exotic Ruddy Shelduck and Egyptian Geese are often present alongside the larger numbers of native Shelduck, which breed and winter out on the Dee saltmarsh in large numbers.

Enormous numbers of ducks, especially dabbling ducks, winter on the Dee Marshes and may visit in large numbers. An impressive range of species breed, including Teal, Shoveler, Gadwall, Tufted Duck, occasionally Garganey, Wigeon and Pintail. American Wigeon and Green-winged Teal are found with some regularity. Garganey are annual. Scaup, Lesser Scaup and Ring-necked Duck have been seen mingling with the Tufted Duck and occasional Pochard, and other diving duck species have been recorded. The reed-fringed scrapes hold Water Rail all year and these have bred. The site is also the most reliable in the county for Spotted Crake. A huge heronry in the wooded areas contains hundreds of occupied nests, including one of the largest Little Egret colonies in the country. Double-figure counts of both Great White and Cattle Egrets have become almost commonplace. Bittern is recorded in winter and there is a recent breeding record. Glossy Ibis and Night Heron have also been recorded and the heronry has attracted Spoonbills.

Unsurprisingly, Burton Mere Wetlands is a superb site to see waders, especially ones that prefer feeding in shallow water such as *Tringa* species. Black-tailed Godwits are more or less ever-present, with a non-breeding flock over-summering. Four-figure counts are frequent. In 2024, the flock attracted a Hudsonian Godwit. Ruff and Spotted Redshank too can be present every month. Redshank, Avocet, Oystercatcher and Lapwing breed locally. The usual range of scarce passage waders occurs, including Little Stint and Curlew Sandpiper, while Pectoral Sandpiper and even Long-billed Dowitcher could be described as rare but expected visitors. Other rarities have included Black-winged Pratincole, Black-winged Stilt, Lesser Yellowlegs, Marsh Sandpiper, Terek Sandpiper, Broad-billed Sandpiper, Wilson's Phalarope and Red-necked Phalarope.

There is a substantial, and noisy, Black-headed Gull colony, which may contain very small numbers of Mediterranean Gulls. Common Terns breeding nearby are regular visitors, and other tern species have turned up, including all three marsh terns and Gull-billed Tern. Marsh Harriers breed and can be seen at any season, Hen Harriers visit frequently from autumn to early spring and Hobby is seen reliably in summer and early autumn. Short-eared and Barn Owls are also fairly regular.

The extensive reedbeds and ditches offer breeding sites for Reed, Sedge and Cetti's Warblers. Bearded Tit has bred in the recent past. The woodland areas hold the expected species, though Wood Warblers are passage migrants only. Whitethroats and a few Lesser Whitethroats have territories in the many hedges and scrubby areas. Water Pipits may be seen from October until early spring, and diurnal passage migrants – especially thrushes and finches – pass over in autumn, following the Dee. A surprising number of scarce passerines and near passerines have been recorded, probably as a combination of this passage line and the very high level of coverage that the site receives. Savi's Warbler, Wryneck, Alpine Swift, Barred Warbler and several Yellow-browed Warblers have also been recorded.

Cheshire

18 INNER DEE MARSHES TO HESWALL INCLUDING PARKGATE

> **Dee Estuary Nature Reserve – Parkgate**
> Old Baths car park, North Parade, Parkgate, Cheshire CH64 6RL
> SJ292777
> tides.rising.wildlife
>
> Website: rspb.org.uk/days-out/reserves/dee-estuary-parkgate
>
> Opening times: Nature reserve open 24 hours; car park open 8am–5pm in winter and 8am–8pm in summer. Entrance is free; donations welcome.
>
> Parking: Shared public car park
>
> OS Landranger Map 117, OS Explorer Map 266
>
> SPA, Ramsar, SSSI

This section encompasses a stretch of the Dee Estuary that is rich in biodiversity. It serves as a crucial habitat for a variety of bird species, especially during migration periods.

SPECIES

Marsh Harrier, Shelduck, Redshank, Oystercatcher, Little and Great White Egrets, Stonechat and Reed Bunting are present year-round.

From December to February, Chaffinch, Brambling, Greenfinch, Linnet and Goldfinch gather along the marsh edge, joined by occasional Twite. Hundreds of Skylark and Reed Bunting are on the marsh; Rock and a few Water Pipits in creeks. Lapland Bunting may be seen in winter but usually only when the marsh is flooded. Snipe, Jack Snipe and Water Rail may be flushed into view by a high tide. Pink-footed Geese, a few wild swans and large numbers of Pintail, Teal and Wigeon are on the marsh. Spotted Redshank and Greenshank are occasionally seen. Peregrine, Merlin, Hen Harrier and Short-eared Owl hunt the marshes, with Kestrel and Sparrowhawk.

Wildfowl mostly depart in March. Wader flocks disperse a little later. A sprinkling of stints and shanks pass through in April or May. Summer-plumaged Water Pipit are present in March and early April.

A few Pintail, Teal and Shoveler linger along with small numbers of non-breeding waders between June and July. Common Terns fly out from Shotton to fish the Dee; Grasshopper, Reed and Sedge Warblers in Neston. There have been Quail in some years.

Wader passage in August may bring Spotted Redshank and Greenshank in large numbers. Wildfowl, Hen Harrier and Short-eared Owl return between August and November.

ACCESS

Neston Station (SJ292777) is in cycling range of all the access points, and buses pass along Parkgate promenade.

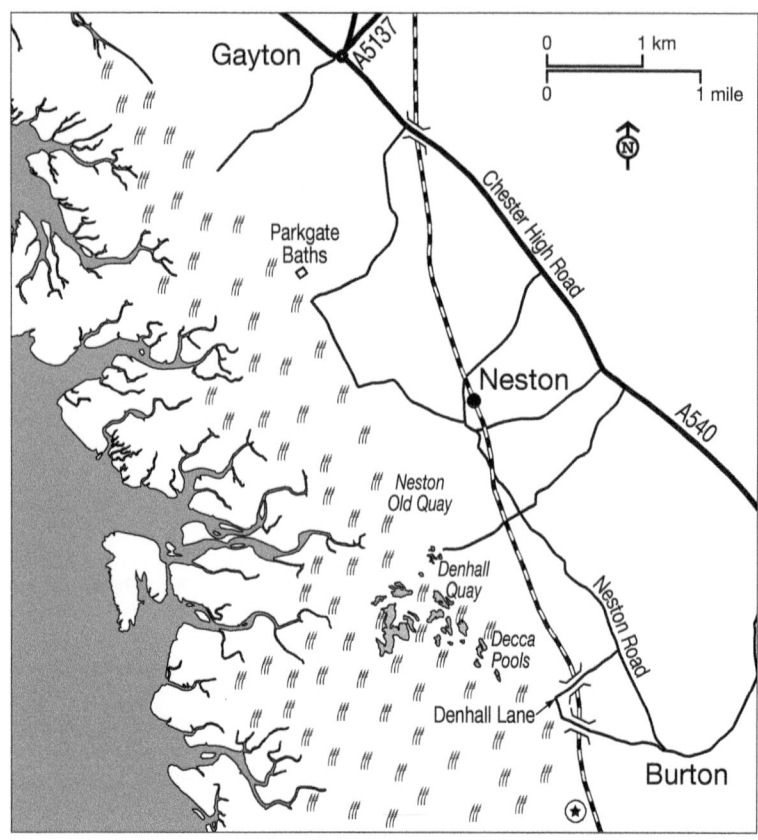

YOUR VISIT

The Dee Estuary has been silting up throughout recorded history. King Henry II left Shotwick Castle with a large army to invade Ireland in 1171. An expanded quay at Shotwick was built in 1278 but is now 2.5km from the water. New quays were built at Burton in 1322, Denhall in 1541 and Neston in 1555, but the last silted up before it was finished! Between 1735 and 1736 a new cut was made, canalising the Dee between Chester and Connah's Quay on the Welsh coast, and much of the Dee flood plain was reclaimed. Tidal creeks and gutters persist within the saltings, and in places shallow brackish pools have formed. Sheep heavily graze the upper parts of the marshes, but elsewhere saltmarsh plants such as Sea Aster, Scurvy-grass and Sea Purslane flourish. The marshes now extend from beyond the Cheshire border at Burton Point seaward as far as Heswall. The RSPB manages much of the saltmarsh, which is internationally important for waders and wildfowl, and protects a large part of the estuary. The whole of the marsh from Burton Point to Gayton is now owned by the RSPB. Pools in this vast marshland attract waders such as Redshank and Lapwing to breed. These are joined by many other waders and wildfowl in winter. From Neston northwards the character of the marsh changes because it is not grazed.

The Wirral Way, a disused railway track, runs parallel to the coastline north-west from Neston. It is lined with scrub and passes through farmland, which includes stubble fields in winter. To the south of Burton Marshes and separated from the estuary by the Burton to Wrexham railway embankment is an area of flat fields which attract wintering geese and wild swans as well as waders such as Lapwing and Golden Plover.

For most of the year the vast Dee Marshes can be viewed distantly from what was once the English shore. Several pools and viewpoints are described below. Once or twice a year, one of the most impressive wildlife spectacles in the north-west occurs when tides at or above astronomical high tide (the highest tide to occur under normal meteorological conditions) flood the outer marsh, forcing thousands of small mammals, mainly voles and shrews, towards the historic shore, which in turn attract a gaggle of predators trying to take advantage of the situation. Wildfowl, raptors and owls, normally far out on the marshes, are concentrated into a narrow strip of land. Normally secretive birds such as crakes and rails are obliged to seek refuge, sometimes under parked cars, and uneasy truces, for example between Weasels and mice, are struck up as they share dry refuges. The optimum location to view this spectacle will vary with the final tide height. A further benefit is the establishment of vast drift lines of saltmarsh vegetation and temporary pools, which attract waders, wet meadow-loving passerines, seed-eaters and insectivores when the decaying vegetation results in an invertebrate hatch. Viewing locations from south (least likely to flood) to north (most likely to flood) are listed here, along with access and species highlights.

The marshes may appear quiet between tides, but activity increases as the river channel floods and waders are forced onto the marsh to roost. Raptors and owls then become more active. It takes a 10m tide (at Liverpool) to flood the marsh. The peak of the tide happens earlier at the north end of the estuary. The large car park at the north end of Parkgate promenade can be standing room only, and even if the tide does not make it all the way to the wall, wildfowl, waders and raptors will be much closer. For the very best experience the tide needs to be the first in a sequence to flood the saltmarsh and displace the small mammals, though the birds will repopulate the marsh between the tides. It is possible to start at Gayton, then, if the tide is going to be a big one, go to Neston Old Quay or Denhall Quay for a second chance. In addition to checking the tide tables, it is important to monitor the predicted storm surge at ntslf.org/storm-surges/surge-model/latest-surge-forecast. A few centimetres higher or lower can make all the difference, and in exceptional circumstances, a 9m tide can flood unexpectedly. For example, Storm Eunice in February 2022 flooded at Neston, but not at Parkgate, because the later tide at Neston coincided with a 1.2m surge – 0.5m higher than the surge at Parkgate. Evening visits to the Old Baths can be good for harriers and Barn Owl. Prolonged easterly or south-easterly winds in May can produce passage harriers or perhaps even an overshooting southern migrant. September or October may bring scarce and rare waders to the pools.

Burton Point and Burton Marsh
Burton Point: Burton, Cheshire CH64 5SB
SJ302735, gurgling.bypassed.device
SSSI, SPA, Ramsar

A foot and cycle path can be accessed from the bottom of Station Road (SJ301745) or from the Welsh side (SJ309718). Burton Point itself is a sandstone outcrop which allows elevated views. The area is particularly good for Wheatear, Whinchat and Cuckoo in spring. The swan herd, which may contain Whoopers in winter, is most likely to be found towards Weighbridge. Burton Marsh, an area with *Phragmites* and Sallow bushes, has a track record of attracting wintering Bittern and breeding Grasshopper Warbler. The many fence posts are worth scanning for birds of prey and owls. Subalpine Warbler and Richard's Pipit have been recorded here.

Denhall Lane
Denhall Lane: Denhall, Cheshire CH64 5SB,
SJ300747, data.cornfield.elevates
SSSI, SPA, Ramsar

It is possible to park in the lay-bys at the end of Denhall Lane and view the marsh (SJ300747). There are similar species to elsewhere in the area, but after heavy rain or a huge tide, pools can form here and attract wildfowl, Spoonbills and waders. This was once the premier location for rarities on the Dee Estuary. The rarest bird there since 2010 was a Buff-bellied Pipit, which took up residence in the tidal wrack after a particularly high tide in 2013.

Decca Pools
Decca Pools: Denhall Lane, Ness, Neston, Cheshire CH64 0TG
SJ294753, grit.perfumed.speeded
SSSI, SPA, Ramsar

A foot and cycle path (Marsh Lane) leads north from the corner of Denhall Lane (SJ300747) towards Net's Cafe. It is also possible to walk south from Denhall Quay Ponds. The area between the track and Ness Gardens has attracted Osprey while the hedges attract migrants. The pools, named after the former wireless telegraphy masts that once stood there, are permanent freshwater pools in a bend of Denhall Gutter. It is traditional to view the pools (SJ293753) from the bank to get a bit of height. The fields have attracted Quail, and the pools themselves can hold duck and waders, with Long-billed Dowitcher, Lesser Yellowlegs and Wilson's Phalarope recorded in the past. Otherwise, birds are similar to other local sites.

Denhall Old Quay
Denhall Old Quay: Denhall Lane, Ness, Neston, Cheshire CH64 0TG
SJ294753, escalated.scoping.walkway
SSSI, SPA, Ramsar

Denhall Quay is accessed from the B5134 to the west of the A540. From Neston take the minor road south for Burton (signposted Ness Gardens); after 0.8km turn right (west) at a mini-roundabout at a modern church onto Marshlands Road. Follow road to the end. Turn left onto Quayside when the marsh is reached. Denhall Old Quay is 300m on the right at the end of the public road. There is an excellent view out over the marsh and a stand of trees that can attract passerines. In winter this is one of the best places to see Short-eared Owl and harriers close without the need of a flood tide. If there is a big tide, the road access can be submerged, but there are worse places to be cut off than sitting outside the Harp public house, watching harriers and pipits. With all the tidal gutters flooded, significant numbers of Rock Pipits (presumed to be migrant Scandinavian *littoralis* birds) may be seen. Do not enter the marsh – there is no need. Richard's Pipit and Pallid Harrier have been seen here.

Neston Old Quay, Neston Reedbed and Neston Sewage Farm
Neston Reedbed: Clayhill, Parkgate, Neston CH64 6TS
SJ282772, interacts.personal.bleaching
Neston Old Quay: Dee Marsh, Little Neston CH64 0SY
SJ285767, pretty.forwarded.crossings
Neston Sewage Farm: Laundry Lane, Little Neston, Parkgate, Neston CH64 6TF
SJ287768, drilled.picnic.reframe
SSSI, SPA, Ramsar

The walk from Denhall Quay to the main access point to Neston Reedbed at Moorside Lane (SJ283773) passes some excellent birding habitat. A stream enters the marsh at SJ285767, in the middle of the Old Quay, but it is very prone to disturbance from dog walkers. Get there early and there is a chance of Water Pipit or perhaps Lapland Bunting. The hedges in the area are excellent for migrants. Neston Sewage Farm is excellent for wintering Chiffchaffs and sometimes a Siberian bird joins them. The site, or more accurately the boggy field next to it (SJ287768, unhappy.candidate.ideas), is the best location in the county for Water Pipit. A footpath to the rear of the sewage farm leads from Old Quay Lane, though a parking space is all but non-existent (SJ288770). Neston reedbed is a large *Phragmites* reedbed adjacent to the Old Quay. It suffered a major fire in March 2022 but is recovering and has breeding Marsh Harriers and typical reedbed species.

Parkgate Promenade, Donkey Stand Flash and The Old Baths
Donkey Stand Flash: The Parade, Parkgate CH64 6SQ
SJ279780, looks.copper.summer
Parkgate Old Baths: North Parade, Parkgate, Neston CH64 6RL
SJ273789, until.monkey.grew
SSSI, SPA, Ramsar

Donkey Stand Flash SJ276786 – a scrape created opposite where the traditional donkey rides set out from for visiting Victorian families – is excellent for waders but is challenging to park at other than very early in the morning. There is a car park at the north end of the promenade (SJ273789), which makes it a very popular spot for birders on the highest spring tides. There is a winter harrier roost off the Old Baths. Tides flood here and north of here more regularly than at the previous sites and there is tidal debris more frequently. In turn, this attracts finches, mostly Linnet, but occasionally Twite visit. The fields inland can attract finches and thrushes in winter.

Gayton
Cottage Lane, Heswall, Wirral CH60 8NU
SJ265800, upper.drop.tame
SSSI, SPA, Ramsar

There is limited parking at the end of Cottage Lane, which is a short walk or ride from the Wirral Way. A footpath from there leads back towards Parkgate, and this allows views over the saltmarsh and the golf course that runs alongside the estuary. The golf course can be good for Wheatears and finches, but otherwise there is a similar range of species to other inner Dee sites. The attraction of Gayton is that it floods before the more popular site of Parkgate, and it is possible to walk along the path with the incoming tide.

19 THE OUTER DEE: HESWALL TO WEST KIRBY

Wirral Country Park Visitor Centre
Wirral Way, Caldy, Wirral CH48 1QX
SJ238834
ranks.punk.squirted

Heswall
Riverbank Road, Heswall, Wirral, CH60 4SQ
SJ262805
pepper.mining.badge

Banks Road
Heswall, Wirral CH60 9JS
SJ254814
bared.gross.models

Target Road
Heswall, Wirral CH60 9LB
SJ249819
sponsors.branching.showering

Thurstaston Cliffs
Wirral Way, Caldy, Wirral CH48 1QX
SJ230839
retail.clips.haggling

Caldy Blacks
Croft Drive, Caldy, Wirral CH48 2JT
SJ222848
upper.tried.scaffold

West Kirby Marine Lake
Dee Lane, Grange, West Kirby CH48 OQF
SJ210867
backhand.nurses.flames

Websites: *https://www.wirral.gov.uk/leisure-parks-and-events/parks-and-open-spaces/find-park-or-open-space/wirral-country-park*, deeestuary.co.uk

> Opening times: The visitor centre has a cafe and toilets and is open from 10am–4.30pm, 7 days a week (closed Christmas Day), while the surrounding parkland is generally accessible.
> Parking: The car park at the visitor centre is £6 for all day.
> OS Explorer Map 266
> SPA, Ramsar SSSI

The Dee coast, from Heswall to West Kirby, starts off as ungrazed saltmarsh, with more open gutters and tidal creeks than at Parkgate or further upstream. Boulder clay cliffs rise between Thurstaston and Caldy, allowing views down over the mud and sandflats of the outer estuary.

ACCESS

The Heswall sites can be accessed by bicycle from Heswall station, and West Kirby Marine Lake is a short walk from West Kirby station. There are regular buses to West Kirby from Birkenhead and to Heswall from Chester. By car, the marsh at Heswall can be accessed from convenient car parks at Riverbank Road or Banks Road. The northernmost extent of the saltmarsh can be reached from Target Road, though parking is restricted. Wirral Country Park visitor centre at Thurstaston provides a useful base from where to explore the Wirral Way. There is very limited parking at the cliff top at Thurstaston, but the views are outstanding and there is access to the beach. The situation is similar at Croft Drive in Caldy. West Kirby Marine Lake is accessed via Dee Lane, which is one way from north-west to south-east, but it is possible to park along its length.

SPECIES

Throughout the year, Shelduck, Oystercatcher and Redshank can be seen, along with Lapwing, Black-headed Gull, Marsh Harrier, Peregrine, Kestrel, Reed Bunting, Skylark, Grey Heron and Little Egret.

From December to February, Peregrine, Merlin, Hen Harrier and Short-eared Owl hunt the marshes. Knot and Dunlin are in huge numbers on the estuary. Goosander, Red-breasted Merganser – and maybe Goldeneye, Shag or Great Northern Diver – are on West Kirby Marine Lake.

From March to May, large numbers of Whimbrel from south of the Marine Lake to Heswall may be seen; passage migrants are in the coastal field.

During June and July, Yellowhammers may still be seen around Thurstaston; breeding warblers along the Wirral Way.

Wader passage in August may include Curlew Sandpiper; overhead passages of pipits, thrushes and finches follow the coast.

YOUR VISIT

The Wirral way runs along the Dee Estuary from West Kirby to Parkgate before turning inland and ending in Hooton. It follows the path of a disused railway line and provides access to a number of different viewpoints over the estuary but is of interest in its own right.

The hedges radiating into the fields are very productive. Heswall Dungeons is a wooded valley, SSSI designated, like Thustaston (Dee Cliffs SSSI), for its geological features. The mussel beds off Caldy formerly held a large flock of Scaup, and may still attract a few Eider, though very distantly. West Kirby Marine Lake is a large saltwater lake which has a good track record of holding sea duck and divers.

Heswall Shore has a similar list of species to the more famous Dee Estuary sites. Whimbrel can gather in large numbers in late April and early May, with counts of over 200 made, whilst Golden Plover and Lapwing flocks are prominent in winter. Up to 1,000 Teal may be present, and a returning Green-winged Teal has been seen in 2018, 2023 and 2024. Curlew Sandpiper and occasionally Little Stint occur on passage and both Red-necked Phalarope and Terek Sandpiper have been recorded. At low tide, the outer estuary holds thousands of Knot, Oystercatcher, Dunlin and Shelduck. The area just south of West Kirby Marine Lake can also be excellent for Whimbrel and Curlew Sandpiper during passage periods. Peregrines hunt the waders over the mudflats while Merlin, Hen Harrier, Marsh Harrier and Short-eared Owl can be seen over the saltmarsh. As with most of the Dee coast, visible migration occurs in both passage periods and scarce and rare birds can turn up, especially at Heswall. Crane, White-tailed Eagle, Gull-billed Tern, Chough, Richard's Pipit and Pallas's Warbler have occurred.

West Kirby Marine Lake can be excellent, but it is prone to disturbance. Red-breasted Merganser and increasingly Goosander are recorded through the winter. Goldeneye and more rarely Scaup can be seen, and in some years a Great Northern Diver can take up residence. Mediterranean Gulls are seen with some regularity, especially in late summer/early autumn, and Shag is seen more reliably than at any other Cheshire site apart from Hilbre. Theake can get very busy, even in winter, so an early morning or evening trip is best. Calm conditions also dissuade windsports enthusiasts.

20 HILBRE BIRD OBSERVATORY

Hilbre Bird Observatory
Hilbre Islands LNR (check tide times and crossing advice before using),
West Kirby, Wirral CH47 1HZ
SJ185878
vacancies.encoding.nips

Website: hilbrebirdobs.blogspot.com
Email: secretary@hilbrebirdobs.org.uk
Opening times: Check tide times and weather conditions before attempting the crossing.
Parking: The nearest parking is a pay-and-display car park on Dee Lane in West Kirby.
Permits: Needed for groups of more than six people (see below)
OS Landranger Map 106, OS Explorer Map 266
LNR, SSSI, SPA, SAC, Ramsar

Hilbre Bird Observatory was established in 1957 and for approaching 70 years has recorded and analysed the Island's birds as well as other aspects of natural history.

SPECIES

Linnet, Meadow and Rock Pipits, Cormorant and Oystercatcher are present for much of the year.

From December to February, vast flocks of estuarine waders – such as Dunlin, Oystercatcher, Curlew and Knot as well as rock-dwellers Turnstone and Purple Sandpiper – are present. Large wintering flocks of Brent Geese, Common Scoter, Red-breasted Merganser and other sea ducks are common. Peregrine and Merlin hunt the waders. Red-throated Diver and Great Crested Grebe are likely on the sea, and scarcer divers and grebes are possible. Severe weather brings westward movements of Lapwing, Golden Plover, Skylark and thrushes. Wandering Pink-footed Geese and Snow Bunting are possible.

Winter ducks can be seen in March to May. Brent Geese and wader flocks disperse. Spring passage brings Sanderling, Whimbrel, Common Sandpiper and other wader species. Purple Sandpipers depart by early May. There is a possibility of Fulmar, Gannet and Kittiwake from March onwards. Terns appear during April. Passerine migration starts in March; there is heavy movement of Meadow Pipits in late March and early April. Passage peaks late April to early May with falls of Willow Warbler, other warblers and chats. White and Yellow Wagtails are regular. There is a steady passage of hirundines. Wheatear passes through in good numbers.

From June to July, Common, Sandwich and Little terns move offshore. Gannet, Manx Shearwater and Storm Petrel are seen in appropriate windy conditions. Small numbers of non-breeding waders linger through the summer months in the estuary; by late July, Sanderling and other waders start to return.

Wader flocks increase in the autumn, and early birds may be in breeding plumage. Purple Sandpiper return in October and November. Westerly gales in September and October, even early November, bring Leach's Petrel, Arctic and Great Skuas, perhaps with Pomarine and Long-tailed Skuas. Kittiwake and Little Gull are regular in these conditions, with the occasional Sabine's Gull. Manx Shearwater are typically seen until September, auks into October and Gannet until November; Common Scoter remain throughout. There are large gatherings of terns in early autumn. Falls of warblers can be seen in August or September, mostly Willow Warbler and Chiffchaff, with the occasional rarity. Diurnal passage in October and November involves winter thrushes, Starling, Chaffinch, Linnet, Greenfinch and buntings, including Snow Bunting.

ACCESS

West Kirby station is a short walk from Dee Lane, the recommended starting point for the crossing to the Islands. Permits are needed for groups of more than six people, in which case apply to the Wirral Borough Council Ranger Service Visitor Centre, Wirral Country Park, Station Road, Thurstaston CH61 0HN. The centre is open daily from 10am until 5pm (tel: 0151 648 4371/3884).

The walk across the sands to Hilbre can be tricky and should not be attempted in foggy conditions. From junction 2 of the M53 follow signs for West Kirby, then the brown signs for the Marine Lake which lead to Dee Lane. It is usual to set out from the end of Dee Lane at least three hours before high tide and head for the left-hand side of the southernmost (left-hand) island, Little Eye, thereby avoiding dangerous gutters. Follow the western side of Little Eye and then the eastern side of Middle Eye to Hilbre. Visitors must then remain on the island until at least two hours after high tide when a return to the mainland is again possible.

YOUR VISIT

At just under 5ha and with cliffs rising 17m on the exposed western side, Hilbre is the largest of a group of three islands lying 1.6km west of the north-west tip of Wirral in the mouth of the Dee Estuary. In the twelfth century, the islands were gifted to the Benedictine monks of St Werburgh's Abbey in Chester, who built a chapel on Hilbre. Ruins of the chapel and traces of medieval fish ponds remain. In the nineteenth century, a lifeboat station was built at the north end of the main island. At low tide the islands are accessible with caution across the sands from West Kirby, but at high tide they are cut off from the mainland. All three islands are composed of red sandstone and topped with rough grassland. Apart from some Blackthorn, a few bushes and trees on Hilbre and some bracken on Middle Eye, there is little cover for birds, and other than the remains of the lifeboat house and a seawatching hide (built by Hilbre Bird Observatory), even less shelter for birdwatchers. The West Hoyle Bank, a large sandbank to the west of the Islands, is a regular haul out for large numbers of Grey Seals and a few Common Seals. Up to 600 Grey Seals have been recorded. At high tide the seals often bob around the main island. At low tide, the sands are exposed, stretching from Hoylake and West Kirby to Hilbre and south down the Dee Estuary, but there are a number of deep gutters that can trap the unwary. These extensive flats are the feeding grounds for thousands of shorebirds.

Some 270 species have been recorded at Hilbre, but only a few breed because of the restricted habitat. Shelducks are seen throughout the year. Cormorants rest on sandbanks at low tide, often in large numbers. Vast flocks of waders winter in the estuary. In late July the first Dunlin, perhaps 1,000 or more, return from the Arctic tundra. They are joined by flocks of Sanderling, particularly in August, on the way to their winter quarters on African beaches. Up to 7,000 Oystercatchers arrive in August. They remain in similar or even larger numbers through the winter. August sees the peak of Ringed Plover passage. Dunlin, Curlew and Redshank increase, and early Knot, some of them red summer-plumaged birds, black-bellied Grey Plover and godwits of either species may arrive. Turnstones reach 80–100 on Hilbre in August and September, flicking over seaweed and pebbles in their search for food. Grey Plover numbers may exceed 200 in September. Knot numbers increase, numbering several thousand by October. A speciality of the island is a flock of Purple Sandpipers that builds up from mid-October into November, peaking at around 20. Winter Knot flocks can exceed 20,000 and there may be 9,000 Dunlin and up to 600 Curlew and 300 Redshank. Hundreds of Grey Plover and a smaller number of Ringed Plover will also be present. Oystercatcher remain abundant and highly conspicuous. Not all of these birds will use the outer estuary at any one time. Many commute between the Dee and the Alt or Ribble estuaries. High-tide roosts may gather on the Middle Hilbre and Little Eye, which should be left undisturbed at high tide for this purpose. Quite often the roosts are disturbed by a Peregrine, several of which hunt the estuary daily in winter. Rather less often a Merlin appears. Numbers of waders in the estuary start to fall by March. In April and May the Purple Sandpiper flock returns north, perhaps to Greenland (Hilbre Bird Observatory had a ringing recovery there). By June only a few Oystercatcher, Curlew and Turnstone remain, with occasional birds of other species.

A conspicuous feature of winter and early spring is a flock of up to 600 Pale-bellied Brent Geese around Hilbre. Small numbers of Dark-bellied Brent Geese may also be present. A favourite haunt for these birds during low tide is around the Tanskey Rocks south of Little Eye, but they may be watched feeding all around the islands. Parties of Pink-footed Geese fly over on their way to or from their Lancashire and Dee Estuary feeding grounds and, especially in late March and early April, they can be seen leaving the estuary and gaining height as they fly back to Iceland when the wind is favourable. Red-breasted Merganser and Goosander are regular outside summer, and Eider may be present in very small numbers. Huge rafts of Common Scoter can be seen near the Burbo Wind Farm from September and other sea duck may be seen with them; Long-tailed Duck and Velvet Scoter are present most years. Divers occur from September to May, although daily totals seldom rise above 20. Red-throated Diver is by far the most numerous; up to 40 can be recorded in a day. Great Northern Diver is seen on a handful of days each year. Black-throated is even less regular, although spring (March through to May) is a good time. Great Crested Grebe is common in winter and the smaller grebes also occur, albeit rarely.

Visitors to Hilbre often go there specifically for seawatching. Two to three days of strong westerlies in autumn are almost guaranteed to produce Leach's Petrels. Hilbre offers an almost unique opportunity to view Leach's Petrels, in that there is always close water, the birds can be very close and there can be an opportunity to look down rather than across wave troughs, which helps the search for small petrels keeping low in a heavy swell. Additionally, birds can be picked up early as they approach, allowing prolonged views and photographs. Ideal conditions are infrequent, but most autumns produce at least a handful of birds and three-figure counts still occur. Most birds pass westward into the wind close by the north end of the island. Birds may settle for a while on the sea. Storm Petrels are rare in autumn but can be seen reliably during summer gales. Conditions that produce Leach's Petrels will also bring other of the region's seawatching specialities, Sabine's Gull and Grey Phalarope. Long-tailed Skuas tend to pass a little earlier, from late August, and Pomarine Skuas a little later, in October. Feeding movements of Manx Shearwaters, Kittiwakes, terns and Gannets can occur through the summer and autumn, especially if there are concentrations of bait fish in Liverpool Bay, and these will draw in skuas. Arctic Skua is scarce in spring but can be seen with persistence through summer and autumn. Great Skuas are currently very scarce, but it is hoped that numbers will recover if the species recovers from the HPAI-induced population crash. All seabirds that breed to the north of Hilbre may be seen in strong westerlies as they pass south. Hilbre offers the best chance in Cheshire to see large shearwaters, but they are still extremely rare in comparison to most other UK coasts.

A small flock of auks moult in Liverpool Bay during summer, though only single-figure counts of Razorbill or Guillemot can be expected. Onshore winds in autumn and occasionally early spring can produce much larger counts. Other auks – Puffin in spring, Little Auk in late autumn/winter and Black Guillemot at any time of year – are rare, though the last is currently increasing, with two summering recently.

There is a reliable passage of Little Gulls from March to April. They may roost on West Hoyle Bank, albeit rather distantly observed from Hilbre, or feed off the north end and out towards East Hoylake Bank. Terns can be seen offshore from April to October. Sandwich and Common Terns are generally the most numerous and occur throughout the summer. Dozens may be counted on spring days, with much larger numbers from July to September. Arctic Terns pass through in small numbers in spring and autumn. There is a Little Tern colony at Gronant, and birds may be seen, especially post breeding. Black Terns may also be seen on passage, particularly after easterly winds, either passing or roosting along with other birds on the east side of the Main Island at low tide. Caspian Tern, and Ring-billed, Laughing and Bonaparte's Gulls have been recorded.

Movements of passerine migrants have been studied intensively at Hilbre. Spring passage begins in early March or even February with Stonechat. The first Chiffchaffs, Wheatears and perhaps a Ring Ouzel arrive in March. A spectacular movement of Meadow Pipits reaches a peak in late March or early April, when more than 700 birds may pass in a day. Migration is at its height in late April or early May when, given suitable conditions (usually south-east winds), the bushes may be full of migrants shortly after dawn. Willow Warbler is the principal species involved. Spring passage at Hilbre is more varied than in autumn. Sedge and Grasshopper Warblers. Whitethroat, Redstart, Spotted Flycatcher and Whinchat are typical spring migrant species. Tree Pipit and Yellow Wagtail are also frequent. Many other summer visitors occur in smaller numbers. The shortage of vegetation cover on Hilbre means that migrants are unlikely to stay for more than a few hours, or a day at most. Considerable numbers of finches pass through in April and May, with Goldfinch, Redpoll, Linnet and Siskin the principal species. Late spring is largely quiet, but Spotted Flycatcher occurs in May and early June and there are occasional rarities, including Bee-eater, Red-rumped Swallow, Melodious, Dartford, Subalpine, Paddyfield and Blyth's Reed Warblers, Little Bunting, Bluethroat, Nightingale, Woodchat Shrike and Red-breasted Flycatcher. Migration of summer visitors picks up in August. when the largest autumn totals of Willow Warblers occur. There is less variety on the island during autumn passage, but Garden Warblers occur more often at this season than in spring, and Pied Flycatcher is sometimes seen. Yellow-browed Warblers make annual appearances several times in late September and October. Yellow-breasted Bunting, Red-flanked Bluetail and Pallas's Warbler have been exceptional west-coast records. Relatively few diurnal migrants pass over compared with the mainland coastline. However, immigrant winter thrushes and Starlings fly over the island in large numbers from October onwards, when finch movements may also be heavy. Lapland Buntings may be recorded at this time and Snow Buntings are seen regularly in some winters but only occasionally in others.

For seawatching, strong winds from the west are ideal. A persistent blow for two or more days in September or October will bring in Leach's Petrels and other seabirds. Consult tide tables and time your visit around high tide but pay particular attention to the state of the tide when visiting this tidal island (see Access). In April and May, the largest arrivals of warblers and other passerines take place with south-easterly winds. Birds arrive shortly after dawn, rest and feed briefly, then move on. Few remain by mid-morning. In autumn, east or north-east winds bring birds. Severe winter weather may bring large westward movements of birds seeking milder conditions. A thaw following prolonged cold induces return movements.

21 RED ROCKS, HOYLAKE AND MEOLS SHORE

Red Rocks
Stanley Road, Hoylake, Wirral CH47 1HN
SJ204882
jogged.glosses.barbarian

Hoylake Shore
North Parade, Hoylake, Wirral CH47 2DQ
SJ215894
mopped.discloses.games

Wader Walk
Wirral Circular Trail, Hoylake, Wirral CH47 6AW
SJ221900
additives.disengage.later

Monkey Woods
Roman Road, Meols CH47 6AG
SJ226901
back.useful.hers

Dovepoint
Meols, CH47 6AW
SJ233906
tracks.proper.lots

Opening times: Sites accessible at all times
Parking: Available at or near each location. Check local signage for specific parking restrictions and charges.
OS Landranger Map 106, OS Explorer Map 266
SSSI, SAC, SPA, Ramsar

With 280 accepted species of bird, Hoylake has the biggest list of anywhere in Cheshire, just pipping Hilbre.

SPECIES

Waders may be present all year, though numbers drop in midsummer. There is barely a gap between the late spring passage of high Arctic breeders and the first return of European breeding species.

From December to February, the saltmarshes are host to wader flocks, scoter, Brent Goose, Linnet, Snow Bunting and Rock Pipit. Peregrines hunt the waders.

Passerine migration starts in March and peaks in late April. Spring brings impressive concentrations of Meadow Pipit and White Wagtail if their passage coincides with a sandfly hatch. Sustained south-easterly winds combined with warm, muggy conditions often bring the chance of rarities, while late spring sees passage of Redpoll.

Terns are seen offshore during June and July, and shearwaters and other seabirds

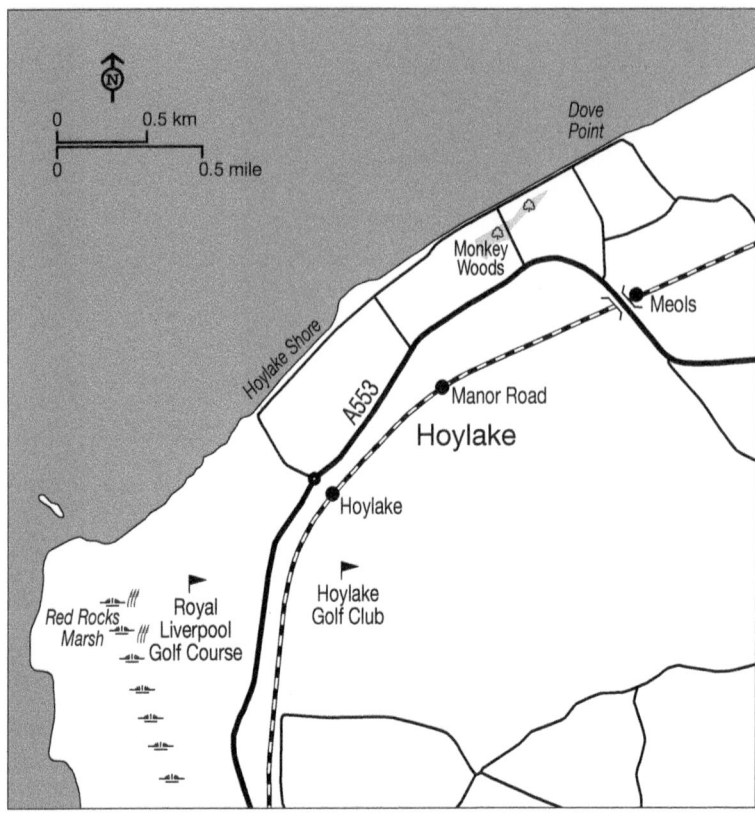

too, if there are bait fish in Liverpool Bay. The first returning waders appear, and Mediterranean Gull are likely in the roost.

Early autumn can bring falls of common warblers, with perhaps the odd rarity, along with impressive dispersal movements of relatively sedentary species. Later in the season, visible migration passage includes winter thrushes and Starling, Chaffinch, Brambling, Siskin and buntings, including Lapland.

ACCESS

Hoylake station is closest to Red Rocks, while Manor Road station is closest to Dovepoint. Buses run regularly from Birkenhead and West Kirby. Red Rocks can be accessed via Stanley Rd, which is off King's Gap, north from the roundabout on the A553. Turn right for the point and left for the reedbed and saltmarsh. The latter can also be viewed from the top of the fixed dunes accessed from the footpath that crosses the Royal Liverpool Golf Course from Pinfold Lane. This also provides an opportunity to scan the fairways for Wheatear, larks, pipits and wagtails. King's Gap also leads to North Parade and Meols Parade, which leads to Dovepoint. Monkey Woods can be accessed from a number of locations, but most easily from Roman Road.

YOUR VISIT

Red Rocks, at Hoylake, is the north-west tip of mainland Wirral and consists, like Hilbre Island, of a sandstone outcrop. The farthest point, Bird Rock, is cut off by high tides and is then popular with roosting waders. A reserve managed by Cheshire Wildlife Trust was established as an SSSI to protect a reintroduced population of Natterjack Toads. There is a cluster of Buckthorn bushes just south of the end of Stanley Road, an Alder bush adjacent to the boardwalk leading south to West Kirby and a few scattered bushes in a dune hollow; otherwise there is now little cover aside from succession vegetation comprising dunes, dune slacks and pioneer saltmarsh. Following management work to remove dune scrub, a poplar stand and all but one clump of reedbed bushes, the attractiveness of the site for migrant and breeding birds is much diminished.

Hoylake Shore is a rare accreting embryo dune and Atlantic salt meadow, which has developed since glyphosate spraying was halted in 2019. It is of international importance for its botany, but is also attractive to birds. The seafront gardens along the north shore of the Wirral and North Parade can hold migrants, and an area of housing-enclosed dunes and trees, known locally as the Monkey Woods, is worth checking in migration periods. Depending on the state of the tide and the time of the year, waders may be feeding or roosting along the shoreline. The area in front of the tennis courts is often very productive, with Baird's Sandpiper and Citrine Wagtail in recent years. Very rapid accretion and vegetation succession means that the optimum location to view will move eastwards as this edition matures. Seawatching is possible from Red Rocks, the RNLI station or various shelters along the seafront, but in conditions where storm-driven pelagic birds are present, there are alternative locations with closer birds and more shelter.

Red Rocks is primarily a site for visible passage, being located at the north-west corner of the peninsula. Regardless of season, nearly all birds observed are moving south. The major paradox of the site is that the conditions most likely to produce a good bird are also most likely to mean it will be a fleeting visit. Its location on the corner of the peninsula makes the site the best place in Cheshire – and arguably in the region – for visible migration. Birds tend to approach from the east, following the coast, regardless of season, then after circling, head south along the edge of the estuary.

The reedbed at Red Rocks is almost deserted in winter, though a few Water Rails may be heard and one or two Cetti's Warblers overwinter. A wintering flock of Linnets is sometimes present in the saltmarsh towards West Kirby, or more frequently at Hoylake beach, with up to 100 recorded. They are worth checking for Twite, but the latter is declining in Cheshire. Snow Buntings are present most years, but they may range all the way from West Kirby to the lifeboat station. Rock Pipits are present in both saltmarshes and they often show signs of being the Scandinavian race *littoralis*.

A movement of Meadow Pipits is the first sign that spring is coming; 300–500 a day are not unusual. It is usually a race between White Wagtail and Wheatear for the first 'proper' spring migrant. The outer edge of the saltmarsh, the exposed lawn garden of the last house on the north side of Stanley Road and the rocks at the RNLI, are the best places to see an early Wheatear, while the wagtails are invariably on the beach. It is not unusual to get counts of more than 50 White Wagtails and the record is more than 300. These large counts coincide with hatches of sandflies and need a wet beach. Yellow Wagtails and Meadow Pipits join them, the latter in large numbers. All the common migrants can be seen, and the area is particularly good for Tree Pipit, but records will mostly be flyovers, including flycatchers. The fences running southwards may collect Whinchats and Redstarts, but they tend to move through quickly. The taller plants in the saltmarsh on Hoylake beach also hold migrant chats. On very rare occasions there can be huge falls (eg, up to 600 Willow Warblers and 250 Wheatears) and these are associated with a clear night, south-easterly winds and rain just before dawn. In late spring, Sedge and Reed Warblers will be on territory at Red Rocks. Although the bulk of the passage is over, it is now that Red Rocks comes into its own. With the right weather conditions, they can be excellent for spring 'overshoots' and drift migrants. Serin, Bee-eater, Hoopoe, Great Reed Warbler, Alpine Swift, Bluethroat, Red-rumped Swallow, Savi's, Dartford, Subalpine, Melodious and Blyth's Reed Warblers, Iberian Chiffchaff, Citrine Wagtail, Red-throated and Tawny Pipits, Red-backed Shrike and Collared Pratincole have all been recorded.

Whereas the bulk of autumn passage migrants move through in late August and early September, it is the later period that can produce rarities. In general, there are far fewer birds on autumn passage than in spring, presumably something to do with the local geography. Once again, almost anything can turn up. The most impressive aspect of autumn birding is the overhead passage at dawn. Counts of 1,000-plus finches are not unusual and occasionally there are enormous thrush and Starling movements, usually on easterly winds. Another feature is an extraordinary passage of less obvious dispersing species. Day counts of hundreds of Blue and Coal Tits have been made, with birds collecting on the point's houses and making exploratory flights out to sea. Nuthatches and woodpeckers on chimneys is a local speciality. Once again almost anything can turn up. Rarities and scarcities have included Great Spotted and Black-billed Cuckoos, Greenish, Marsh, Icterine, Melodious and Barred Warblers. Yellow-browed Warbler is regular. Richard's and Red-throated Pipits, Red-rumped Swallow and Cirl Bunting have also been seen at Red Rocks; and Arctic, Yellow-browed, Pallas's and Barred Warblers, Pallid Swift, Richard's Pipit, Citrine Wagtail, Wryneck and Red-breasted Flycatcher have been seen in or over the Hoylake seafront gardens and Monkey Woods.

Seawatching is possible from the point or the RNLI station, though there needs to be a big tide (9m or more) at Red Rocks. In stormy weather, you are probably better off going somewhere with shelter. In calm weather, there can be vast rafts of thousands and sometimes tens of thousands of scoter, with Long-tailed Duck and Velvet Scoter often present. However, the distances involved are huge and a trek out to the top of East Hoyle Bank two hours either side of low water is the best way to get acceptable views with a telescope. The ducks tend to stay out at the same range when the tides come in. Changes to the nature of the shore – more sand, less mud and a filling of the channels – have made walking out a practical option in the last few years, though you are still advised to be aware of the tide times. Internationally important numbers of Great Crested Grebe are also present, occasionally a Surf Scoter is recorded and a Black Scoter was present in 2025. Brent Geese are seen with increasing frequency on both West Kirby and Hoylake beaches.

Perhaps the best reason to visit in winter is the huge numbers of shorebirds: 7,000–30,000 Knot, 10,000–20,000 Dunlin and good numbers of all the other common waders. As the feeding grounds on East Hoyle Bank and Mockbeggar Wharf flood, huge flocks of birds will fly over the point at Red Rocks and attempt to roost on the shore at West Kirby or Hilbre, though in recent years they may just stay at Hoylake. At low tide, the Knot often feed or roost close to the sea wall at Dovepoint, or Roman Road. Of course, the local Peregrines know this and often try and pick out their breakfast. The sound of the whole flock preparing to fly on a false alarm is worth the wait, but that is nothing compared to the twisting turns of the flock if it's not a false alarm. Watching the tide flood in from the Meols end is also worth doing if you have the chance. Most birds fly, but up to 8,000 Oystercatchers walk ahead of the incoming water, sometimes all the way to the lifeboat station from Dovepoint. Curlew Sandpipers and Little Stints are regular on autumn passage but scarce in spring. White-rumped Sandpipers are seen every few years, while Baird's, Western, Semipalmated and Broad-billed Sandpipers have been recorded. Locating and relocating a rarity in the huge flocks is a serious challenge, but high tide roosts or feeding along the wader walk in front of the tennis courts at low tide offer your best chance. Dotterel and American Golden Plover have been tempted down by the new salt meadow, while Snipe and Jack Snipe winter in both saltmarshes.

Towards the end of summer, an impressive tern roost can build up. The Little and Sandwich Terns at Gronant and the Common Terns at Shotton and Seaforth all bring their fledged young out to the mouth of the Dee to feed them. They roost at high tide, and rarer species may be present. Arctic and Black Terns are annual, Roseate Tern is much rarer, and Caspian, White-winged Black and Gull-billed Terns have also been recorded. There is also a substantial gull roost, which usually contains Mediterranean and Yellow-legged Gulls.

Peregrines are present on most tides, and multiple Ospreys are seen on passage most years, often catching and eating fish out on the tide edge. Ravens at Hoylake beach have become bold enough to try and steal your sandwiches if you sit in the shelter near Clydesdale Road.

For migrant passerines and near passerines, birds present at dawn move off within the first hour or two, and birds that have made landfall elsewhere on the Wirral coast will move through within three hours of dawn. For spring rarities, the wind needs to be in the south-east, or better still east-south-east, for at least three days. It needs to be hot and muggy, with clear skies but poor visibility – so birds that are moving get to the point but can't see to cross to Wales. For autumn visible migration, a south-westerly wind is often better, largely because birds stay low enough to see or hear as they avoid the stronger headwinds at higher altitudes; otherwise first light and about an hour afterwards. Seaduck are best seen at low tide on calm days from late autumn until spring. Waders are dependent on the state of the tide. Looking for a single bird in 13 square kilometres of sand and mudflats is not for the faint-hearted, and the roost location will depend on the height of the tide. A tide in the 8.6–9.3m range maximises the chances of a close but undisturbed tern roost in the vicinity of Bird Rock or by the RNLI station.

22 HOYLAKE CARRS AND GILROY ROAD NATURE PARK

Hoylake Carr Wetlands
Hoylake, Wirral CH48 6DE
SJ221880
startles.cartoons.hobble

Cheshire

> **Gilroy Road Nature Park entrance**
> Gilroy Road, Grange, Wirral CH48 6DP
> SJ224874
> rewarded.burden.lakes
>
> Website: https://www.wirral.gov.uk
> Opening times: Accessible at all times
> OS Landranger Map 106, OS Explorer Map 266
> LNR

The Hoylake Carrs or Langfields are an area of former willow and Alder carr, cleared and drained in the Georgian period and now used as low-grade pasture, lying between Hoylake, West Kirby, Newton and Meols.

SPECIES
Buzzard, Raven and Peregrine are present throughout the year. In winter, from December to February, seasonal floods attract waders, especially Black-tailed Godwit and Snipe and wildfowl on seasonal floods. As spring unfolds between March and May, passage migrants include chats, wagtails and Ring Ouzel. By June and July, small numbers of breeding warblers settle. From August into

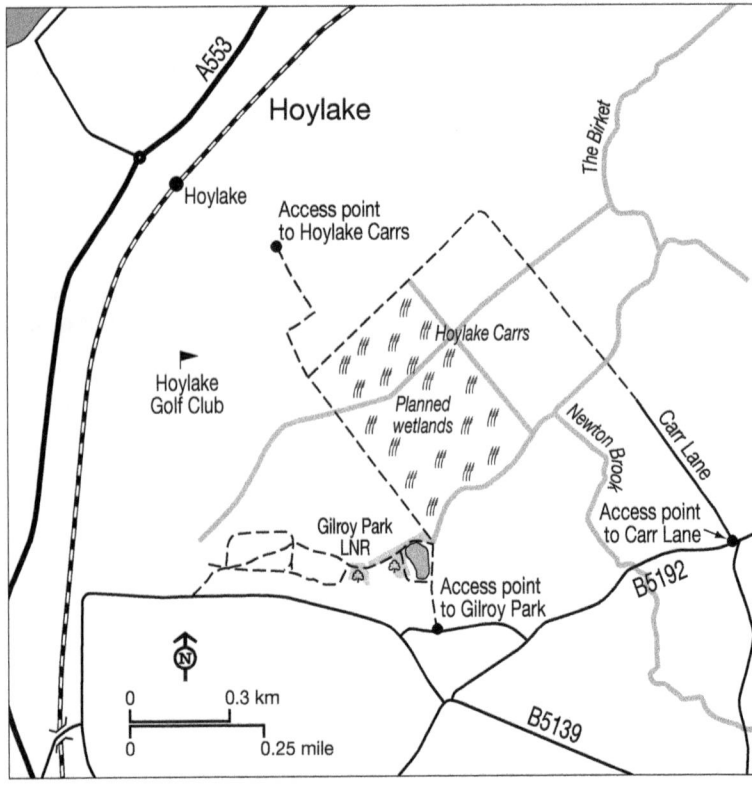

November, returning waders, especially Black-tailed Godwit, include thrushes and finches on passage.

ACCESS

The New Hall Lane entrance is just behind Hoylake train station. Two tracks cross the Carrs, one between New Hall Lane in Hoylake (SJ219886) and Gillroy Road in West Kirby (SJ224874) and another from Newton (SJ233877) back towards Hoylake (SJ222887). Both are known locally as Carr Lane and the latter requires wellingtons in winter. A public footpath runs from Oldfield Road (SJ240880) to Fornall's Green Lane in Meols (SJ232891). Oldfield Lane offers a clear, if distant, view overlooking the whole area. The other surrounding roads are unsuitable for stopping.

YOUR VISIT

The south-west corner of the area is a grassed-over tip and the north-west corner adjoins a now disused municipal golf course. Though drained by numerous ditches, and with the River Birket straightened, the old meandering path of the river and its floodplain is still visible in the fields and the area is prone to flooding in periods of heavy rain. When this happens, it acts as a magnet for wildfowl and waders. Gilroy Road Nature Park is a small pond and wooded area created at the site of the former council tip. The patchwork of arable fields and 200-year-old hedges can be excellent for passage migrants.

Following a failed attempt to establish a golf resort on the site, Environment Agency Flood Alleviation funds were awarded and there are plans to establish a wetlands reserve, managed by Cheshire Wildlife Trust. This will involve the creation of scrapes within the original floodplain and leaky dams to push water into them in times of heavy rain. It is expected that the birds that currently appear in extremely wet conditions will become a regular feature. Work is planned to begin with improved access and hides/screens.

When the fields flood due to seasonal storms, large numbers of wildfowl take advantage of the temporary pools. Once, 5,000 Icelandic Black-tailed Godwits were a regular feature of the flooded field opposite Gilroy Park before a blocked drain was cleared. Wood, Green and Common Sandpipers were also attracted. Godwits reappear quickly when the carrs flood, along with up to 150 Snipe, 50 Teal and other ducks. After Storm Christoph in 2021, 8,000 Pink-footed Geese moved in. In 2024, two Hen Harriers appeared in the area within a day of flooding. Cattle and Great White Egrets occasionally join the frequent Little Egrets, and Bittern has appeared in cold snaps. The location is easily visible from the air from the entrance to the Dee and Mersey estuaries. The rapidity with which birds investigate it when it floods suggests that there is a great potential to attract passage waders and wintering wildfowl.

Gilroy Pond is largely populated with domestic Mallard, but Little Grebe has bred, and small numbers of Reed and Sedge Warblers occupy the surrounding ditches. The wooded area attracts Bullfinch, breeding warblers and finch flocks, including Siskin and Redpoll.

The arable fields and their hedges and the trees surrounding the disused municipal golf course can be excellent in passage periods. Falls of Wheatear are frequent, and Whinchat and Redstart can be seen. Ring Ouzels have become regular in April, when White and Yellow Wagtails are also present. Corn Bunting is recorded occasionally and Lesser Whitethroat breeds. Peregrines frequently roost on the pylons,

Kestrel and Barn Owl nest on the golf course and Tawny Owl is present. Ravens and Buzzards are frequent visitors. Great Grey Shrike has wintered, whilst Yellow-browed Warbler and Hoopoe have been recorded.

23 LEASOWE LIGHTHOUSE AND NORTH WIRRAL COUNTRY PARK

Leasowe Lighthouse
Lingham Lane, Moreton, Wirral CH46 4TA
SJ251913
passes.asserts.cities

Lingham Lane Bridge
CH46 4TB
SJ251910
asleep.text.freed

The Gunsite
Green Lane, Leasowe CH45 8NA
SJ274925
rides.probe.sharp

Park Lane
Great Meols CH47 8XX
SJ242907
fires.needed.backed

Ditton Lane
Moreton, Wirral CH46 3SW
SJ261912
horn.deaf.sailor

Website: https://leasowelighthouse.com
Phone: 01513 530861
Opening times: Sites accessible at all times; the lighthouse is open to the public twice a month.
Parking: Gunsite can be accessed via Green Lane with car parks along the way and a main car park by the sea defence; Leasowe Lighthouse has a car park off the A551 (SJ251913); Meols end is accessible via Dovepoint Road and Meols Parade, with limited parking on Park Lane.
Phone: 01513 530861
OS Landranger Map 108, OS Explorer Map 266

The name Leasowe is derived from the Anglo-Saxon word 'Leasowes', meaning 'Meadow Pastures', reflecting its pastoral origins. A prominent feature is Leasowe Castle, constructed in 1593 by Ferdinando, the 5th Earl of Derby, possibly as an observation platform for the Wallasey races, considered forerunners of the famous

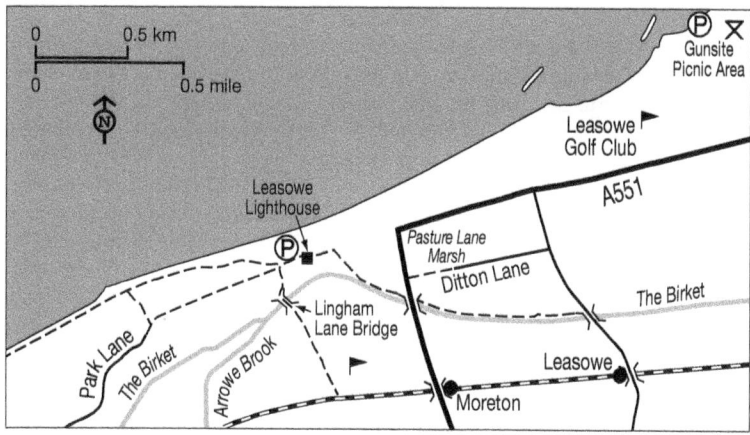

Derby races. Another significant landmark is Leasowe Lighthouse, built in 1763. Much of the land is at or below sea level, necessitating protection by a coastal embankment. Additionally, local tradition holds that the legendary demonstration by King Canute to show the limitations of royal power against nature took place on the north Wirral shore, encompassing areas like Meols, Moreton and Leasowe.

SPECIES

Stonechat, Skylark, Reed Bunting and Little Egret can be seen throughout the year, and a few non-breeding Knot remain through the summer.

During the winter months, from December to February, the area attracts shore waders, Snow Bunting, Linnet, wildfowl and Snipe if the fields flood.

Wheatear arrive in March, followed by other migrants. The paddocks are particularly good for Ring Ouzel, Redstart and Whinchat and in late spring Cuckoo. White and Yellow Wagtails and Meadow Pipits feed in the fields, and overhead passage will include Tree Pipit and Redpoll. Little Gull passage in March and early April and terns may be seen.

Between June and July, Whitethroat, Grasshopper, Reed and Sedge Warblers breed, while Terns feed offshore, and the first waders return.

From August to November, migrant passerines move through in smaller numbers than in spring; and there is overhead passage of finches and thrushes. Yellow-browed Warbler and Siberian Chiffchaff are recorded most years. Stormy weather can bring Leach's Petrel and other storm-driven pelagic close inshore.

ACCESS

Meols, and especially Moreton station, are conveniently close. For the Gunsite take the A554 from Junction 1 of the M53 towards New Brighton. After the second slip road there is a sign for North Wirral Coastal Park. Turn left here into Green Lane. Follow the lane all the way. passing two car parks on the left until eventually arriving at the car park by the sea defence. Leasowe Lighthouse car park (SJ251913) is accessed from the A551 directly from Junction 3 of the M53. Access to the Meols end of the site is via Dovepoint Road and Meols Parade. Park Road leads to Park Lane, but there is limited parking.

YOUR VISIT

The site comprises a large area of coastal fields, footpaths and market gardens stretching from Meols to Wallasey. It is defended from the sea by a fixed dune system and an artificial concrete embankment. The whole area is attractive to migrants and can be covered on foot from the car park at Leasowe Lighthouse or Moreton station. From the corner of Pasture Road and Leasowe Road, Lingham Lane leads down towards the lighthouse car park. The field opposite this corner is permanently flooded, though access is restricted. Kerr's Field, inland of the track to the lighthouse, is excellent for ground-feeding migrants and, when it floods, wet meadow-loving species. There is a track down the side of this field that leads to a footpath along the river Birket. Trees and bushes at Stone Cottage also attract migrants. At the lighthouse car park, Lingham Lane takes a sharp left turn and there is nowhere to park past this point. The hedges and fields adjoining this lane are excellent. The lane crosses the Birket, then continues inland past Lingmere fishery towards a former brick pit, now filled and scrubbing over apart from a fishing lake at the west end. Leading west from the lighthouse car park are two footpaths towards Meols which allow a circular walk around the horse paddocks. The start of the inland path runs between two lines of hawthorn and Sycamore trees, which can be excellent for migrants. The paddocks themselves, and the many fences, are always worth scanning, and the wet ditches and a small reedbed have breeding Reed and Sedge Warblers. The inland path continues onto Park Lane, Meols, and offers further views over rough pasture and paddocks as well as market gardens and mature willow trees. The fields here are prone to flooding if the water level in the Birket rises rapidly and can be excellent, attracting wildfowl and waders. The fixed dunes at the Meols end of the area can be viewed from the path which leads away from Dovepoint car park. Huge sand- and mudflats can be seen from the top of the seawall and a permanent gutter at the base of the wall can hold waders and egrets. A groyne (SJ241911) holds a wader roost at high tide and can be excellent for Wheatear. Ditton Lane, which leads east off Pasture Road, gives access to a dauntingly large scrub and tree patch between here and the Birket. The Gunsite car park is an excellent place to seawatch from a car, and in westerly autumn gales the car park can be full of birders.

In migration periods the coastal fields immediately behind the seawall are particularly good for Wheatear, Whinchat, Redstart, Ring Ouzel, wagtails and pipits. Stonechat and Skylark breed, as did Lapwing until very recently. The hedges and ditches are attractive to warblers, with Reed, Sedge and Grasshopper Warblers being regular breeders. The trees around Stone Cottage and the inland path leading west from Lingham Lane provide cover for the usual summer migrants. A wintering Linnet flock can occasionally attract Twite, and Snow Buntings may be found along the seawall. Visible migration occurs in both migration periods, and the area has a good track record for producing scarce and rare passerines and near passerines. These have included Serin, Bluethroat, Red-throated Pipit, *iberiae* Yellow Wagtail, Tawny and Richard's Pipits, Short-toed Lark, Wryneck, Little Bunting, Barred, Melodious, Subalpine, Radde's and Pallas's Warblers; Yellow-browed Warbler is annual. Though the Meols end of the area has attracted fewer rarities, Desert Warbler, Pied Wheatear and Hoopoe have been seen.

A few freshwater waders may be seen on the paddocks and fields after heavy rain, but the real wader action is out on Mockbeggar Wharf, where thousands of Knot, Dunlin and Oystercatcher and hundreds of Bar-tailed Godwit, Ringed Plover and Sanderling may be present in winter. As the tide floods, most fly to roost sites at Hoylake, Hilbre or the Alt, but a few will roost on the groyne at Meols, as do wintering Lapwing. At low tide, the gutter at the base of the seawall can be good for Whimbrel in spring, and there is often a wintering Greenshank. Collared Partincole, Dotterel and American Golden Plover have been seen, though briefly, and a Stone Curlew once spent a day in one of the inland fields.

The Gunsite is rightly famous for close view of seabirds. A good blow in July and August should produce Storm Petrel, terns and Arctic Skua. The same conditions in September and October are good for Leach's Petrel, while Great Skua, Sabine's Gull and Grey Phalarope are possible. At Dovepoint seabirds are further away, but there is more chance of seeing stronger-flying species such as shearwaters, assuming the tide is in. Balearic, Sooty and Barolo Shearwaters have been recorded here.

Migrants tend to stay longer than at Red Rocks, so an early start is less essential, though waders on the inland fields tend to leave at the appearance of the first dog walker. South-easterly winds are most likely to result in migrants in spring, and anywhere from the easterly quarter is best in autumn. Autumn overhead passage is best with the wind from the southerly quarter and low cloud. See New Brighton for the conditions required for seawatching, though the tide state is more critical at the Gunsite and especially Dovepoint, where low tide means you will be sand-watching!

24 NEW BRIGHTON AND WALLASEY

Marine Promenade
New Brighton, Wirral CH45 2JS
SJ310942
crew.lock.still

Wallasey Coastguard Station
Kings Parade, Wallasey, Wirral CH45 3PZ
SJ286935
rush.video.tile

Derby Pool
Bay View Drive, Wallasey, Wirral CH45 3PZ
SJ284932
drift.melon.upgrading

Opening times: Accessible at all times
Parking: Seafront parking at New Brighton; Derby Pool pub also has a car park.
OS Landranger Map 108, OS Explorer Map 266

New Brighton is located on the north-eastern tip of the Wirral peninsula on Liverpool Bay, where the River Mersey meets the Irish Sea. New Brighton was born in 1826 and by the turn of the twentieth century was a bustling seaside resort with hotels, bathing pools, a pier, the promenade and a ballroom, as well as many

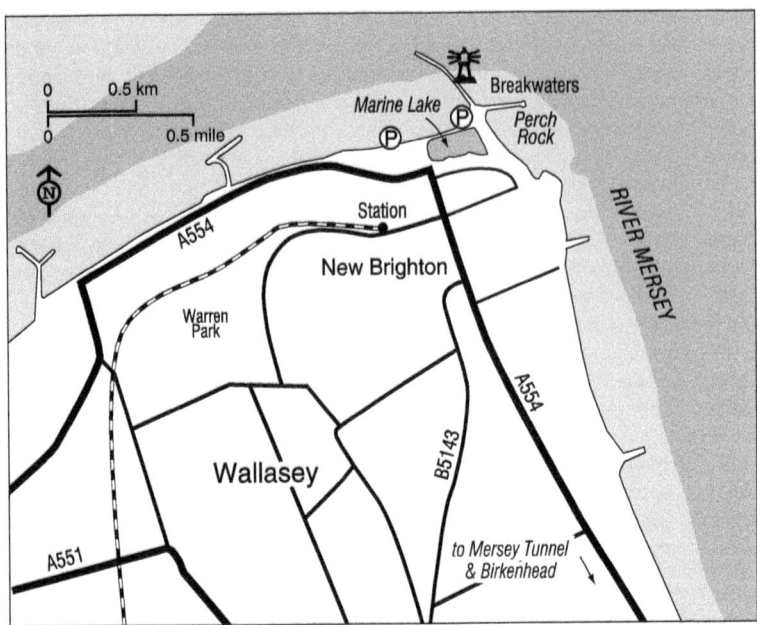

other attractions and amenities. Much of the tourist infrastructure is decayed or removed but it still hosts the longest promenade in the UK. Fort Perch Rock was built in 1826–29 as part of a series of permanent defences on the Mersey. It is now a museum.

SPECIES

From December to February, Purple Sandpiper and Turnstone in the high tide roost on Marine Lake pontoon. Mediterranean Gull are likely, and Little Gull may be present after gales.

From March to May, Purple Sandpipers leave for their Arctic breeding grounds, but passage terns may be seen feeding in the river, including Black Terns, especially after easterly winds. Little Gull numbers may build up in late March and early April.

During June and July, Kittiwakes and terns, breeding nearby, feed in the river. Storms may bring Storm Petrels.

From August through November, prolonged storms will bring pelagic species close inshore. Purple Sandpiper will feed on rocks and breakwaters at low tide and roost on Marine Lake pontoon at high tide. In some years, Snow Buntings arrive in late winter.

ACCESS

New Brighton station is a few minutes' walk from the seafront shelters. There are also frequent bus services from Birkenhead. During storms and in the winter months, there is seldom any difficulty in parking along the seafront at New Brighton. View from the shelters near the fort or from a parked car. Wallasey Coastguard Station provides shelter for anyone wanting to brave the elements, whilst the Derby Pool public house and its car park offer an alternative view of

Liverpool Bay when the tide is in, though offshore groynes tend to push birds a little further out.

YOUR VISIT

New Brighton Marine Lake is more disturbed than West Kirby Marine Lake and prone to algal blooms. The primary source of the disturbance, a water-sports centre, has seen the introduction of pontoons, which host high tide gull and wader roosts and are a reliable site for Purple Sandpiper, which feed with Turnstone on the breakwaters around Fort Perch Rock and leading out to the lighthouse when the tide is out. The 'Dips' are grassy hollows along the seafront which often hold roosting gulls. Vale Park contains mature trees and can attract passerines. However, it is the views of the mouth of the Mersey that attract most birders to New Brighton, especially after westerly gales in autumn, when seabirds can be almost within touching distance. New Brighton is best known for the shelters on its promenade that provide cover for seawatching during autumn storms. It is also possible to seawatch from a vehicle in the extensive car parks and causeway to Fort Perch Rock.

The Irish Sea is largely land-locked and shallow so contains fewer seabird species than more exposed Atlantic or North Sea coasts. Westerly gales can produce birds in any month, but the chief attraction is in autumn when hundreds of Leach's Petrels may reliably be seen. Strong westerlies drift southward-migrating seabirds that would otherwise pass west of Ireland into the Irish Sea, and once there, further strong winds will bring them into the Mersey mouth. Poor visibility and squalls will increase the chances of this. At night birds may even carry on up the Mersey and may be seen upriver as far as Seacombe Ferry. Birds often shelter under the seawall on the Wirral side of the river and may pass very close. Other storm-driven seabirds, including auks, seaduck, shearwaters, Gannets, Kittiwakes and skuas, may be seen in similar conditions. Notably, Sabine's Gull, Grey Phalarope and Long-tailed Skua may linger in the mouth of the river. Stronger-flying species may also be seen in the river but are more likely to stay in Liverpool Bay and be seen on the seaward side of Fort Perch Rock or from Wallasey Coastguards at Harrison Drive. Summer gales can be good for terns and Storm Petrels. Away from gales, terns can be seen on passage and Little Gulls may be seen, especially in March and April, when 50 or more have been recorded in a day. Mediterranean Gulls can be seen in any month but are scarce from April to mid-July and rarer gulls recorded include Glaucous, Iceland, Caspian, Ring-billed, Bonaparte's and Laughing Gulls. Turnstone return in late summer and Purple Sandpipers may be seen from late September, though rarely more than 10 are present.

Though New Brighton might not be the first-choice location to look for migrant landbirds and near passerines, it is on the passage line for visible migration and notable finds have included Little Swift, Hoopoe and Yellow-browed Warbler. In some years, Snow Buntings are present in New Brighton. More often they are present in the dunes west of Wallasey Coastguards. Ospreys are recorded most years, while Peregrines breed nearby and may be seen overhead.

For autumn seawatching at least two days of strong westerly winds (force 7 or more) are required for pelagic birds. Once birds are present in the Irish Sea, shorter periods of strong winds may be enough to bring them close to land. Poor visibility also helps. Keep an eye on reports of pelagic species on the west coast of Scotland. It can be worth paying close attention to the river from first light after an overnight gale, when reorienting birds that flew upriver overnight make their way back to sea.

Cheshire

There is always water in the river, but at low tide the tide edge can be surprisingly distant on the coast, so it's worth checking tide tables. Visits at high tide, when the extensive rocky shore and breakwaters are underwater, make finding Purple Sandpipers a great deal easier, since they will be roosting on the Marine Lake pontoons.

25 NEW FERRY SHORE AND PORT SUNLIGHT RIVER PARK

New Ferry shore
New Ferry, Wirral CH62 1DB
SJ341856
shift.option.notice
Parking: Limited street parking available

Port Sunlight River Park
Dock Road North, Bromborough, Wirral CH62 4TQ
SJ349849
swaps.honest.fired

Website: *thelandtrust.org.uk/space/port-sunlight-river-park*
Phone: 07587 550060
Opening times: Port Sunlight River Park car park is currently open Monday to Saturday 9am–4pm
Parking: On-site car park at the River Park
OS Landranger Map 108, OS Explorer Map 266
LNR, SSSI

New Ferry Shore and Port Sunlight River Park are important coastal and riverside habitats on the Wirral, offering a mix of mudflats, saltmarsh and landscaped wetlands. The area supports a rich variety of waterfowl, waders and breeding passerines throughout the year.

SPECIES

Shelduck and Oystercatcher are present throughout the year on the mudflats, while Cetti's Warbler and Reed Bunting frequent the wetlands.

From December to February, Curlew, Redshank and common shore waders feed along the estuary, joined by Pintail and other dabbling ducks on New Ferry Shore.

Skylark sing from early spring and warblers return to breed from late April. Wheatears and other passage migrants may be seen at the River Park, with Whimbrel and Greenshank possible on the shore.

Post-breeding Mediterranean Gull may be on the shore between June and July, while breeding warblers feed fledged young.

Icelandic Black-tailed Godwits arrive from late summer, and wintering waders and wildfowl return.

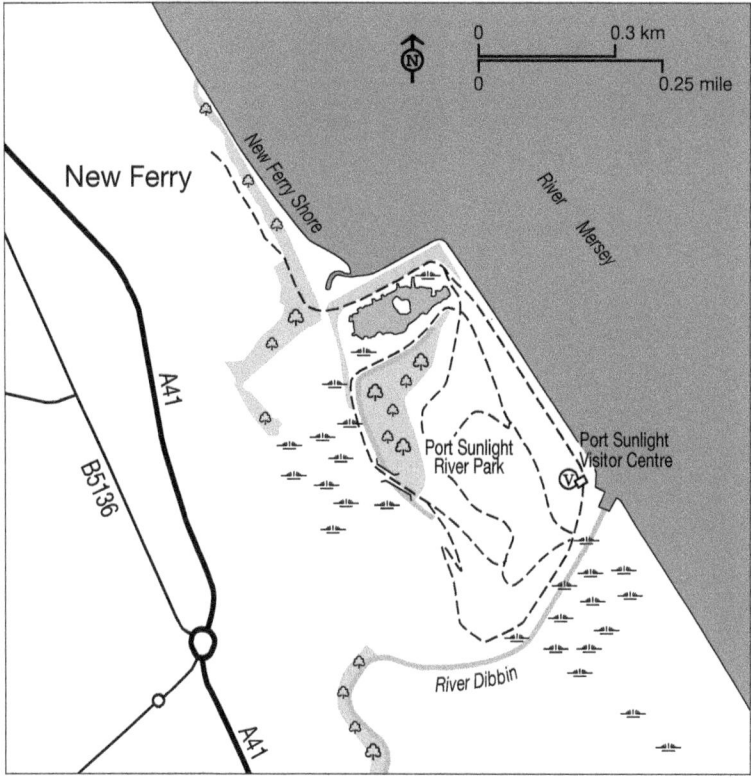

ACCESS
Port Sunlight River Park is about 10 minutes from Port Sunlight station, while New Ferry shore can be reached from Rock Ferry station. To view the New Ferry wader roost and wintering wildfowl, take suburban roads east from New Ferry village and look down from the clay cliff at SJ343855.

YOUR VISIT
Between Rock Ferry oil terminal and New Ferry, there is a stretch of muddy beach with some rocky areas and developing saltmarsh that attracts waders and wintering wildfowl. Port Sunlight River Park was opened in 2014 following the reclamation and landscaping of a former landfill site. Walks offer fantastic views over the river and a section of wetland to the north of the site, which covers 23ha. The River Mersey mudflats are an important site for large populations of waterbirds.

The river mudflats attract waders in winter and at passage periods. Curlew, Redshank and Oystercatcher are usually present on the mud, and rocky areas attract Turnstone and very occasionally Purple Sandpiper. Black-tailed Godwit, sometimes in the hundreds, may appear in early autumn through winter. Shelduck, Mallard, Pintail and Teal are frequent on the shore in winter. Reed Bunting, and Reed, Sedge and Cetti's Warblers breed in the wetland area, Whitethroats and sometimes Lesser Whitethroats hold territories in the scrubby areas and Skylarks sing over the grass. In spring and autumn, both overhead passage and grounded migrants can be seen.

26 WOODLANDS ON THE WIRRAL

Arrowe Park
Arrowe Brook Lodge, Arrowe Brook Road,
Woodchurch, Upton CH49 1AB
SJ265869
packet.spine.forces

Stapledon Woods
King's Drive North, Grange, Caldy CH48 1LH
SJ229860
insiders.silks.lavished

Royden Park and Thurstaston
Hill Bark Lodge, Hill Bark Road, Newton, Frankby CH48 1NR
SJ245862
broth.canoe.sticky

Storeton Woods
Red Hill Road, Storeton, Bebington CH63 6HQ
SJ315840
loss.dawn.soils

Eastham Woods
Green Lane, Heathfield, Bromborough CH62 3QQ
SJ363818
noon.metals.plan

Burton Woods
Mill Lane, Burton Point, Ellesmere Port, Burton CH64 5TA
SJ310745
blatantly.chain.newlywed

Bidston Hill
Wilding Way, Bidston CH43 7QZ
SJ289898
undulation.digs.bring

Brotherton Woods and Dibbinsdale
Spital Road, Heathfield, Bromborough CH62 2AF SJ346828
lush.frames.rubble

Riveacre Valley
Riveacre Road, Overpool, Ellesmere Port CH66 1LJ
SJ383777
option.mint.looks

Cheshire

> Opening times: Most sites are open year-round, from dawn to dusk, with no formal entry restrictions.
> Parking: Each site has on-site or nearby public parking.

To avoid repetition of species from wood to wood, this section gives a general overview of the species to be found in the Wirral's woodlands. The peninsula's geographical position causes it to miss out on a number of species typical of other woodlands. Pied Flycatcher, Wood Warbler, Lesser Spotted Woodpecker, Turtle Dove and Marsh and Willow Tits have been lost as breeders in recent years.

ARROWE PARK

From junction 3 of the M53 take the first exit and continue on this road past Asda on your right up the hill. The park can be accessed on foot via the gate opposite the Arrowe Park public house. Alternatively, there is a car park (SJ265869) with access to a footpath following Arrowe Brook. Woodcock are present in winter, especially in the surrounding fields, and Mandarin may be present on the small lake within the country park at the head of Arrowe Brook.

STAPLEDON WOODS

Take the A540 from West Kirby heading south. Proceed past the boys' grammar school on the left and Kings Drive North is about 275m on the right. Park and enter the woods here. Migrant Wood Warblers are less than annual and Green Woodpecker records have dried up, but for its size Stapledon is a productive piece of woodland. There is a lower path with mature deciduous trees and a central path with birches. The field immediately south of the woods can be good for Wheatear, and a pond at the bottom of the hill on the other side of the A540 attracts 1,000 or more Black-tailed Godwit in early autumn.

ROYDEN PARK AND THURSTASTON

Continue along the A540 until you reach a roundabout, then turn left up Montgomery Hill and right opposite the Farmer Arms public house to bring you to the entrance for Royden Park. Alternatively, continue straight over the roundabout through the sandstone cutting and access Thurstaston Common on the left where there is ample parking (SJ246845). This is quite a large area to cover. Green Woodpecker is still present, Redpoll may be seen in the open, heathy areas, Nightjar has been recorded recently and an Iberian Chiffchaff held territory for two months in 2018.

STORETON WOODS

Take the junction 4 Birkenhead exit off the M53 and continue along this road until you reach Red Hill Road on the left. Park on this road and enter the woods where all the expected species may be seen.

EASTHAM WOODS

Take junction 5 off the M53 and follow the A41 in the direction of Birkenhead until you reach Eastham village road on the right. Follow the road all the way, continuing along Ferry Road until you reach the country park car park at the end. The site is good for Tawny Owl and has occasional Firecrest records.

BURTON WOODS

Take Mudhouse Lane off the A540 at Burton. This continues into Burton village and park. Enter the wood via Mill Lane. Crossbill is seen occasionally and Wirral's last proven breeding of Lesser Spotted Woodpecker was here.

BIDSTON HILL

Take junction 1 off the M53. Follow the Bidston link road towards Birkenhead then the A553 Hoylake Road towards Birkenhead. Turn right by the church up Worcester Road then right into Vyner Road North, where the wood can be accessed at a number of points. Firecrest is annual, while both Yellow-browed and Pallas's Warblers have been recorded.

BROTHERTON WOODS/DIBBINSDALE

From junction 5 of the M53, take the B3157 towards Bromborough along Spital Road, turning right at the mini-roundabout. The entrance is on the right. In addition to a wide range of woodland species, the presence of the River Dibbin adds extra interest. Garden Warbler is frequent here and Cheshire's first Iberian Chiffchaff was next to the car park in 2004.

RIVEACRE VALLEY

From junction 7 of the M53, take the B5132 Netherpool Road and back on yourself onto Riveacre Road to the country park, which has a large car park. In addition to woodland species, Kingfisher is a regular visitor.

LANCASHIRE

No fewer than 381 species have been recorded in Lancashire, a respectable tally on the west coast of England. Given the large estuaries of the Ribble and Morecambe Bay (shared with Cumbria) it's not surprising the shorebird list is particularly high, but thanks to the nature reserve in Liverpool's Seaforth Docks there have even been several American passerines.

In terms of breeding birds the reedbed at Leighton Moss in particular ensures these include Bittern and Bearded Tit. In lowland farming areas there are widespread if declining populations of Corn Bunting and Yellowhammer. Wood Warbler, Pied Flycatcher and Ring Ousel occur in upland woodlands and cloughs. Red Grouse harvested by shooting and several nationally scarce raptor species share the moorland in the east of the county, though the interaction between them is a difficult one.

As well as large wintering wader populations the estuaries are favoured by

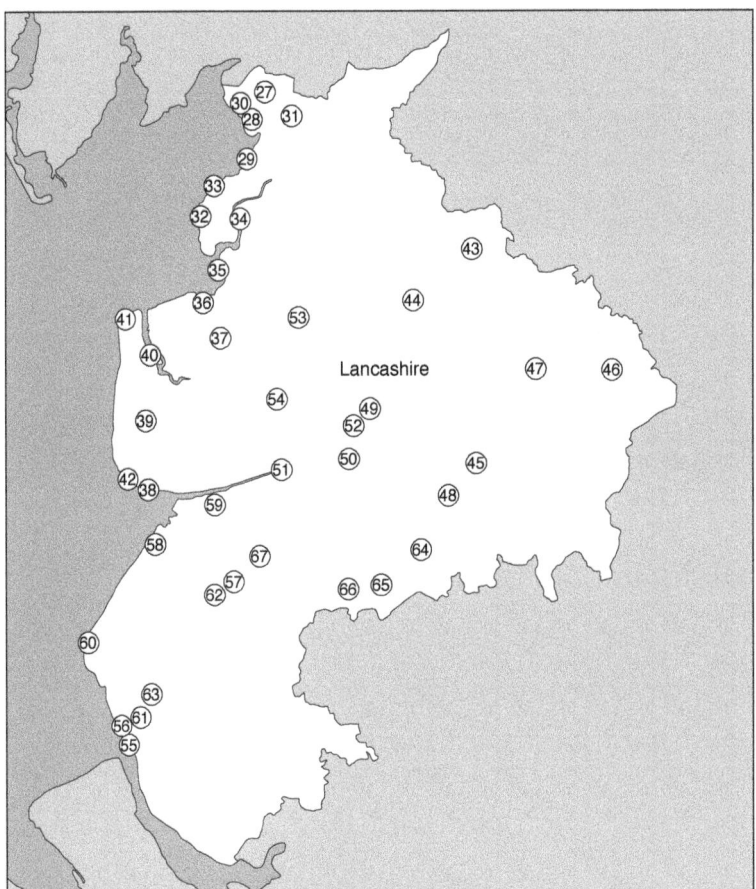

wildfowl for feeding and roosting. These include internationally important numbers of Wigeon, Pink-footed Goose and Whooper Swans. 'Wild goose chasing' is a popular winter birding activity and over the years all the rare goose species and subspecies on the British List have been recorded.

The Irish Sea off the Lancashire coastline is favoured by Red-throated Divers, and particularly Common Scoter that occur in huge flocks. Skuas are seen at passage times, whilst in autumn Leach's Petrels can be 'wrecked' along the county's shores. Many birders from elsewhere have seen their first Leach's in Lancashire.

Inland reservoirs are a magnet for migrating Black Terns and Little Gulls in spring and autumn. Several of them also attract gull roosts which are easier to work than those on the coast, and regularly turn up scarce species including Glaucous and Iceland and increasingly Caspian Gull.

Clifford Oakes' 1953 *Birds of Lancashire* remains a fascinating read on the ornithology of the county. A more recent update on status is provided by the 2008 publication *The Birds of Lancashire and North Merseyside*, edited by Steve White, Maurice Jones and the late Barry McCarthy. A breeding bird atlas edited by the late Peter Golborn and Robert Pyefinch was published in 2001.

NORTH LANCASHIRE

27 LEIGHTON MOSS RSPB

> **Leighton Moss RSPB**
> Myers Farm, Storrs Lane, Carnforth LA5 0SW
> SD476750
> trespass.orange.crumble
>
> Website: rspb.org.uk/days-out/reserves/leighton-moss
> Phone: 01524 701601
> Email: leighton.moss@rspb.org.uk
> Opening times: Nature reserve open dawn-dusk, visitor centre 10am–4pm
> Parking: Reserve car park on Storrs Lane
> OS Landranger Map 97, OS Explorer Map OL7
> Ramsar, SPA, SSSI

For over half a century, Leighton Moss has been one of the RSPB's flagship reserves nationally. The reserve is a key stronghold for Bitterns, Bearded Tits, and a wide variety of other waterbirds, waders and passerines, and its carefully managed reedbeds, meres and woodland areas provide year-round interest for birdwatchers and wildlife enthusiasts.

SPECIES

Bitterns have bred at Leighton Moss for many years, and currently around four nests are occupied each year though pairs would be a misnomer as males mate with up to five females in a season. Satellite reedbeds have been created nearby to provide refuge for dispersing young which in the past lacked suitable habitat when encouraged to leave in the autumn. Birds are most easily recorded from the spring 'booming' calls of the males, like the sound of blowing in a bottle. This can be very far carrying and heard anywhere on the reserve and beyond. Patience can be rewarded with birds feeding in the open or flighting over the reedbed, particularly in harsher weather conditions. In some years winter migrants engage in pre-migration flights at dusk in spring as they prepare to depart, uttering gull-like calls as they circle the reedbed.

Bearded Tits have also become a reliable fixture of the reserve, though in some years many will irrupt and leave the site in late autumn, particularly when high water levels remove access to some food sources. As with Bitterns calls may be the first way to pick them up, distinctive 'pinging' notes unlike those uttered by any other species on site. Post-breeding, and particularly in the months of October and November, the birds use specially provided trays of grit in cut areas of the reedbed alongside the main causeway path. They use this grit to help them ingest food and this is often when the best views can be had, with excellent photographic opportunities of an otherwise active and fairly elusive bird.

Other reedbed species include several pairs of Marsh Harriers, a sizeable Water Rail population, Reed Bunting and Reed, Sedge and Grasshopper Warblers. Cetti's Warbler is a relatively recent but now firmly established colonist, and their strident calls can be heard even if getting a good view is often much more challenging. Water Pipits are scarce regular winter visitors, sometimes showing very well in open areas in front of the hides, and Yellow-browed Warblers have become regular autumn visitors among tit flocks.

A large egret roost in trees on the east side of the reserve regularly includes several Great White and occasional Cattle Egret as well as larger numbers of Little Egret. Spoonbill, Little Gull and Black Tern all drop in on passage, as does Osprey, which may colonise soon as they have started to breed elsewhere in the county. There has been an upsurge in Purple Heron records in recent years in line with general increases in formerly vagrant heron species.

Waterfowl include several breeding duck species with numbers enhanced in winter by immigrants. Wigeon, Shoveler, Teal, Gadwall and Goldeneye are all regular. Tufted Duck and reduced numbers of Pochard breed; the diving duck flock has attracted several Ring-necked Duck over the years and a pair of Lesser Scaup in 2024. Leighton Moss is one of the best county sites for spring passage Garganey, and they have stayed to nest in several years.

Raptors include Hobby, attracted to hunt hirundines in the summer months, Peregrine from nearby nest-sites, Sparrowhawk, Kestrel and fairly regular Hen Harrier. Birds of prey can be attracted to Starling roosts in the reedbeds as well as to the flocks of Swallows and martins feeding over the open water of the various meres. Feeding stations at the visitor centre attract a range of passerine species. Of particular interest is Marsh Tit, with the Silverdale area one of the remaining strongholds in Lancashire. Other birds that may be seen include Nuthatch, Treecreeper, other tit species including Coal Tit and finches such as Chaffinch, Greenfinch and occasional Brambling in winter.

As would be expected at a well-watched site of prime habitat many rare and scarce species have been recorded. These include White-tailed Plover, Squacco Heron, Pied-billed Grebe, Black-headed Wagtail, Penduline Tit, Ross's Gull, Red-footed Falcon, Woodchat Shrike, Montagu's Harrier, Night Heron, Sabine's Gull, Great Reed Warbler, Caspian Tern, Savi's Warbler, White-winged Black Tern, Glossy Ibis, Green-winged Teal and American Wigeon.

ACCESS

Leighton Moss is well signposted from the A6 at Yealand Redmayne, with the reserve car park being 3.5km west of here at the far end of Storrs Lane, just before Silverdale station.

The reserve is free to enter for RSPB members. Non-members need to pay an entrance charge (currently £9). The causeway that goes through the middle of the reserve from SD479752 is a public footpath, so entry to the public hides and viewing of the Bearded Tit grit trays along here can be enjoyed without paying admission or being a member, though all birding visitors are of course encouraged to contribute to the upkeep of this important habitat. There are toilets at the visitor centre.

A walk to all the hides from the visitor centre is around 5km in length, with birding opportunities throughout on the paths as well as from the hides and other viewpoints. Stout footwear is recommended; in winter paths may flood, and wellingtons are then advisable.

The reserve visitor centre is a short walk from Silverdale station on the Lancaster-Barrow railway line. Visitors who are not RSPB visitors but who have arrived by train or on bikes can take advantage of half-priced admission.

YOUR VISIT

Leighton Moss was first leased by the RSPB in 1964 before they went on to buy it and subsequently the associated shooting rights.

The site was acquired for its reedbed habitat, which remains important for key species including Bittern and Bearded Tit as detailed above. Reedbed is by nature a temporary habitat, so ongoing management work involves preventing succession into scrub and maintaining the areas of open water that benefit birds and birders alike. Over time the quality of the reedbed was found to still be degrading and the RSPB has taken steps to address this as well as creating several other similar habitats in the wider Silverdale area (see additional sites below).

Other habitat on the reserve includes woodland around the paths to the Lower Hide, with species including Ash, Oak and Yew. In the wider area there is extensive limestone pavement and in summer a visit to the reserve can be combined with looking for several nationally scarce butterfly species that occur nearby at locations including Gaitbarrows for several fritillaries and Arnside Knott for Scotch Argus.

The reserve visitor infrastructure has been developed in a variety of ways over the years. The former Myers Farm is now a high-quality visitor centre, including a shop on the ground floor and an extensive and fully accessible café above. The five main hides overlook open water and in July 2015 were augmented by the 9m-high Skytower viewpoint which gives panoramic views across much of the reserve. The Skytower adds another dimension to the experience of birding at the reserve, and a vigil on here – whilst exposed to the elements – is recommended other than in inclement weather. At the time of writing the Lower Hide is closed whilst a completely new replacement is built, which is due to open in late 2025.

In terms of seasons the best birding at Leighton Moss is arguably in spring and early summer. As well as booming Bitterns the Marsh Harriers are very active, often engaging in spectacular aerial food passes from the male to the female. The reedbeds are also alive with warbler song at this time, particularly early in the morning from dawn, and there is a good chance of scarce passage visitors.

During the winter an impressive Starling murmuration often occurs; it is worth checking the reserve website and local bird club news for updates on numbers and where on the site they are roosting. This is also generally the best time to see rather than hear Bitterns as their numbers are at their highest, and when dykes and channels are frozen birds may come into the open more often.

NEARBY SITE: BARROW SCOUT FIELDS
SD477736, shiver.jetliner.rocks

The area between the main reserve and the saltmarsh reserve was acquired a number of years ago and has been carefully managed to create satellite reedbed. It can be a particularly good area to see Pochard during the breeding season, as well as scarce migrants including Garganey and Spoonbill. The area can be viewed with a telescope from the elevated vantage points off the Warton Crag Road.

NEARBY SITE: WOODWELL
SD463744, nourished.poster.fountain

The Woodwell area has long been the most favoured site for Hawfinch in north Lancashire, and probably the whole of the county. Numbers had dwindled but more recently a welcome resurgence in fortunes has been reported, including up to four together in early 2024. The site is also, like the wider Silverdale area, a stronghold for Marsh Tit, which has declined significantly elsewhere in Lancashire.

Park carefully at the entrance at SD463744. A public footpath traverses the length of the wood and both species can be seen anywhere along the route. Marsh Tits are vocal and generally easier to locate; Hawfinches are more elusive but can be easier to see in winter when there is reduced leaf cover.

The Lancaster and District Birdwatching Society do regular walks in spring for Hawfinch at other sites in north Lancashire around Gaitbarrows, which are sometimes open to non-members for a small fee. In early 2025 some very good flocks of up to 50 Hawfinches were seen in the Gaitbarrows area. Details are available on the club website and social media pages in the run-up to these events (lancasterbirdwatching.org.uk).

NEARBY SITE: SILVERDALE MOSS
SD471773, fermented.otters.found

As with Barrow Scout this is a satellite reedbed created by the RSPB, partly to create buffer sites for dispersing young Bitterns but they also attract a wide variety of species similar to the main Leighton Moss reserve. Viewing is limited but some of the site can be seen from the minor road north of Waterslack. Hawes Water (SD478766, stuck.claims.concerned) is also potentially worth a visit for waterfowl but is only accessible from public footpaths.

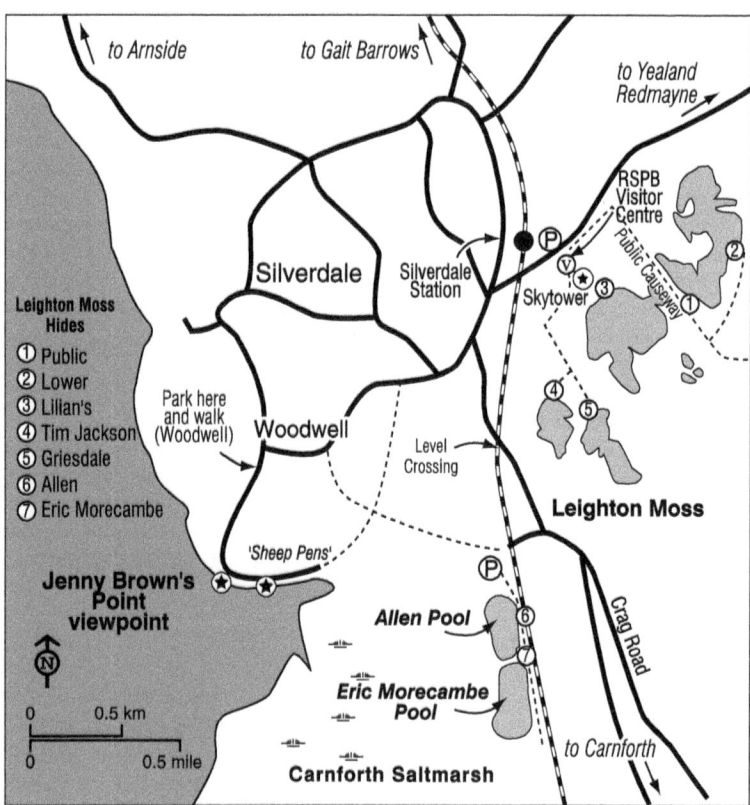

28 CARNFORTH MARSH (MORECAMBE BAY RSPB)

Carnforth Marsh
Off New Road, Warton, Carnforth LA5 9RZ
SD475737
width.duck.definite
Phone: 01524 701601
Email: leighton.moss@rspb.org.uk

Opening times: Accessible at all times
Parking: The car park and entrance paths are accessed at the end of the track from New Road under the Lancaster-Barrow railway line.
OS Landranger Map 97, OS Explorer Map OL7
Ramsar, SPA, SSSI

Carnforth Marsh, managed by the RSPB, is a network of controlled pools, salt-marsh and surrounding fields. The site supports a wide range of passage and wintering waders, wildfowl and other birds.

SPECIES

Regular passage and wintering wader species include Ruff, Green, Wood and Common Sandpipers, Greenshank, Black-tailed Godwit and, in autumn, Curlew Sandpiper and Little Stint. Spotted Redshank numbers have reduced but wintering birds still occur as well as passage individuals.

In terms of scarce wader species Pectoral Sandpiper is recorded in most years. Rare shorebirds that have occurred over the years include Lancashire's first White-tailed Plover and Black-winged Pratincole as well as White-rumped Sandpiper, Long-billed Dowitcher, Lesser Yellowlegs and Wilson's and Red-necked Phalaropes.

Spoonbills are frequently seen in summer and early autumn, and all three regular species of egret visit and roost at nearby Leighton Moss. Teal, Wigeon, Gadwall and Shoveler all use the pools, and Green-winged Teal and American Wigeon have both been recorded with carrier species. Other duck species noted include Scaup, Long-tailed Duck, Garganey and Ruddy Shelduck.

Gravel islands built in the two lagoons hold breeding Avocets, and a Black-headed Gull colony attracts Mediterranean Gulls. The site has held Caspian Tern in the past, ranging between here and Leighton Moss.

The bushes and fields flanking the path between the car park and hides are worth checking for finches and passerines, whilst Woodchat and Red-backed Shrikes have both been recorded.

Marsh Harriers from Leighton Moss hunt the area. Kingfishers can be seen on the fence posts around the pools. The Greylag Goose flock in the area was formerly mostly wild birds at the limit of their wintering range; the situation is a lot less clear-cut these days but ringing studies have shown the flocks do still include some wild birds. As with any goose flock these birds are always worth a closer look in winter for potential scarcer visitors, with Tundra Bean, Snow and White-fronted Geese having been present on occasion.

Lancashire

ACCESS

By car leave New Road at SD477738, head west under the Lancaster–Barrow rail line and then left to the car park. It is a walk of around 500m from the car park to the hides. There is a 2m height restriction accessing the car park.

If travelling by public transport it is a walk of about 1.6km from Silverdale station to the hides. Follow Slackwood Lane from the station, then bear right onto New Road as above.

YOUR VISIT

The RSPB created pools on the edge of Carnforth saltmarsh with sluices to control water levels in the 1980s. Other than when high flood tides have inundated the area it is one of the best sites for passage saltwater waders in Lancashire. Two hides enable close viewing opportunities. One of the hides is named in memory of Eric Morecambe, the famous comedian who was a keen birdwatcher himself.

As well as the Allen Pool and the Eric Morecambe Pool there are also more natural floods further from the hides that sometimes attract additional species. One of the benefits of visiting by train is that close if sometimes relatively brief views can be had of these other floods, and larger scarce species such as Spoonbill and Great White Egret may be identified on the approach to/departure from Silverdale station.

April to early June and late June to early October are the best times for passage waders. As these dates suggest, there is only usually a couple of weeks of 'summer' when birds are not moving through. High tide is generally the best time to visit with more birds pushed onto the pools as saltmarsh is covered, but on very high tides the site can be flooded out and unattractive to waders until feeding areas gradually become exposed again.

Avocets and the gull colony are both present from April, and Avocets have generally left by August and are rare in winter. A visit during or after a thunderstorm or other rainfall can be profitable as overflying waders and terns can be forced down. On hot sunny days heat haze can be a problem.

Visitors to the site should be RSPB members or have purchased a day ticket for Leighton Moss. For visitors not arriving by train the wider flood cannot be seen from the hides but may be viewed distantly, along with Barrow Scout Fields, from viewpoints off the elevated road over Warton Crag.

29 HEST BANK

Hest Bank
The Shore Road, Hest Bank, Lancaster LA2 6HW
SD468665
vegans.greyhound.explores

Website: rspb.org.uk/days-out/reserves/hest-bank-at-morecambe-bay
Phone: 01767 693690
Email: wildlife@rspb.org.uk
Opening times: Accessible at all times

Lancashire

Parking: Available at Shore Road, Hest Bank, LA2 6HN (not an RSPB car park)
OS Landranger Map 97, OS Explorer Map 296
Ramsar, SSSI

Hest Bank is an important coastal site on Morecambe Bay, which is renowned for its large winter and high-tide roosts of waders and wildfowl. It offers excellent birdwatching opportunities from the shore.

SPECIES

Hest Bank is one of the best sites from which to see concentrations of shorebirds in north Lancashire. Whilst to some extent numbers have been diluted by small wader species utilising sea defence groynes in Morecambe, high tides can still generate a large roost at Hest Bank. Knot, Oystercatcher, Redshank and Curlew are among the most regular species here. Whimbrel flocks have been recorded in several springs.

Twite flocks have occurred on the saltmarsh in some winters and may be under-recorded given the amount of available habitat. Skylark and Scandinavian Rock Pipit are also seen when flushed by high tides. A variety of duck species may be

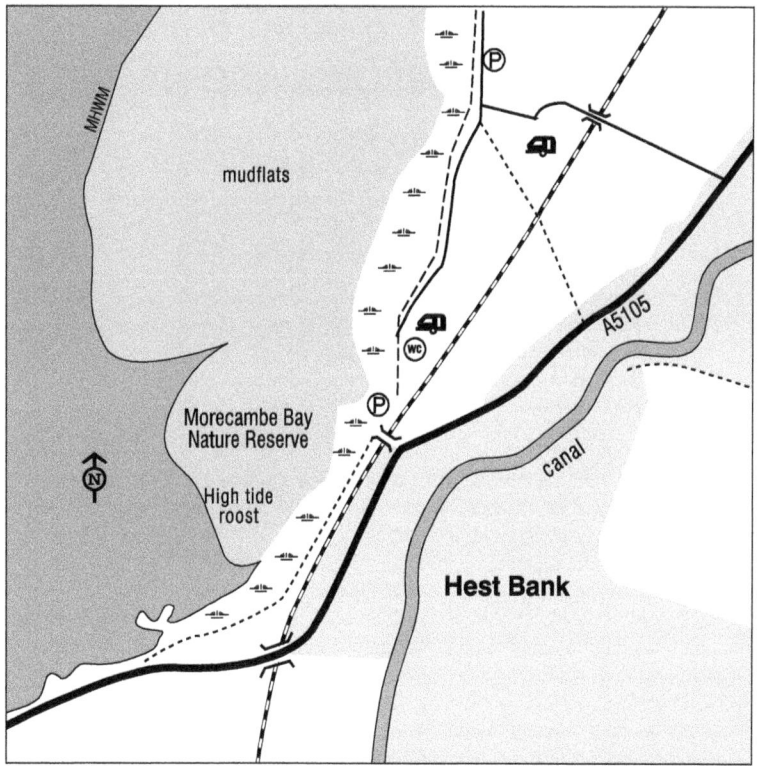

seen at high tide or in low tide channels including flocks of Eider, Pintail, Wigeon, Teal and Mallard, as well as small numbers of Red-breasted Merganser.

ACCESS

From the A5105 north of Morecambe enter Hest Bank and turn left going over the level crossing at SD468665. The railway is the busy West Coast Main Line so there may be delays on arrival or departure for trains heading in both directions. There are a couple of rough car parks for which there is no charge. There are no public toilets, but the Shore Café has facilities for the price of a tea or coffee.

If arriving by public transport there is a half-hourly bus service between Carnforth and Lancaster; at Carnforth the bus stop is at the railway station so convenient for onward connections. The bus stop is a short walk north of the railway crossing.

YOUR VISIT

High spring tides outside the breeding season are the most productive, with the largest numbers of waders in winter. If visiting by car, it can be possible to enjoy the shorebird roosts without leaving your vehicle if visiting at high tide.

It is probably the closest site to the M6 where there is a good chance of seeing Eider. If visiting other than at high tide, the footbridge over the level crossing provides some elevation to allow better scanning of the bay for more distant ducks and other waterbirds; these potentially include divers and grebes.

30 JENNY BROWN'S POINT

Jenny Brown's Point
Lindeth Road, Silverdale, Carnforth LA5 0UA
SD462734
husbands.fleet.cabbies

Opening times: Accessible at all times
Parking: Roadside parking available
OS Landranger Map 97, OS Explorer Map OL7
Ramsar, SAC, SOA, SSSI

Jenny Brown's Point is a small but productive seawatching site on the edge of Morecambe Bay. It offers excellent views of seabirds, passage terns and occasional rarities, as well as low-tide waders.

SPECIES

Jenny Brown's Point is on the face of it an unlikely seawatching venue, being far into Morecambe Bay and not a particularly pronounced headland. A number of dedicated patch-watchers have however shown it can be extremely productive in the right conditions. It is particularly good in spring when skuas advance well into Morecambe Bay, presumably in some cases then gaining height and continuing overland towards the East Coast. All four regular British species have occurred

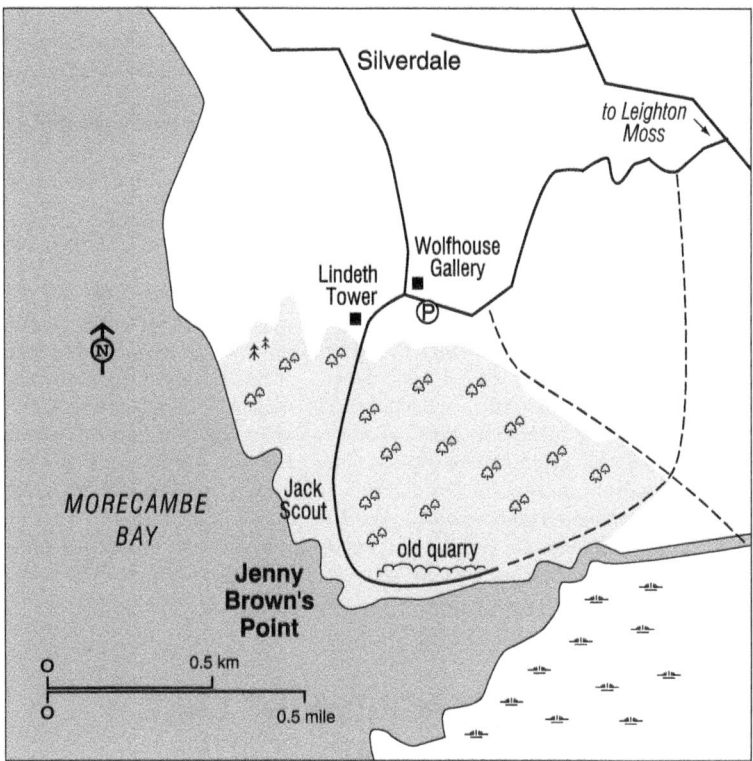

and Pomarine Skuas have been seen in many springs. At other times in the spring Kittiwake and tern passage may be marked, with Arctic and Black Terns regular.

In south-west winds seabirds can also be blown this far into the bay, particularly in autumn storms. Leach's and Storm Petrels have been recorded, and there have been a couple of occurrences of Sabine's Gulls, including a lingering adult in July 2022. In calmer conditions Red-throated Diver can be seen and regular vigils have produced several records of Scaup as well as Long-tailed Duck, Velvet Scoter and Slavonian Grebe. In May 2007 Lancashire's only Black Scoter was found on the railway nearby and was subsequently released here.

As a small headland, it can produce passerine migrants. The most surprising of these by some considerable margin was Britain's first 'Caspian' Reed Warbler, which was picked up dead in December 2011. There have been several records of Yellow-browed and Pallas's Warblers, and Common Rosefinch, Chough and Firecrest have occurred. It is fair to say that there has been relatively little coverage of the area for passerines in recent years, and if more time were devoted to this some rewards would follow.

Species that are regular at the nearby RSPB lagoons can also be seen particularly at low tide, including Greenshank, Spotted Redshank and in autumn Curlew Sandpiper, as well as scarce herons and allies including Great White Egrets and Spoonbills.

ACCESS

The site is accessed by vehicle by turning off on the minor road south at the junction of Lindeth Road and Hollins Lane in Silverdale. Head for SD462734 and park wherever it is safe to do so, given that other vehicles are present.

Although there is no cover in the event of inclement weather there is access from Crag Foot crossing by public footpath, approximately 1.6km away. This gives options to incorporate Jenny Brown's Point into a green birding trip, arriving at Leighton Moss by train.

YOUR VISIT

Arrive at least a couple of hours before high tide and if time allows stay until the tide begins to ebb. A telescope is essential to get good seawatching results. Most prolonged sea vigils are undertaken in spring, partly because overland migration is more pronounced at this time, and partly because there is a lack of cover and average conditions are challenging in autumn at this exposed site. It is worth checking for suitable weather conditions in advance, and even on the day of a planned visit itself; monitoring for news coming out from Heysham and Morecambe Stone Jetty will indicate whether passage is under way. If visiting for autumn seawatching it is advisable to wear layers to keep the cold off.

South-easterly winds are also the best conditions in which to look for passerine migrants. Reports of Yellow-browed Warblers from better-watched sites in the region, such as Heysham and Hilbre, are a useful indicator of when to look.

31 DOCKACRES/PINE LAKE WATERS, CARNFORTH

Dockacres/Pine Lake Waters
Pine Lake Resort, A6 Warton, Carnforth LA6 1JZ
SD509720
paler.driftwood.clues

Opening times: Accessible at all times, but see below for considerate usage of the site
Parking: There is no dedicated on-site parking (see below)
OS Landranger Map 97, OS Explorer Map OL7
Ramsar, SAC, SPA, SSSI

Dockacres and Pine Lake Waters are a series of flooded gravel pits near Carnforth that provide important habitat for wintering and passage wildfowl, and diving ducks.

SPECIES

Gravel extraction to the north-east of Carnforth has created several now-flooded pits that are attractive to wildfowl and Coot in particular. All the sites have extensive leisure usage currently; those to the west of the M6 are given over to holiday

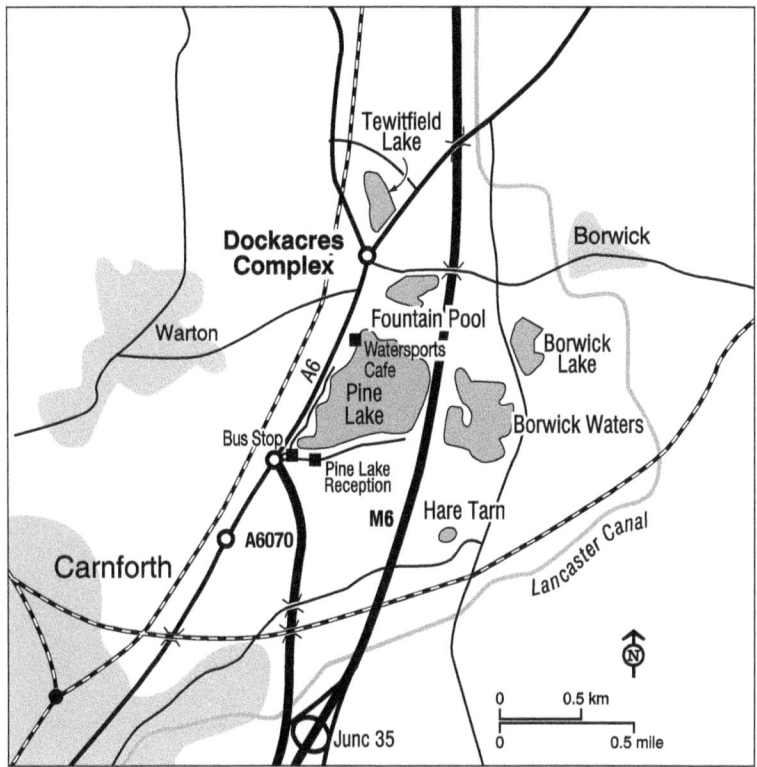

cabins and those to the east are fisheries. As a result, Pine Lake (the largest expanse) is the only one that generally holds significant waterbird gatherings. Scarcities can turn up on any of them, though, and when disturbed from the larger site (e.g. by speedboat activity), birds will temporarily move to the others.

Although numbers of diving duck have dwindled due to the increased disturbance, Pine Lake still attracts reasonable numbers of Tufted Duck and Pochard outside the breeding season. These have a proven track record of pulling in scarce *Aythya* species, including regular Scaup (even double-figure flocks), multiple Ring-necked Duck and Lesser Scaup, Red-crested Pochard and Ferruginous Duck. The area has also been very productive for Smew in the past and, in hard weather, birds could still occur. Long-tailed Duck occasionally occur in winter whilst Common Scoter drop in on passage.

Whilst the site is not particularly productive for divers, Red-throated can turn up and a Great Northern lingered for several weeks in the 2024/25 winter. There have been several occurrences of rare and scarce grebes, notably Lancashire's first Pied-billed in 1997 and also Black-necked and Slavonian. In storms, more maritime species can take refuge; there have been multiple Gannets and Shags and there has been a single record of Storm Petrel. There isn't generally a significant gull roost or pre-roost, but birds drop in to bathe and as well as Mediterraneans among the Black-headed, both Glaucous and Iceland have been seen in the past.

The position of the site on the Keer Valley flightline helps the area to attract passing terns and Little Gulls, particularly in spring. The list of tern species seen includes Whiskered and White-winged Black as well as Sandwich, Arctic and Common. Similarly, in the right conditions large numbers of Swallows and martins, particularly Sand Martins, can swarm over the water of Pine Lake, and whilst Red-rumped Swallow hasn't occurred to date, it feels like only a matter of time before this species is recorded as it has been at nearby Leighton Moss.

In terms of passerines other than hirundines, Siskin flocks can occur in the trees around Pine Lake. Pied Wagtails are regular around the lake edges and may be joined by White Wagtails in the spring and less regularly Yellow Wagtail. Great White Egrets have been regular in the area in recent years, including alongside the Lancaster Canal.

ACCESS

General vehicular access for birdwatchers is not currently permitted for Pine Lake. The 555 bus service between Lancaster and Kendal actually stops on site, however, with a half-hourly service for much of the day. There are cafes at the southern end in the main complex, and at the north end in the water-sports centre, so there is no reason why birders can't visit on the bus and use the facilities whilst doing so. It is important to respect the privacy of residents, but there are several benches alongside the lake and an open vantage point beyond all the chalets on the western shore towards the water-sports centre.

Limited viewing of the other sites is possible from the A6070 (Twin Lakes: SD514732, fades.soggy.offstage), Borwick Lane (South Lakeland Leisure Village: SD515728, stickler.tucked.united) and Kellet Lane (Clearwater Fisheries: SD520722, songbook.glitz.regard). A public footpath runs south from the canal near Borwick (SD524729, layover.exhaling.deflate) from which another footpath branches south-west, offering views of Birchwood Lake.

YOUR VISIT

Winter is the best time for diving ducks. The best numbers occur when other sites in the area are frozen over. There is also less water-sports activity at this time of year. When birds disperse from Pine Lake they may go to neighbouring waters but can also relocate to Leighton Moss.

For Little Gulls and passage terns late April and May are the peak times. Any showery conditions with light winds can deliver, but thunderstorms and south-easterly conditions are optimal. Tern passage is often most productive in the afternoon; on the other hand wildfowl numbers may be greater early in the day before any potential disturbance.

NEARBY SITE: HOLMERE

A seasonal flood known as Holmere is adjacent to the A6 east of Yealand Conyers (SD517744, shipwreck.starch.darting). There is parking in a lay-by on the north-bound side of the A6 just to the north of the site; cross the road carefully to view, as traffic can be both intensive and fast. This site can attract good numbers of Black-tailed Godwit and other waders when water levels are conducive, as well as Gadwall, Shoveler, Tufted Duck, Wigeon and Little Grebe.

NEARBY SITE: HARN TARN

Another seasonal water known as Hare Tarn is shown on the OS map at SD517717 (scarecrow.spellings.revives). This can also be a good site for Black-tailed Godwit,

Little Ringed Plover and hirundines when water levels drop to produce muddy edges. Access is from the minor road that runs north-west out of Carnforth – park carefully to avoid blocking roads – or by walking south from Borwick Lake on public footpaths.

32 HEYSHAM AREA INCLUDING MIDDLETON

Heysham Nature Reserve
Moneyclose Lane, Morecambe LA3 2UW

Middleton Nature Reserve
Main Avenue, Morecambe LA3 3NT
SD407601
inferior.lilac.laminate

Websites: lancswt.org.uk/nature-reserves/heysham-nature-reserve; also heyshamobservatory.blogspot.com for regular sightings updates.
Opening times: Heysham Nature Reserve 9am–5pm or dusk; other sites in the area open at all times.
Parking: Available at the North Harbour Wall, Half Moon Bay Cafe, Moneyclose Lane and Middleton Nature Reserve
OS Landranger Map 97, OS Explorer Map 296
Ramsar, SPA

The Heysham area, including Middleton Nature Reserve, is a diverse coastal and industrial landscape that supports a rich variety of birds. Despite the presence of a freight terminal and power stations, the area provides an important habitat for waders, seabirds and migrants.

SPECIES

The Heysham area includes a major freight ferry terminal and two nuclear power station buildings but still attracts an excellent variety of birds in several different habitats. Whilst not officially accredited by the Bird Observatories Council, a bird observatory is operated from offices at Heysham nature reserve and undertakes regular mist-net ringing of passerines as well as wider recording of the area, including seabird passage. There is another nature reserve to the south of Heysham, adjacent to Middleton Industrial Estate.

The bushes, trees and open ground around the nature reserve can attract good numbers of grounded passage migrants, particularly in easterly winds and especially south-easterlies. To some extent increased tree and bush cover on the peninsula has made it more difficult to find the birds away from the two nature reserves, and many migrants probably pass through undetected. As well as Whinchats and Wheatears, Ring Ouzels, Redstarts, Black Redstarts, Pied Flycatchers and several species of warbler are annual in spring. Autumn as well as producing migrant thrushes invariably sees several Yellow-browed Warblers and usually some Siberian Chiffchaffs.

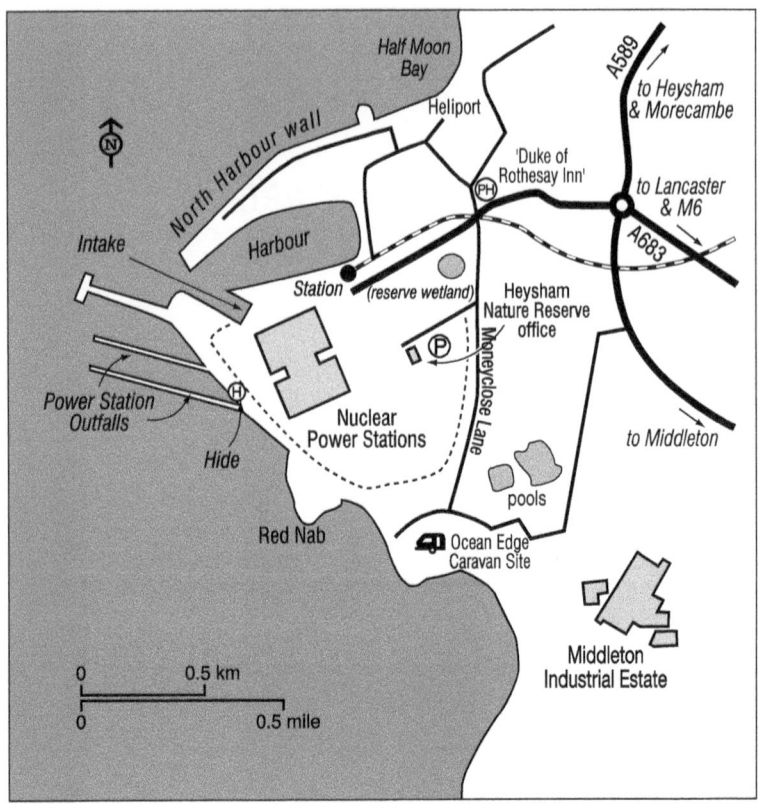

An impressive range of rarities has been recorded over the years, including Tawny Pipit, Thrush Nightingale, White-throated Sparrow, Wryneck, *blythi* Lesser Whitethroat and Icterine, Melodious, Dusky, Marsh and Barred Warblers. A Subalpine or Moltoni's Warbler was recorded in June 2018. Visible migration can also be productive with good numbers of pipits and wagtails passing overhead, and Grey Wagtail is a long-term migration study species, with birds attracted down and colour-ringed on site. Unusual species seen and heard going overhead have included Serin, Woodlark and Bee-eater.

The grassed area of the heliport and the north harbour wall here provide normally undisturbed high tide roosting for waders. Knot occur in their thousands, with Redshank, Oystercatcher, Turnstone and Dunlin also to the fore. It is a good site to read colour rings and leg flags, viewing from the southern edge of the adjacent Half Moon Bay beach.

The power station outfalls attract gulls and terns. The installation of grilles to reduce fish deaths has accordingly diminished the numbers of birds but flocks do still feed in the area. Species seen include Common and Arctic Terns, Kittiwake and Little Gull whilst rarities have included Sabine's Gull and a Bonaparte's Gull in 2013. Mediterranean Gulls gather in large numbers with Black-headed Gulls in late summer and early autumn and have adapted to taking worms from the mudflats.

The harbour itself can be worth a check, particularly after sustained onshore winds. Guillemots and Kittiwakes occur after storms, and the Cormorant roost at the derelict former jetty is also one of the best Lancashire sites for Shag. Black Guillemots have summered in the harbour in the past but are most likely to be seen on a seawatch and have generally become scarcer in the county of late.

Seawatching is these days largely undertaken from vehicles parked carefully on the North Harbour Wall road. The best strategies vary with the season. Spring seawatching peaks in April and May and can be as productive in easterlies as in onshore winds. Early morning watches on a rising tide are often the most fruitful. Species include Red-throated Divers, Kittiwakes, terns and auks and the site is particularly good for skua passage. Birds that continue into Morecambe Bay often then gain height and move overland, and these include annual Pomarine as well as Arctic and Great Skuas. Passage seaduck are often seen at this time in the offshore channels, including Long-tailed Duck, Scaup and Common and Velvet Scoters with the possibility of Great Northern Diver and Slavonian and Black-necked Grebes.

Autumn seawatching is more dependent on westerly winds, particularly sustained south-westerlies, which can push Leach's Petrels into Morecambe Bay; these then battle their way past the harbour as they reorientate. Sabine's Gull, Long-tailed Skua and Grey Phalarope have all been recorded several times at this season. Summer is generally quiet but Manx Shearwater and Sandwich Tern are regular. Little Gull can appear in winter gales and is still annual although numbers have dwindled; Little Auk is a rare visitor at this time. Eider is present in the area all year round.

Middleton Nature Reserve has several pools that attract ducks, including Tufted Duck, Wigeon, Shoveler, Teal and Gadwall, and loafing or bathing gulls. Cetti's Warbler and Reed Bunting are resident here and Grasshopper Warbler is a summer visitor.

In recent years Brent Geese wintering in south Cumbria have increasingly made visits to Heysham to feed on eel grass, and it is now comfortably the most regular site for them in Lancashire. Numbers of birds and the particular areas they favour vary, so it is worth checking the excellent Heysham Observatory blog before visiting; this is usually updated daily in some detail. Rock Pipits nest at a number of sites along the coastal fringe and on the harbour wall, and Snipe and Jack Snipe can be seen departing from Ocean Edge saltmarsh on higher tides.

ACCESS

Follow the A683 bypass from the M6 junction 34 to the traffic lights (SD408602) by the Duke of Rothesay pub. For the North Harbour Wall and heliport roost turn right here, park on the Harbour Wall road for in-car seawatching; there is also a pay-and-display car park by the Half Moon Bay café. For the nature reserve turn left at the same lights down Moneyclose Lane and the car park is on the right after 300m. For Ocean Edge and the power station outfalls walk south from here to pick up the seawall in front of the power station at SD403591, then head north-west.

For Middleton Nature Reserve from the A683 turn left at the roundabout for Middleton Lane at SD414602. Turn right for the industrial estate road (Main Avenue) at SD420591 and there is a car park on the right by the largest pool. There are paths to view other parts of the reserve.

Several bus services from Lancaster and Morecambe serve Heysham. The best stop for covering the main Heysham sites with the least amount of walking is

Combermere Road (SD413603), which is served by the 2X service half-hourly from Monday to Saturday and hourly on Sunday.

YOUR VISIT

Early mornings are generally best for spring seawatching and looking for migrant passerines before dog-walking activity and other disturbance increases. If visiting for the North Harbour Wall wader roost, timing your visit for high tide is essential. Seawatching is generally better on an incoming tide but scanning low tide channels can be surprisingly productive.

The more coastal parts of the recording area are usually most productive in winter. Birds to look for include feeding Brent Goose, Rock Pipit and Eider with at high tide a large Knot roost and other shorebirds on the harbour wall. Kittiwake, Little Gull and Shag are possible in the harbour, on the old jetty or around the outfalls, particularly after strong onshore gales for the former two species.

In spring seawatching from the North Harbour Wall can be productive for terns, divers, skuas, Kittiwake, Manx Shearwater and auk species including, increasingly, Puffin. Heysham nature reserve can produce Wheatear, Whinchat, Ring Ouzel and passage warblers, and visible migration may include Whooper Swan, Redpoll, pipits and wagtails. Migrating Osprey can occur over land and offshore, particularly on calm days.

In summer Cetti's Warblers breed; listen for their strident song at Middleton Industrial Estate, and in early autumn birds from elsewhere may pass through. Large numbers of Black-headed Gulls gather and attract locally significant numbers of Mediterranean Gulls. Terns can be seen feeding on the outfalls. Passage wader species occur off Ocean Edge, and there is a chance of Storm Petrel in strong onshore gales.

In late autumn, visible migration includes flocks of Pink-footed Goose, Meadow Pipit, Fieldfare, Redwing, wagtails and finches. Other passerine migration includes Chiffchaff, Blackcap and Yellow-browed Warbler as well as Wheatear and the occasional Black Redstart.

33 MORECAMBE PROMENADE

Morecambe Promenade
Stone Jetty, Marine Road Central, Morecambe LA4 4NJ.
SD425647
stone.baked.smashes

Opening times: Accessible at all times
Parking: Several public car parks are available along Marine Road
OS Landranger Map 97, OS Explorer Map 296

Morecambe Promenade is a coastal stretch that offers birdwatching along the bay, with opportunities for seabird passage and wintering waterfowl. Its groynes, sea defences and adjacent parks provide a mix of habitats within easy reach of the town.

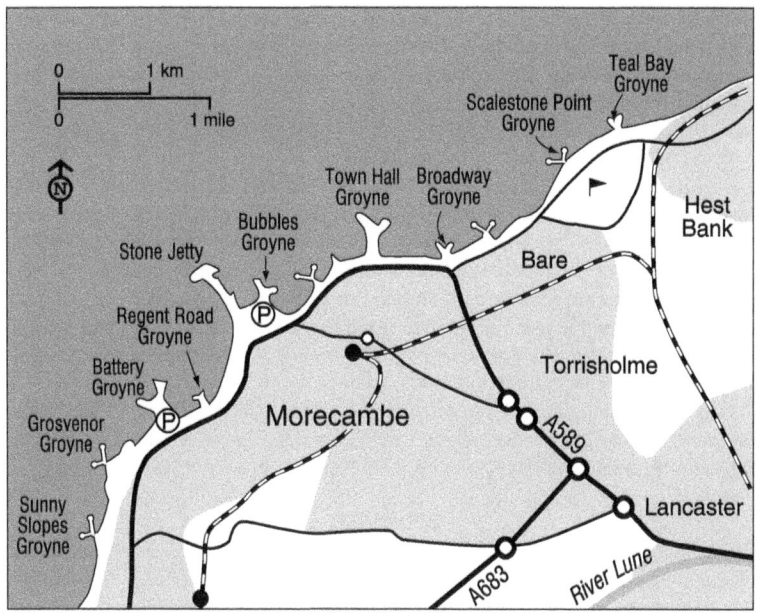

SPECIES

Although less well covered than Heysham to the immediate south, good birding can be had along Morecambe Promenade, particularly during seabird passage seasons and in winter. Although the birdlife has changed significantly with the closure of the Sandylands sewage outflow, and roosting sites have adapted to sea defence works, several interesting species can still be seen.

The groynes between the Stone Jetty and Happy Mount Park provide refuge for waders at high tides. Very small numbers of Purple Sandpipers can still be seen in the winter months among the Oystercatchers, Dunlins, Turnstones and other waders. Curlew Sandpiper is occasional in autumn. Eider and Common Scoter can be seen offshore, and Scaup are more or less annual, whilst Long-tailed Duck and Velvet Scoter can occur. The sea defences have proved attractive to Black Redstart, and Snow Bunting also occurs along the Promenade on occasion.

The Stone Jetty was in the past a good place to watch large numbers of birds drift into the bay on the tide in winter. Numbers of Red-throated Diver and Great Crested Grebe have declined in recent years, but both can still be seen in calm conditions and Great Northern Diver and Slavonian Grebe occur sometimes. It is in spring and autumn that watching from here is now most productive. All the seabird species that occur off Heysham also pass here, including skuas and terns in spring and Leach's Petrel, Grey Phalarope and Sabine's Gull in autumn. As with other coastal watchpoints calm spring days can also produce passage raptors, including Osprey and Hen Harrier. Auks occur offshore and can include Black Guillemot and Puffin. A Forster's Tern passed close inshore in March 2025.

Like any seafront town there are frequently concentrations of gulls on the seafront and the sands. These can reward close attention, since in the past Glaucous, Iceland and Yellow-legged Gulls have all been seen and Little Gull and particularly Mediterranean Gull can be regular.

The amount of cover in the area, including on Morecambe Golf Course, disperses landfalling passerine migrants but Yellow-browed Warbler is regular in autumn. Happy Mount Park can be productive, as well as small areas of cover near the seafront further south.

ACCESS

The end of the Stone Jetty (SD425647) is accessed on foot only from the Promenade near the lifeboat station. Groynes are obvious in the field and on an OS map and can be viewed from SD435647, SD440649 and SD445649. Scalestones Point (SD445657) and the area around Bare can be good for Scaup. Happy Mount Park is accessed off Marine Road East at SD453653.

One of several car parks on Marine Road can be selected depending on the areas being checked. Bay Arena car park (217 Marine Road, beam.cared.winner) is the nearest public car park to the Stone Jetty.

Public transport is a good option with the railway station 300m from the Stone Jetty and regular bus services along the Promenade; the latter run to both Hest Bank and Heysham if combining either of these areas in a day's birding.

YOUR VISIT

Spring seawatching is generally best early morning, ideally on incoming tides. Winter calm-weather seawatches are best started three or four hours before high tide. Autumn seawatching is more weather dependent and is best after prolonged south-westerly gales. Watching waders on the sea defences is best done at or very close to high tide. Passerine migrant searches are most likely to be successful in easterly and particularly south-easterly winds.

On Saturday mornings a parkrun along the Promenade currently attracts several hundred runners from 9am in the area around the Stone Jetty, so this time may be best avoided for that part of the Promenade.

34 LUNE ESTUARY

Lune Estuary
Aldcliffe Hall Lane, Lancaster LA1 5AS
SD459601
talking.louder.remarked

Website: birdingaldcliffe.blogspot.com
Opening times: Accessible at all times
Parking: Several small car parks are available, including Aldcliffe Lane, Glasson, Conder Green and Sunderland Point
OS Landranger Map 102, OS Explorer Map 296
Ramsar, SAC, SPA, SSSI

The Lune Estuary is a rich coastal system stretching from Lancaster down to Sunderland Point and Glasson. Its mix of saltmarsh, mudflats, tidal channels and freshwater pools supports large numbers of waders and wildfowl.

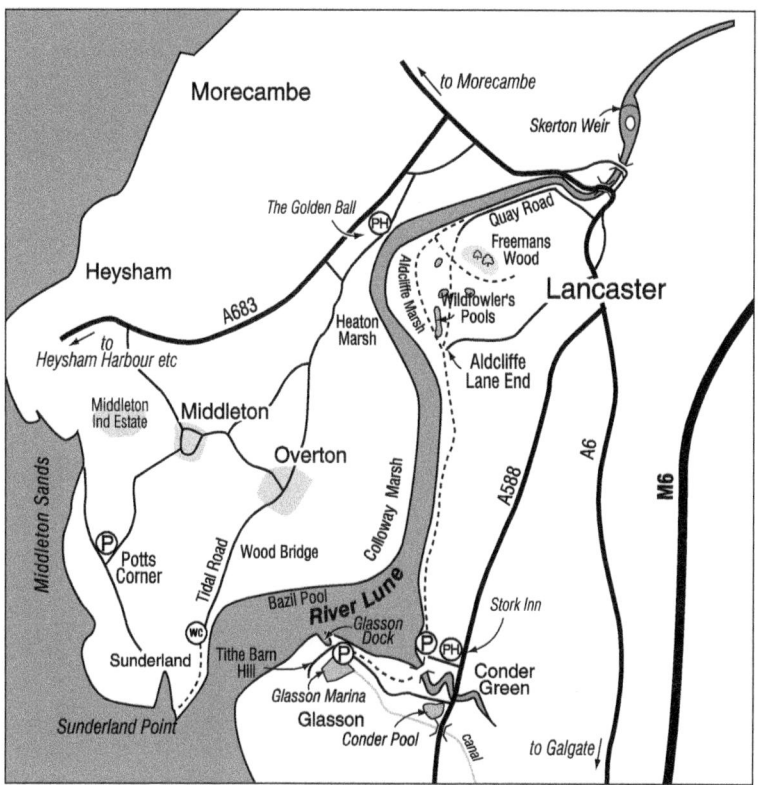

SPECIES

This account follows the previous edition of the book in defining the Lune Estuary as extending from downriver of Skerton Weir in Lancaster to the mouth of the estuary at Middleton Sands and Sunderland Point on the west shore, and Glasson on the east side. There is increasing disturbance on the estuary from jet skis, drones and even hovercraft, but on the other hand several new habitats have been created around the estuary that can offer very productive birding.

The River Lune is bordered by extensive saltmarshes as it leaves Lancaster, which can attract large numbers of waders including Curlew, Lapwing, Redshank, Snipe and Golden Plover throughout the year and Dunlin, Ringed Plover and Black-tailed Godwit on passage, with Curlew Sandpiper and Little Stint on occasion. Bar-tailed Godwit is regular, though the summering non-breeding flock seems to have ebbed away. Avocet is now a summer fixture with several pairs breeding. Wigeon and Teal both occur in numbers, and Eider is now an established visitor to the outer estuary. Increasingly, Pink-footed Geese use the area in winter, particularly at Aldcliffe. The egret population is also increasing, with large numbers of Little and smaller numbers of Great and Cattle Egrets, many of them roosting at Ashton Hall on the east shore north of Conder Green. A number of pools attract freshwater waders including Common, Green and Wood Sandpipers and Little Ringed Plover. Aldcliffe in particular can be good for migrant passerines.

Although the gull interest of the area has diminished with the closure of the landfill site at Salt Ayre there can still be good numbers of birds on the estuary, even on occasion in Lancaster itself but more often between Glasson and Colloway Marsh. Scandinavian Rock Pipits occur regularly outside the breeding season, and when they are pushed off the marshes by high tides Water Pipits are also increasingly picked out among them. The tern populations on the estuary itself may have dwindled away but there is now an established breeding colony of Common Terns at Conder Pool.

ACCESS

GLASSON
Excellent views of the estuary can be had both looking west from near the junction of Bodie Hill and Tithe Barn Hill (SD442560, talking.louder.remarked) across Glasson Marsh towards Sunderland Point, and north from behind the bowling green and toilet block in the village (SD446561, sadly.sofas.condensed) towards Colloway Marsh. The latter usually gives a wider variety of species, both in the channel and on the saltmarsh. Highlights can include Greenshank, Eider, Mediterranean Gull and Yellow-legged Gull. Rarities have included Ring-billed Gull, American Golden Plover and White-winged Black Tern.

In addition, Glasson Basin (SD445559, prop.foremost.grasp) holds diving ducks and grebes in winter with Goldeneye regular in small numbers and Scaup occurring most years. It can be viewed from the car park off the B5290 entering the village, or from the south off School Lane by going through a gate next to the school at SD444557 (avoid school pick-up and drop-off times).

CONDER GREEN/CONDER POOL
The River Conder meanders through saltmarsh at Conder Green between the A588 and the old railway line footpath alongside the estuary. The area is attractive to waders including Redshank and Greenshank, and Ruff and Spotted Redshank are regular. Teal flocks can be seen, and Green-winged Teal has joined them on occasion. Rarities have included Lesser Yellowlegs and Broad-billed Sandpiper. The site can be viewed from the Glasson Dock road, the A588 or Corricks Lane leading to Conder Green Picnic Site.

The picnic site itself offers an alternative vantage point for viewing the estuary to Glasson, with better views of birds in the Conder channel. The footpath along the railway line here can be used to get to Ashton Hall to view birds heading over to roost on the lake there (SD457575, proved.pulled.steadily).

Conder Pool is an area of land reserved by the Environment Agency (EA) for flood protection. The EA has allowed the Fylde Bird Club and others to undertake some management work including the installation of tern rafts. There is a viewing platform and screen next to the lay-by on the B5290 at SD457557 (tooth.corrupted.manicured). The site has nesting Common Terns, Avocets and Black-headed Gulls. Outside the breeding season it hosts a variety of ducks and shorebirds, including Wigeon and Greenshank, as well as regular Kingfisher. Rock and Water Pipits can be seen on flood tides, and Short-eared and Barn Owls hunt the surrounding grassland.

ALDCLIFFE
The Aldcliffe area has a range of habitats and is one of the best sites on the Lune Estuary to see a wide variety of species. There are freshwater pools and some

woodland. Floods on the saltmarsh here attract wildfowl and waders, and the nearby pasture can also hold wild geese in the winter.

There is a small car park at the end of Aldcliffe Lane (SD459601, talking.louder.remarked). The embankment here gives views across the saltmarsh, and it is then possible to take a circular walk north along the Lancashire Coastal Way and back along the old railway line. This route covers the main habitats, including brackish floods on Aldcliffe Marsh, whilst a number of pools at SD456615 attract good numbers of duck, including Wigeon and Goldeneye with occasional scarcer species. The Freemans Wood area is worth checking as a relatively isolated patch of cover; some of it is now managed as Fairfield Nature Reserve with feeding stations.

The favoured areas for waterbirds in the area vary with the timing and extent of tidal flooding, and it is worth checking the recent sightings of the Lancaster and District Birdwatching Society before choosing which location to visit. In many years the main Lancashire Coastal Way footpath north of the car park can be impassable in winter due to flooding.

HEATON MARSH

The western side of the estuary is generally much more difficult to work but it can be viewed from the road known as Lancaster Road in the vicinity of the Golden Ball pub at Snatchems (SD448615, sprint.rival.expand). This area is best at high tide, when birds are pushed closer to the road, and species like Snipe and Jack Snipe may be flushed, as well as saltmarsh pipits.

SUNDERLAND POINT

Sunderland Point is a charming hamlet accessed by a tidal road from Overton. Birding from the vehicle when crossing avoids disturbance. There is rough parking for several vehicles in the vicinity of the toilet block at the end of the road (SD426562, winter.coasting.desk) from where the main Lune channel can be scanned. Following the public footpath south to the point can be productive for migrant passerines in spring and autumn, as well as shorebirds. There are also pools off The Lane near the Sambo's Grave monument. There is a bird hide at SD421560 (surprise.reports.bloomers), oriented to view Morecambe Bay rather than the Lune Estuary.

The approach road is impassable for several hours some days, and care is required to avoid being cut off. A dedicated website sunderlandpoint.net gives more details and includes tide times for the following few weeks. If intending to visit on a date when tide conditions are not favourable one option is to drive to Potts Corner car park (SD413572, gullible.jolt.twins) then follow the coastal footpath south-east for 1.6km, birding Middleton Sands en route.

PUBLIC TRANSPORT

For those able to walk reasonable distances it is possible to take the coastal way from Lancaster railway station along the river to the Aldcliffe sites mentioned above. Lancaster is a West Coast Main Line station with regular train services. A limited bus service (every two hours) runs between Lancaster and Knott End. This can be taken to access Conder Green and Glasson Dock, and one option then is to walk back to Lancaster, birding Aldcliffe on the way back.

YOUR VISIT

A rising or ebbing tide is generally best for shorebirds at most of the saltmarsh sites above, whilst wildfowl can be best at high tide. At Conder Green the creeks can be rewarding at all states of the tide other than the highest high tides when they are generally deserted by birds in favour of the safe roosting at Conder Pool. Spring tides are best for getting good views of Scandinavian Rock and Water Pipits. February and March are the best times to observe wild geese.

35 COCKERSAND (INCLUDING BANK END)

Cockersand
Moss Lane, Cockerham, Lancaster LA2 0AZ
SD430530
belief.rail.retiring

Opening times: Accessible at all times
Parking: Limited roadside parking is available near Bank End (SD449527) and near the entrance to Cockerham Sands Holiday Park for Cockersand
OS Landranger Map 102, OS Explorer Map 296

Cockersand, including the Bank End area, is a coastal and estuarine site. Its diverse habitats, including saltmarsh, tidal flats, flooded fields and small inland pools, support a wide range of waders, wildfowl, seabirds and passerine migrants throughout the year.

SPECIES

Although a little off the beaten track, the Cockersand area has been covered regularly by a series of local patchers and has proved very productive. The variety of species to be seen reflects the diversity of habitats in the area.

The intertidal area attracts waders, and at high tide a roost forms at Plover Scar, which can be viewed without disturbance to the birds from the Lancashire Coastal Way north of Cockersand Abbey. In spring flocks of Dunlin and Ringed Plover occur as well as smaller numbers of Grey Plover and Bar-tailed Godwit. In the autumn Curlew Sandpiper and Little Stint are more or less annual. Ruff and Spotted Redshank occur, though both are generally easier at nearby Conder Green. Rarities have included Broad-billed Sandpiper, Long-billed Dowitcher and Kentish Plover. The Golden Plover flock has proved to be very productive for American Golden Plover, with several individuals over the years making it one of the best sites for the species in north-west England.

The fields inland of the coastal path are also productive. When there is stubble this can attract Lapland Buntings, particularly in autumn. Corn Buntings occur in the area in small numbers, though the species is generally in decline locally. Pastures are frequented by wild swans and geese. In recent years the area by the approach road around Clarkson's Farm and Gardener's Farm has been particularly favoured by swans, with flocks of several hundred Whoopers and one of the few remaining small herds of Bewick's Swans in the region. Wintering Pink-footed Geese occur in

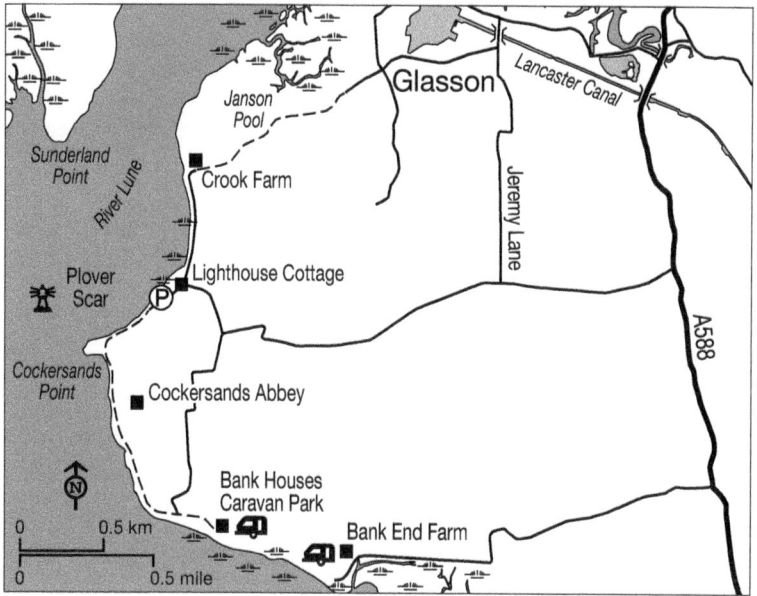

this area in good numbers each winter and will usually carry Tundra Bean Geese, Eurasian and Greenland White-fronted Geese and Dark-bellied and Pale-belled Brent Geese as well as rarer species. Also in the winter months flocks of Twite and one or two Snow Bunting frequent the fields closest to the estuary.

Although overshadowed by Heysham to the north, as a headland of sorts the area can be good for passerine migrants. Black Redstart and Whinchat in particular are recorded regularly. Scarce migrants in the area have included Red-backed Shrike, Wryneck and Yellow-browed Warbler. Lancashire's only confirmed Buff-bellied Pipit to date was seen in a ploughed field here in May 2014. The fields also attract hunting harriers at times, with both Hen and Marsh occurring, whilst in 2018 a juvenile Pallid lingered for a day. Dotterel have occurred on passage and singing Quail have been noted on occasion. Cattle Egrets are increasingly frequent with livestock and Short-eared Owls hunt the fields most years.

Eider breed locally and are usually easily seen on the estuary. Flocks of Wigeon and Teal are regular and can repay close scrutiny with American Wigeon and Green-winged Teal both having occurred, the latter on multiple occasions. When the inland fields are flooded Gadwall and Shoveler also occur, and Garganey may be seen. There have also been interesting seaduck records over the years, with small flocks of Scaup and occasional records of Long-tailed Duck and Velvet Scoter. Black-necked Grebe has occurred offshore more than once in recent years. On high tides pipits can be flushed off the saltmarshes; Rock Pipits (mostly Scandinavian) generally predominate but Water Pipits do occur.

Cockersand arguably punches above its weight in terms of seabird records. Whilst it would never be as productive as sites projecting further into the bay such as Heysham the species list is surprisingly extensive, including Arctic and Great Skuas, Leach's and Storm Petrels, Manx Shearwater, Kittiwake, Sabine's Gull, Fulmar, Arctic and Black Terns and Grey Phalarope.

The south of the area at Bank End is one of the best locations locally to see Yellow Wagtail, with several birds together on the sheep-grazed saltmarsh most autumns. As with Cockersand, Twite flocks can occur and Wheatear is regular in spring and autumn. The same area also hosted a couple of Shorelarks for several weeks in the winter of 2016/17. A high-tide wader and gull roost forms here at the mouth of the River Cocker off Bank Houses Farm. This can attract Mediterranean Gulls and in autumn Curlew Sandpipers and Little Stints.

ACCESS

The area is accessed by minor roads from off the A588 between Cockerham and Glasson Dock as follows.

Bank End: leave the A588 and head west at SD462352 down Hillam Lane. Parking for a couple of cars is available just before the seawall at SD449527. If time permits it can be worth stopping at the start of Hillam Lane in the lay-by on the left to view the flooded pits of Cockerham Quarry (SD460530, lists.breakfast.transit), which can attract waders and wildfowl. In winter the Mute Swans roosting on the water here are often joined by Whooper Swans. Sand Martins have nested in the area and these attract Hobbies.

Cockersand: leave the A588 at Brigg's Brow, joining Moss Lane (SD460545, corrosive.plunge.degrading), and head east to the junction at SD433541. Here you can go right along Slack Lane to a small parking space at Lighthouse Cottage (SD430543, heat.lightly.squirted) though this car park has deteriorated and is not recommended, or preferably left to park on the upper foreshore near the entrance to Cockerham Sands Holiday Park. From both parking areas, the Lancashire Coastal Way footpath can be followed to Cockersand Abbey. If combining with birding at Glasson Dock/Conder Green instead approach from the north down Jeremy Lane, which can be productive for wild swans and geese.

The roads to Cockersand, in particular, are narrow and can be relatively busy for their size with holiday park traffic so care is advised. If stopping en route to view fields, park carefully to avoid blocking entrances to farms and fields.

Although there is a bus route to Cockerham from Lancaster, public transport is not a great option for birding Cockersand. If already birding the Glasson area by bus, walking the Lancashire Coastal Way to Lighthouse Cottage would be an option, and passes some good wildfowl wintering and breeding habitat at Janson Pool (SD435552, gosh.patio.rainwater).

YOUR VISIT

Before and over high tide is best for waders as birds feeding in Morecambe Bay get pushed inshore. There are high-tide roosts both at Bank End and – more easily viewed – on Plover Scar off Cockersand Abbey. If visiting in winter for Whooper and Bewick's Swans and wild geese check Lancaster and District Birdwatching Society or Fylde Bird Club sightings pages for currently favoured locations. The lay-by on the northbound side of the A588 at SD460543 offers some elevation to scan the fields for groups of geese and swans. For passerine migrants in spring, and Twite and Snow Bunting in winter, the Lancashire Coastal Way footpath gets very busy around the Abbey as the day progresses and early morning visits may be more profitable.

Lancashire

THE FYLDE COAST

36 PILLING MARSH/COCKER'S DYKE

Pilling Lane Ends Amenity Area
Backsands Lane, Pilling PR3 6AU
SD415495
apart.buddy.evolving

Opening times: Accessible at all times
Parking: There is a car park just off the A588 on Backsands Lane
OS Landranger Map 102, OS Explorer Map 296
Ramsar, SPA

Pilling Marsh and Cocker's Dyke both offer sweeping views of a large part of southern Morecambe Bay. Pilling Marsh is particularly productive for Pink-footed Geese and wildfowl, whilst Cocker's Dyke attracts roosting gulls.

SPECIES

Both these adjacent sites give extensive views of the southern expanse of Morecambe Bay, but offer different habitats. Pilling Lane Ends has a slightly elevated perspective over the saltmarsh of Pilling Marsh, whereas Cocker's Dyke gives a vantage point over open sands. This gives a different balance of key species, and a visit to both is often the best option if time permits. Lane Ends Amenity area also has a couple of pools and some tree cover, both of which have attracted interesting species over the years.

Pilling Marsh is particularly known for the large roost of Pink-footed Geese that can be viewed from the embankment by the car park in the winter months, departing to feed in fields early morning and returning to roost at dusk. On higher tides it

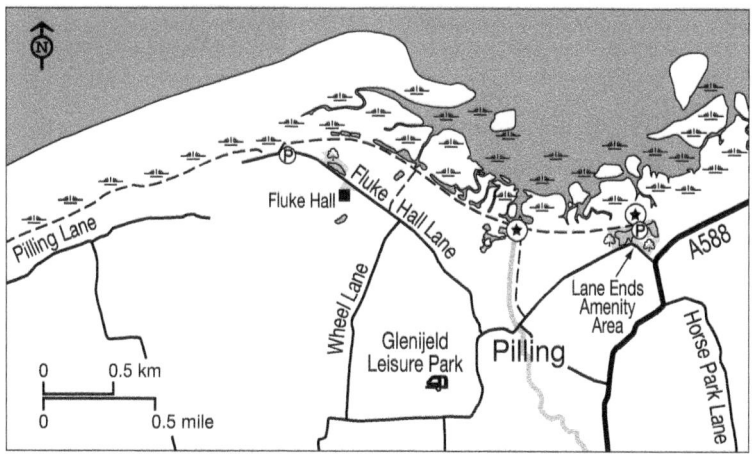

137

can also be a good site to view wildfowl and waders pushed close in to the embankment, particularly from where Pilling Water flows into Morecambe Bay at SD406495. Large numbers of Teal and Wigeon are often joined by Pintail and, more unusually for a saltwater location, Shoveler. The pools around the car park have breeding Little Grebe and Reed Warbler and a nocturnal egret roost, and they attract passage ducks. The fields inland of the embankment can be attractive for passage passerines including Wheatear and Whinchat in spring and autumn.

Pink-footed Goose flocks of several thousand birds often attract wild geese of other species. Just about every goose species that occurs in Britain has been seen here including Barnacle, Taiga and Tundra Bean, Ross's, Snow, Lesser White-fronted, Red-breasted, Eurasian and Greenland White-fronted, Black Brant and other races of Brent Geese. The origin of some of these birds is open to debate but many are clearly wild.

Other rare or scarce visitors on or over the marsh have included Pacific Golden Plover, Buff-breasted Sandpiper, Montagu's Harrier, American Wigeon, Glossy Ibis and Richard's Pipit. Great and Cattle Egrets are increasingly frequent. Perhaps surprisingly all four regular skua species have been seen on the marsh, including wintering Pomarine. The trees around the pools held a long-staying Iberian Chiffchaff in 2019 and have attracted increasingly regular autumn Yellow-browed Warblers. Slavonian Grebe, Long-tailed Duck and Garganey have occurred on the ponds.

Cocker's Dyke flows out into Morecambe Bay between Knott End and Pilling Marsh, and the fresh water attracts loafing large gulls. As well as several Caspian Gulls there have been regular Yellow-legged Gulls, with both species at the north-western edge of their normal range, and a number of Glaucous and Iceland Gulls. Little Stint and Curlew Sandpiper are seen here most years among Dunlin flocks, and a Semipalmated Sandpiper was present in November 2013. A Pied or Eastern Black-eared Wheatear was on the sea wall here in September 2019. The fields behind the sea wall can be productive for passerine migrants including Wheatear, Whinchat, Cuckoo, Yellow Wagtail and Ring Ouzel.

ACCESS

Lane Ends Amenity area: there is a two-tier car park just off the A588 on Backsands Lane. The higher level, furthest from the road, affords views of the marsh from vehicles but only holds five or six cars and can be full.

Cocker's Dyke: view from the sea wall where the dyke enters the sea. Park sensibly on Pilling Lane at the start of the track (SD374493, king.sometimes.spans) and walk north.

Between the two sites Fluke Hall is another option, though there can be a lot of disturbance on the sands here since there is an access ramp off the car park (SD389500, booth.pockets.wriggled). In addition to the species mentioned above, birds seen here have included Red-breasted Flycatcher, Firecrest, Twite feeding in fields south of the car park, and Blue-headed Wagtail in spring.

In terms of public transport, Cocker's Dyke can be reached on regular Blackpool to Knott End buses. The Lancaster to Knott End bus service passes fairly close to Lane Ends with a stop at the opposite end of Backsands Lane.

YOUR VISIT

For viewing the goose roost early morning is best. Be aware that wildfowling may be taking place on the marsh, and this may be upsetting to some birders.

Alternatively visit at dusk to see birds returning from inland though views are generally less satisfactory at this time. For waders and wildfowl arrive before peak tide, view from the car park or alternatively where Pilling Water enters the marsh below the embankment at SD406494. Early morning is best for migrant passerines along the sea wall before disturbance from dog exercising increases.

At Cocker's Dyke low or ebbing tide is better to view gulls or feeding waders. There is often a smaller Pink-footed Goose roost here, which is best viewed in early morning, but other species are more likely to be encountered in the larger roost off the amenity area.

37 OVER WYRE FIELDS

Bradshaw Lane, Eagland Hill, Pilling PR3 6SN
SD430452
digit.scarecrow.courts

Feeding station webpage: fyldebirdclub.org/fylde-feeding-stations
Email: fyldebirdclub@gmail.com
Opening times: Accessible at all times
Parking: Limited roadside parking is available along Bradshaw
OS Landranger Map 102, OS Explorer Map 296
Ramsar, SPA

The Over Wyre fields around Eagland Hill and Pilling provide winter feeding grounds for wild geese and support some of Lancashire's most threatened farmland birds.

SPECIES

The agricultural fields around Pilling and Cockerham, and in particular those around the hamlet of Eagland Hill, have two main attractions for birders. Throughout the winter Pink-footed Geese that roost on Pilling Marsh use the area for feeding. Numbers fluctuate as some birds leave for Norfolk before returning in the New Year. From late winter into early spring birds also range more widely including onto the marshes around the Lune Estuary. The flocks attract regular waifs of other goose species, the most regular being Barnacle, White-fronted of both subspecies, Light-bellied and Dark-bellied Brent and Tundra Bean Geese. Rarities have included Ross's and Lesser White-fronted Geese in the past (considered to be wild), with more recent highlights of Red-breasted, Todd's Canada, Snow and Taiga Bean Geese. Whooper Swans also still use the fields around Eagland Hill, with Bewick's occasionally joining them, but numbers have reduced as the area around Glasson has become more favoured.

The other main interest is farmland breeding birds, which are increasingly scarce locally with changes in agricultural practices. In particular, these include Yellowhammer and Corn Bunting, whilst Tree Sparrow, which is faring rather better, occur in very good numbers. All three species may visit farmland bird feeding stations provided in the Eagland Hill area by local birder Bob Danson and the Fylde

Lancashire

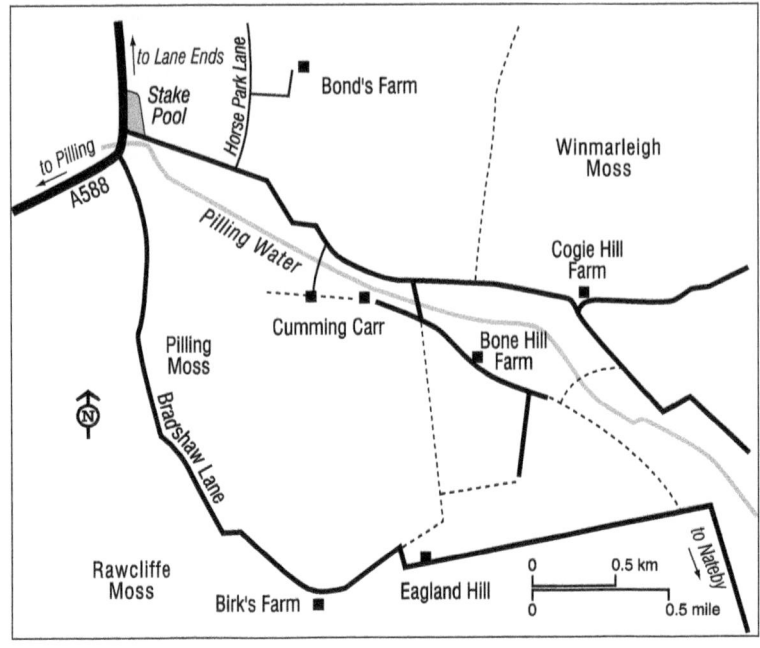

Bird Club. Currently the feeding stations are off New Lane at SD439452, and on a track in the hamlet itself at SD430451, whilst there is sometimes an additional feeding programme at Bradshaw Lane Head at SD417456. It is worth checking the Fylde Bird Club website's dedicated feeding stations page as locations may change. Always view feeding stations from the car to avoid disturbing the birds and give way to farm traffic and local residents at all times.

Most years Quail can be heard calling in the summer. Raptors include Barn Owl, Sparrowhawk, Kestrel and Peregrine all year round whilst Short-eared Owl and both Marsh and Hen Harriers can be seen in winter. On occasion flooding in the stubble fields can attract gulls, waders and wildfowl, and depending on the time of year birds may include Little Gull, Kittiwake, Little Stint, Curlew Sandpiper, Ruff and Spotted Redshank. Rare and scarce birds over the years, in addition to the geese above, have included Lesser Yellowlegs, Buff-breasted Sandpiper, Crane, Dotterel and Great Grey Shrike, whilst the feeding stations hosted an Ortolan Bunting in May 2009, and a juvenile Purple Heron unexpectedly wintered in Eagland Hill in 2019/20.

ACCESS

By car, in Pilling Village turn off south at SD412477 towards Scronkey, then after 600m when the road heads west continue south along Bradshaw Lane to cover the Eagland Hill sites. On reaching the T-junction at Copthorne Farm (SD454443) head east into Nateby and proceed back along Black Lane and Garstang Road to cover more habitat at Bone Hill, Cogie Hill and Cumming Carr. There are several other sites in the area that can reward attention including Cockerham Moss, Sand Villa, Braides and Great and Little Crimbles. Winmarleigh Moss (SD454478) is lowland heath which, whilst only accessible from public footpaths, regularly attracts Cuckoo, Hobby and Tree Pipit. It is also the only site in Lancashire with Large Heath butterflies.

Note that some of the roads in the area, including the one through Eagland Hill, are narrow with passing places and in some cases very uneven so care is required. When parking, always show appropriate consideration to residents and stick to public footpaths at all times when leaving roads. Bear in mind that shooting activity can be taking place in the area.

'Wild goose chasing' is not really suited to public transport birding, but it is possible to look for farmland birds on foot using the 88 and 89 bus services, which stop in Stakepool at the Elletsons Arms, with some services also stopping at Nateby.

YOUR VISIT

Geese will use different fields to feed in during the season, and even during a single day if accidentally disturbed or deliberately moved on. One option is to watch the birds depart the roost at dusk from Pilling Lane Ends Amenity Area and see where they head. If you do find wild geese, ideally view from some distance with a telescope to avoid disturbing them, though they will sometimes tolerate close approach when shooting activity is less intensive.

For farmland birds in winter visit the feeding station sites as covered above. Otherwise for singing birds, visiting in late evening before dusk in May–June is recommended. Unless there is significant farm traffic, singing Corn Buntings, Yellowhammers and even Quail can be heard at relatively long range on calm evenings. Driving through Eagland Hill and stopping considerately in passing places is perhaps the best strategy for listening for these species.

38 RIBBLE ESTUARY NORTH SHORE

Fairhaven Lake Visitor Centre
Inner Promenade, Lytham Saint Annes FY8 1BD
SD341274
silver.putts.elections
Opening times: Accessible at all times
Parking: Limited parking is available at Fairhaven Lake Visitor Centre, along Seafield Road (free), at the RNLI lifeboat station near Lytham Jetty
OS Landranger Map 102, OS Explorer Map 286

The north shore of the Ribble Estuary is a quieter alternative to the more intensively watched south bank, but it supports a wealth of wintering wildfowl, waders and breeding birds.

SPECIES

The Ribble Estuary is one of the most important sites for shorebirds and wildfowl in Britain. Key wintering birds include Wigeon, Pink-footed Goose, Oystercatcher, Bar-tailed Godwit and Icelandic Black-tailed Godwit. Spring brings large numbers of Dunlin and Ringed Plover en route to Arctic breeding grounds. Breeding birds

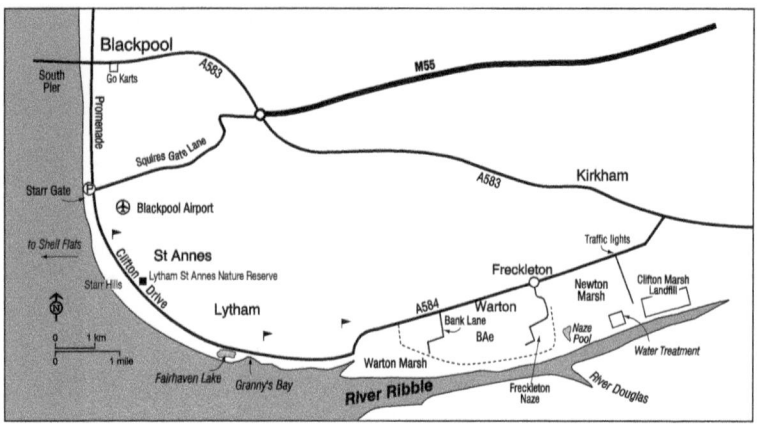

include Lesser Black-backed Gull and declining wader species including Lapwing, Redshank and Snipe. Continental Black-tailed Godwit breed most years.

The north bank of the Ribble Estuary lacks some of the 'mod cons' of the south in terms of nature reserves with hides and visitor facilities. There have also been more habitat management works for birds on the south bank. As a result, the north shore area is considerably less well watched, but can be very rewarding. Regular recent coverage of the Lytham–Warton stretch has shown some of the rarities, and scarcities on the south bank can be viewed from across the river or move between both sides of the Ribble.

ACCESS

NEWTON MARSH
SD455292, onions.paint.safe

This grazed freshwater marsh is managed to help sustain the important wader species nesting here. It has been the northern outpost of nesting Continental Black-tailed Godwits and also holds good numbers of Lapwing, Redshank and Snipe. Most of the site is not easily viewed but the Entrance Pool along the minor road to Clifton Marsh is most accessible and produces most of the reports. Rarities have included Citrine Wagtail, Collared Pratincole and Red-necked Phalarope, the Wigeon flock in winter has held American Wigeon, and Green-winged Teal has also been recorded among Eurasian cousins. Ruff and Garganey are regular passage migrants. The site is the best on the Fylde for Temminck's Stint; late May is the best time for this species.

By car, leave the A583 at Clifton and follow the A584. After 1.6km or so there is a set of traffic lights and a small garage on the left. The Entrance Pool is down this road on the right. Birders are advised not to leave their car if possible as the pool is very close to the track. The garage may park cars along this road, in which case viewing may be difficult without leaving your vehicle. There are also lay-bys further west along the A584 towards Freckleton from which the rest of the marsh can be scanned distantly. The nature of the site makes public transport less of an option than other nearby sites.

FRECKLETON NAZE
Naze Pool: SD434276, shred.decoder.tigers
Naze Point: SD434272, refilled.paler.flaked
These two sites are accessible from the start of the Lancashire Coastal Way. Freckleton Naze Pool is viewed from the elevated footpath above Freckleton Pool. The pool attracts wildfowl and waders, and there is also a large egret roost in the trees on the island. Viewing of the pool has become increasingly difficult because of vegetation growth, but some of the birds that occur here can also be seen at Naze Point at other states of the tide.

Continuing to the estuary along the Lancashire Coastal Way there are benches at Naze Point, near the trig point overlooking the confluence of the Douglas and Ribble rivers. As well as Greenshank, Spotted Redshank and sandpipers that occur on the Naze Pool large groups of geese can be viewed on the marshes across the river. Very large numbers of gulls can be observed passing west in late afternoon as they head to roost. Although the Lancashire Coastal Way follows the estuary from here to Warton care is advised as the footpath is frequently extremely wet immediately west of the Naze.

Rarities and scarcities here have included Collared Pratincole, Cattle and Great White Egrets, Ruddy Shelduck, Honey Buzzard, Spoonbill, Hobby, Sabine's Gull, Blue-headed Wagtail and Great Grey Shrike. In the past a sewage farm attracted scarce and rare waders but this has been replaced by a more efficient but less ornithologically rewarding water treatment works.

In terms of public transport, the 68 bus service from Blackpool to Preston stops regularly in Freckleton and also Warton (see Warton Bank below).

WARTON BANK
OS SD403274, redeeming.crashing.balconies
This area of marsh on the estuary viewed from the end of Bank Lane includes areas that are only flooded on the highest spring tides. As a result it is a surprisingly reliable site for Water Pipit, and breeding species have included Gadwall and Water Rail. As with other sites on the outer estuary it can be a good site from which to look for Hen Harrier, Marsh Harrier and Merlin in winter, particularly towards dusk. In some years Short-eared Owls hunt the marsh, and when present can be flushed by the spring tides that also afford the best views of saltmarsh pipits.

Get off the bus in Warton or if driving park sensibly at the northern end of Bank Lane. The car park at the east of the estuary end of Bank Lane is owned by wildfowlers and is private. Alternatively, park at the east end of Lytham and join the sea wall at SD396279.

LYTHAM
Lytham Quays: SD379271, galleries.thread.convinces
Lytham Jetty: SD367266, ombudsman.summer.smelter
Seafield Road: SD354269, huts.pronouns.translate
There are several options for viewing the estuary in Lytham. The embankment behind the housing development at Lytham Quays has become increasingly popular with local birders viewing the saltmarsh and river channel. This can afford views of birds on the far side of the estuary including regular Spoonbill in recent years. Hen Harrier and Merlin can also be seen from here in winter. Park sensibly

on adjoining roads and pick up the footpath overlooking the estuary to the rear of the Danbro offices.

West of Lytham Windmill the car park by the RNLI lifeboat station overlooks Lytham Jetty (SD367266). The jetty is public and protrudes well out into the estuary, giving good views both up and downriver. This is the best site on the north bank from which to view Eider. Low tide is best. Take care when the tide has ebbed as the wooden slats of the jetty can be slippery when wet.

Finally Seafield Road (SD354269) has free car parking with convenient access to view Church Scar, which is a good area for feeding shorebirds such as Black-tailed and Bar-tailed Godwits and duck including Wigeon and Pintail.

There is a regular train service to Lytham from Preston and Blackpool. Frequent buses run from Blackpool and also the 68 route from Preston, as above.

FAIRHAVEN

Although something of a tourist trap and heavily disturbed, Fairhaven offers several birding habitats and options in close proximity. Granny's Bay can hold thousands of shorebirds, particularly Dunlin and Ringed Plover in spring. An Ivory Gull was seen here in 2002. Fairhaven Lake has proved to be a good site for seaduck, particularly Scaup, as well as having several records of Grey and Red-necked Phalaropes and Red-necked Grebe and a regularly returning Ferruginous Duck.

Scattered trees along the edge of the lake and others near the pumping station to the west attract passerines in spring and autumn. Highlights over the years have included Nightingale, Melodious Warbler, Woodchat Shrike, Ortolan Bunting, Dusky and Barred Warblers and regular Yellow-browed Warbler. This area is also on a particularly productive visible migration flightline, and in autumn in particular morning vigils on calm days have produced good totals of many species; these have included local scarcities including Richard's Pipit, Lapland Bunting and Crossbill.

There is a train station at Fairhaven (Ansdell and Fairhaven), which is a 10-minute walk from Fairhaven Lake. For many years, there was an RSPB interpretive centre between the lake and the Inner Promenade, though its future is currently under review.

YOUR VISIT

Newton Marsh is most productive during spring passage. Early morning visits are best to reduce the likelihood of disturbance to birds near the road. Take care when the recycling centre is open as there is regular HGV traffic along a relatively small road.

The Ribble from Naze Point is best viewed on the incoming tide, or towards dusk as gulls head past to roost. Warton Bank comes into its own on spring tides in late winter and early spring, when Water Pipits and Short-eared Owls are flushed by the advancing tide.

Lytham Quays can be productive at all tide states and in winter is a good area to view harriers hunting before dusk.

Lytham Jetty is often covered by the river and is best visited at lower states of tide when it gives a viewpoint upriver and across to outer Banks Marsh. Fairhaven can be productive at all times. The often large numbers of waders feeding in Granny's Bay may also roost in view if not disturbed too much. If visiting for birds other than shorebirds, storms can be good for dropping terns and hirundines onto

Fairhaven Lake, and early morning is best for passerine migrants with visible movement most pronounced and grounded birds least disturbed by visitors.

39 MARTON MERE

Marton Mere
Mythop Road, Blackpool FY4 4XN
SD339353, third.overnight.spell

Website: blackpool.gov.uk/Residents/Parks-and-community-facilities/Parks,-playgrounds-and-community-facilities/parks-and-sites/marton-mere-local-nature-reserve.aspx
Phone: 01253 478478
Email: martonmere@blackpool.gov.uk
Opening times: Accessible at all times, but the visitor centre hours vary
Parking: Unavailable on the reserve. There is nearby parking at Blackpool Zoo.
OS Landranger Map 102, OS Explorer Map 286
LNR, SSSI

Marton Mere is an urban wetland in Blackpool where waterfowl, reedbed species and passerines can be observed. Its range of habitats makes it a hotspot for common and scarce birds.

SPECIES

Marton Mere is a reed-fringed water adjacent to Blackpool Zoo. It is reputedly the 'black pool' after which the town is named. Inspite of its relatively urban location and being bordered on the entirety of the south side by a busy caravan park it attracts a diverse range of birds and has a long track record of turning up rare and scarce species.

The main interest is naturally on the water and the reedbed around it. Numbers of waterbirds can be variable but significant flocks of Teal and Wigeon can occur in winter, whilst Goldeneye, Gadwall, Shoveler and Tufted Duck are all regular and Garganey are annual passage migrants. The reedbeds hold Bittern every winter, and birds can be relatively easy to see despite the skulking habits of this species. In some years birds have been seen to clamber up the reeds at dusk to snatch Starlings out of the air as the latter come into roost! Cetti's Warbler now occur all year round and are easy to hear if trickier to see, whilst Reed and Sedge Warblers also visit. Little Gull, and Black, Common and Arctic Terns are fairly regular passage visitors.

Hirundines are regular on passage, and there is a Sand Martin breeding wall though the species has stubbornly preferred sandbanks nearby for nesting. The scrub to the north of the mere and part of the nature reserve is a good area for warblers in summer and has held roosting Long-eared Owls in past winters. The fields to the east, whilst not part of the reserve, can repay observation with geese and waders including flocks of Pink-feet in winter and occasional other species

Lancashire

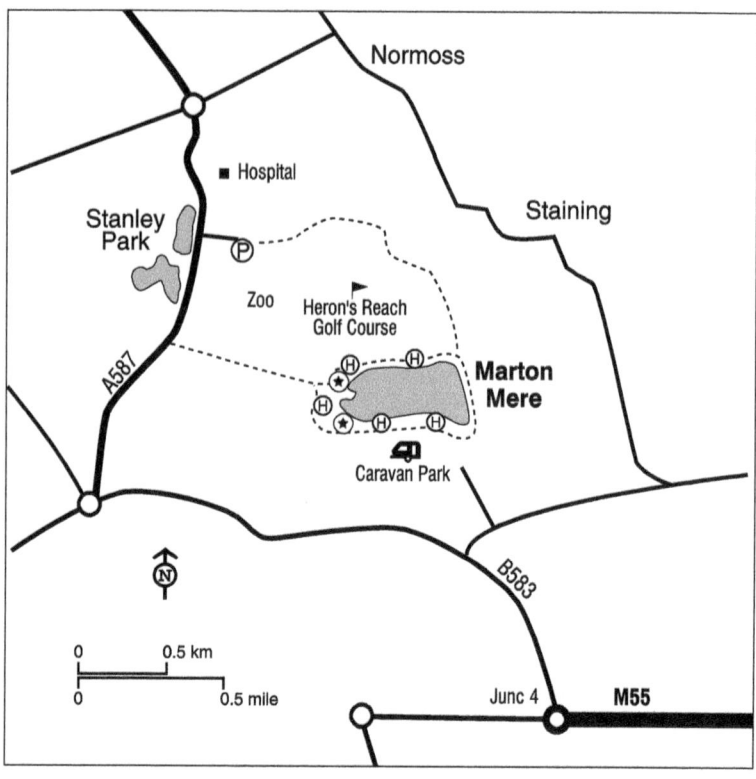

among them. A free-flying population of Barnacle Geese that spends the day in Blackpool Zoo sometimes roosts on the Mere, at other times favouring fields at nearby Staining Nook. The barn to the east remains a good site for the locally declining Little Owl. Barn Owls often hunt over the reserve and wider area. Kestrel, Sparrowhawk and Peregrine are regular, with occasional Hobby in late summer and Marsh Harrier in all seasons.

Four hides overlook the mere itself. One on the south bank is restricted to members of Fylde Bird Club but the others are open to all at all times. At either side of the inflow at the west end there are also strategically placed benches giving a good view of much of the expanse. The elevated causeway along the eastern edge is also an excellent viewpoint and the best place on site to listen for Grasshopper Warbler in spring. A fifth hide overlooks a bird feeding area which attracts garden birds and has in the past produced Brambling and Mealy Redpoll.

Rare and scarce birds seen over the years include American and Little Bitterns, Night Heron, Ross's, Laughing and Ring-billed Gulls, American Wigeon, Green-winged Teal, Collared Pratincole, Pectoral Sandpiper, Lesser Yellowlegs, Glossy Ibis, Montagu's Harrier, Red-footed Falcon and White-winged Black Tern. Regular scarcities include Osprey, particularly in spring, and in irruption years Bearded Tits have occurred in small flocks.

ACCESS

Parking by the mere is restricted to people staying at the caravan site or on nature reserve work party business. Three alternatives provide other birding opportunities: use the Blackpool Zoo car park and combine with checking Stanley Park Lake and Woodland Gardens (north of the zoo entrance road); park on Lawson Road near SD334351 and follow the footpath north past Lawsons Field nature reserve and east to the Mere; or park at Staining Nook SD345358 and check the marsh to the west before following the path round Herons Reach Golf Course to the mere.

There are bus services to all three options mentioned for car parking above (currently 3A, 5A and 5B respectively). If visiting from outside Blackpool the 61 bus service from Preston stops on Preston New Road at the Metropolitan Business Park; this is 500m away from the path down to the mere at the end of Cornwall Place (SD337348).

There is a visitor centre for Marton Mere in the caravan park (gentle.shady.icons); its opening hours have varied greatly over the years, so it is probably worth checking current times before going into the caravan site to visit it. The centre has a library of donated bird and other nature-related books. There are weekly working parties and regular guided walks run by the Friends of Marton Mere, details of which can be found on their Facebook page.

YOUR VISIT

There is access at all times. Early mornings are best, particularly in spring, since the reedbeds are then alive with singing warblers and disturbance by dog-walkers is at a minimum. In winter a late afternoon visit is often best as the Starling murmuration may be spectacular and this is the best time to see Bittern, Barn Owl and possibly Woodcock in the scrub areas. School holidays have more disturbance as the caravan park is busier; this includes Scottish holidays which may be on different dates to those locally. On summer evenings in particular there can be some anti-social activity in hides and motorbiking on footpaths can be an issue, so care is advised.

NEARBY SITE: STANLEY PARK

SD330359, ankle.tulip.places

Although Stanley Park is ornithologically in the shadow of Marton Mere nature reserve virtually just over the road, it can be rewarding. The main areas worth attention are the boating lake and the trees adjacent, around the boathouse.

The northern part of the lake is no longer accessible for pleasure boats and with weed growth attracts increasing numbers of ducks. These include Tufted Duck, Shoveler and Gadwall, whilst occasional visitors have included Garganey, Smew and Scaup. The large island in this half of the lake holds a long-established Grey Heron colony in the summer.

The southern half of the lake is more disturbed and has a traditional urban park feel with a herd of Mute Swans, various feral geese and usually reasonable numbers of gulls. Mediterranean Gull is occasional and Iceland Gull has occurred. Kingfisher is occasionally seen in early mornings, and Grey Wagtail nests in the area.

Although some distance from the coast, Stanley Park is one of the best sites on the Fylde for passerine migrants that nest in mature woodlands inland, including Wood Warbler, Pied Flycatcher and Spotted Flycatcher.

Ring-necked Parakeets have colonised, and the park has surpassed Lytham

Crematorium as the best place on the Fylde to see them. The golf course adjoining the park occasionally gets Wheatear, and a Chough fed on the short turf here in March 2017.

Between Stanley Park and Marton Mere, former football pitches off Lawson Road have been converted into a small scrub and wetland nature reserve. By car, this can be accessed by parking at the junction of Lawson Road and Lancaster Road at SD333350 and joining the path east of the allotments.

In terms of public transport access the 18 bus runs from Blackpool town centre hourly and stops at the park gates. In summer there is also a tourist service 22 which as well as going to the park gates stops at the zoo, nearer to the best parts of the park for birdwatching.

NEARBY SITE: SALISBURY WOODLAND
SD331361, crab.hook.edges
If visiting Stanley Park, Blackpool Zoo or Marton Mere it can be worth popping into the small park north of the zoo entrance road known both as Salisbury Woodland and Woodland Gardens. This offers a further chance to look for owls, woodpeckers and other woodland species. Kingfisher and Little Egret occur on the water courses alongside the woodland.

NEARBY SITE: HERONS REACH GOLF COURSE
SD339361, photo.piano.settle
There is a public right of way around the periphery of the golf course, and also a footpath that crosses the fairways. Cetti's Warblers are resident, passerine migrants include Wheatear, Whinchat and Redstart, and Tufted Duck and Little Grebe are regular on the various pools. A pair of Lesser Scaup visited in spring 2024. Do not wander off the public right of way that bisects the northern part of the golf course, and on the footpath be alert to the possibility of stray golf balls.

NEARBY SITE: STAINING NOOK
SD345358, charm.slick.tunnel
This small wetland has the advantage of being quickly accessed off Broad Oak Lane in Staining. If time permits it is a pleasant walk south from here past Mere View Farm to Marton Mere. The site holds Snipe and Water Rail in winter, with Cetti's Warbler resident and Grasshopper Warbler present in summer. The trees at the east of the site can hold migrants including regular Spotted Flycatcher.

40 WYRE ESTUARY

Wyre Estuary Country Park
River Road, Thornton-Cleveleys, FY5 5LR
Stanah SD335431
rockets.streaking.degree

Website: wyre.gov.uk/WyreEstuaryCountryPark
Email: countrysideservice@wyre.gov.uk

Lancashire

> Opening times: Wyre Estuary Country Park is open daily from 8am to dusk
> Parking: Wyre Estuary Country Park car park
> OS Landranger Map 102, OS Explorer Map 286/296
> SSSI

The Wyre Estuary has a rich mix of estuarine and farmland species, from saltmarsh-nesting waders to wintering geese and waterfowl, as well as occasional rare vagrants. Access points along the river provide views of both feeding and roosting birds, with opportunities for high-tide wader counts, gull roosts and wintering wildfowl.

SPECIES

The Wyre Estuary is relatively underwatched in Lancashire terms, partly because of access difficulties on the east side and partly because there aren't many easily watched shorebird roosts. Excellent birding can be had, though.

Several species of wader nest on the saltmarshes including Lapwing, Redshank and Oystercatcher. These are augmented outside the breeding season by flocks of Curlew, Black-tailed Godwit, Dunlin, Ringed Plover and Golden Plover. Whimbrel roosts form on spring passage in late April and early May. Greenshank are regular

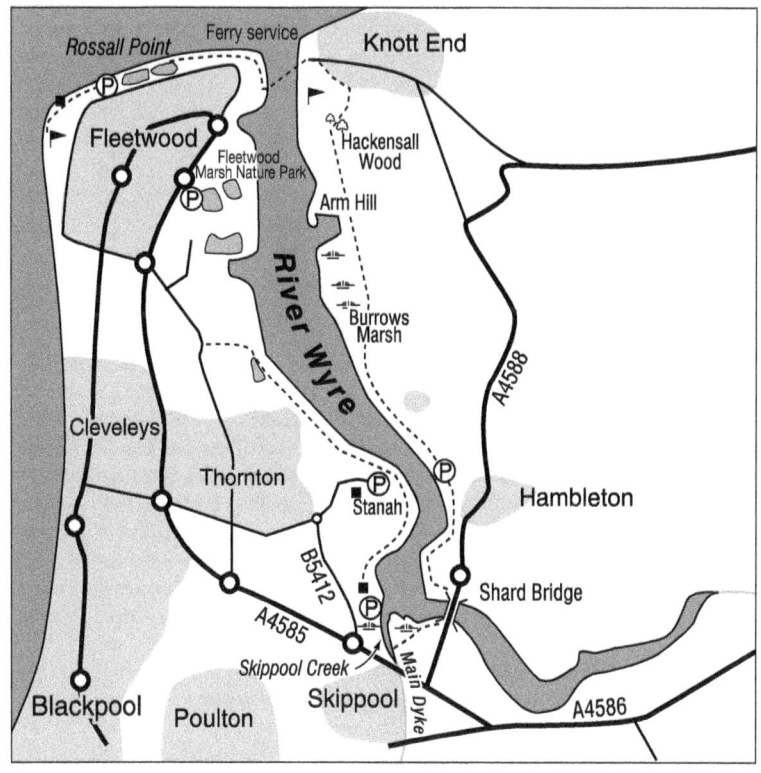

on passage and can overwinter, whilst autumn in particular sees Spotted Redshank, Curlew Sandpiper and Little Stint between Skippool and Ramper Pot. Ducks include flocks of Wigeon and Teal.

Water Pipits are increasingly regular among the Scandinavian Rock Pipits on the saltmarsh in winter, whilst Pink-footed Geese roost on the marshes in the middle reaches of the estuary and can feed in fields either side of the river with other species occurring among them regularly. Further upriver the large numbers of Greylag Geese at Little Singleton are less exotic in origin but an equally impressive sight in flight. Raptors include regular Peregrine and Kestrel, with Merlin in winter and Marsh Harrier particularly in autumn.

Rare waders have included Britain's third Great Knot, Semipalmated Sandpiper, White-rumped Sandpiper and American Golden Plover. American Wigeon has occurred a couple of times, off Burglar's Alley in 2002–3 and more recently well upriver at Little Singleton in January 2025. Regular scrutiny of loafing gulls at Skippool has produced Caspian, Glaucous, Little, Ring-billed and Yellow-legged Gulls, whilst it is probably only a matter of time before a Bonaparte's Gull is found among the large Black-headed Gull assemblies.

Visits to the sites mentioned below are generally better on a rising tide, though high-tide roosts form at Burrows Marsh.

ACCESS

STANAH
Park at the Wyre Estuary Country Park car park (free) at the end of River Road. There are good views of the estuary from the footpath north. The creek immediately adjacent to the car park regularly holds Teal in particular and some waders. If you go to the end of this path the fenced-off pool at SD340445 (still known locally as the 'ICI Reservoir') can hold Little Grebe and ducks including Tufted Duck, Wigeon and Goldeneye; the first Ring-necked Duck for the Fylde occurred here.

There is a café here and toilets. An access ramp from one of the car parks can mean the estuary is frequented by powerboats and jet skis.

SKIPPOOL CREEK
Wyre Road, Thornton Cleveleys FY5 5LF
SD356410, defaults.economies.united
At the River Wyre pub roundabout on the A585 turn off onto Skippool Road, then turn first right and drive carefully down Wyre Road to access the free car park on the left, just beyond the cricket ground. Walk between the jetties to view the extensive bend in the river between Shard Bridge and Ramper Pot. Do not be tempted to get an elevated viewpoint from any of the jetties because they are private, and many are in an unsafe condition. Skippool is arguably the best site on the estuary, partly because of the large areas of feeding exposed at low tide and partly because of the relatively easy access and wide views across a 'U-bend' in the estuary. As well as feeding waders there is usually a gull roost at dusk here, with large numbers of Black-headed Gulls in autumn drawing in Mediterranean Gulls.

Blackpool and Fleetwood Yacht Club is based at Skippool. Their vessels are usually afloat around high tide when birding is least productive. They have a café, open to non-members including birders.

LITTLE SINGLETON
SD377398, jogged.imprints.pegs

Park considerately at the end of Poolfoot Lane at SD376395 (pepper.thus.porridge), walk back along the A585 toward Skippool then follow the footpath down Occupation Lane to the riverbank. The area upriver is private but there is access downriver to Shard Bridge. There are often feral geese in the river channel and on the marsh, and Pink-footed Geese are increasingly frequent here, apparently roosting. Goldeneye are regular on the river in winter.

SHARD BRIDGE
SD368412, pimples.rebounder.producing

Park sensibly, off the road and not obstructing field gates across the bridge either side of the river, or along the western end of Old Bridge Lane where the parking restrictions stop. Do not use the car park of the Shard Riverside Inn unless you are eating or drinking there, but on a nice day the outside seating does give good views upriver, with Lapwing flocks and other waders at most tide states.

WARDLEY'S CREEK
SD367427, hardly.amicable.rewarding

Park sensibly off road where possible along Kiln Lane before it becomes Wardleys Lane; view the creek here or walk out to the estuary. This part of the estuary is often favoured by Greenshank, including occasional wintering birds.

BURROW'S MARSH
SD355450, snacking.nerves.gender

Head west from Stalmine along Grange Lane and past Little Height o'th'Hill to where the road meets the estuary at the above grid reference. There are places to park on the road here and to the south-east of The Heads. You can view wader roosts on Burrows Marsh from here. On very high tides fields at The Heads flood and are then the best place on the Wyre to view Rock and Water Pipits.

YOUR VISIT

The sites furthest upriver (Little Singleton, Shard Bridge and Skippool) are generally best at low tide with feeding waders on view. That said, Little Singleton can be worth watching on very high spring tides when birds downriver get pushed further up the estuary. Skippool can be worth visiting towards dusk for the gull roost, particularly in autumn. Burrows Marsh is probably best visited on an incoming high tide, as Jack Snipe and Snipe leave the marsh and a roost of other wader species forms. Note that on some high tides recreational sailing occurs at Skippool, and speedboats may launch from the ramp at Stanah, which can cause disturbance from the estuary mouth to Little Singleton.

Lancashire

41 FLEETWOOD AREA

Rossall Point Tower
Outer Promenade, The Esplanade
Fleetwood FY7 8PG
SD314479
dollars.payer.acrobats

Website: wyre.gov.uk/rossallpoint
Phone: 07976 650803
Email: countrysideservice@wyre@gov.uk
Opening times: Most sites in the Fleetwood area are accessible at all times.
Parking: Available at multiple sites (see below)
OS Landranger Map 102, OS Explorer Map 296
SSSI

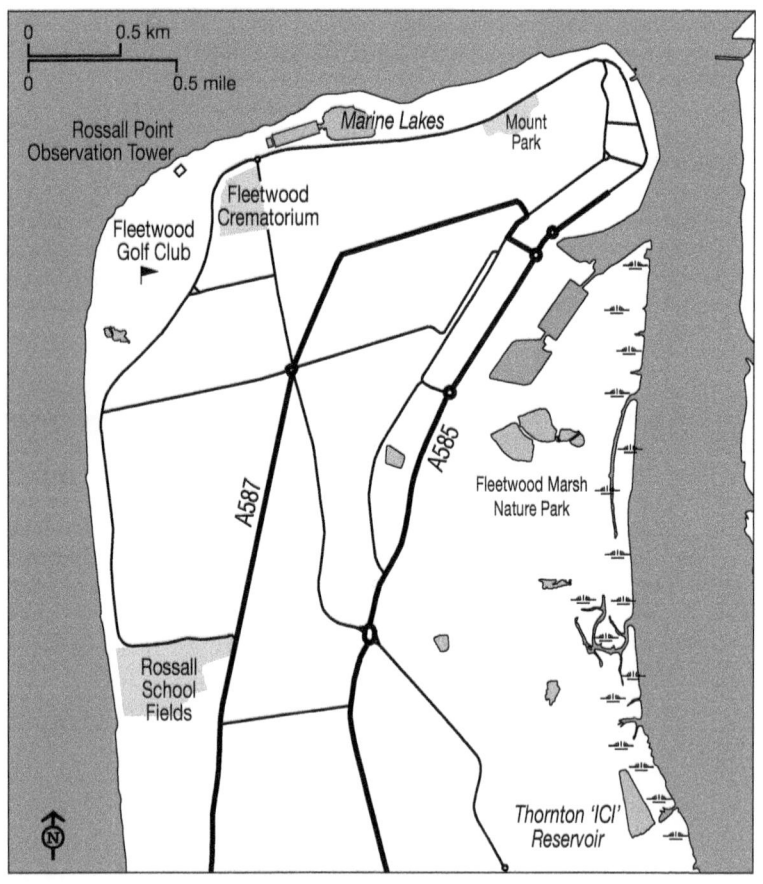

Fleetwood sits at the northern edge of the Fylde coast and the southern lip of Morecambe Bay. It's a strategic position for seawatching, estuary birding and passage migrant observation. The area supports waders, waterfowl and passerines.

SPECIES AND ACCESS

The town of Fleetwood is well placed to be a good birding location, being on the southern lip of Morecambe Bay and the northern extremity of the Fylde coast. As a result it can be a productive area for migration and also offers opportunities to see a range of species all year round. The town also has a wetland nature reserve, and the fields on its southern fringes are frequented by Pink-footed Geese and other wildfowl and waders in winter.

ROSSALL POINT AND FLEETWOOD GOLF COURSE

The coastguard station at Rossall Point has traditionally been the favoured seawatching location in the area, and birders once had access to view from it before the former building was demolished. The new Rossall Point Tower (completed in 2013) does have an external viewing platform used occasionally for seawatching events, but it is only open for limited hours on Saturday and Sunday (currently 11am–4pm) and observers often therefore watch from beside the building, using the building walls as shelter from the wind. In very strong winds, particularly where birds are blown into the mouth of the Wyre Estuary, an alternative seawatching site is the area around the Lower Light on the main Promenade at SD338485.

In winter Red-throated Diver, Great Crested Grebe, Eider and Red-breasted Merganser are all regular whilst the Kings Scar sandbank that has formed offshore in the last few years holds Cormorants and shorebirds. Dunlin, Ringed Plover and Sanderling also roost on the beach here, offering excellent views when present, but they are susceptible to disturbance by dog-walkers and to a lesser extent anglers. Snow Bunting occurs along the stretch from here to the Marine Hall.

Spring seawatches in calm weather can be very productive. Skuas are regularly recorded and include Pomarine, which are more frequent here than at other Fylde seawatch locations as they come into the bay from the Irish Sea. Auks can include Black Guillemot and increasingly Puffin in spring. Visible migration can comprise Pink-footed Geese and Whooper Swans as well as passerines. Yellow Wagtail and Tree Pipit are regular, though the former are declining significantly. Hooded Crow and Ring Ouzel are annual passage migrants. Stonechats pass through and a pair usually breeds also.

In autumn visible migration can include large numbers of Meadow Pipits. Grounded migrants including Wheatear and Whinchat with the possibility of scarce species. In gales Leach's Petrel, Sabine's Gull, Grey Phalarope and Long-tailed Skua are all possible. Strong south-westerly winds pushing birds into Morecambe Bay are best as Leach's Petrels can then show well at relatively close range as they battle their way out again.

Past rarities and scarcities have included Ross's Gull, Caspian and Elegant Terns, Hoopoe, Tawny Pipit, Desert Wheatear and Icterine Warbler.

Several car parks can be used when visiting the area, both to the north and the south of Fleetwood Golf Course. The most convenient for the coastguard station is accessed off Princes Way at SD319479. Alternatively, park on the road at SD317478 and walk along the path by the northern golf course boundary to the Promenade.

FLEETWOOD MARINE LAKE
SD322480, general.lawful.pedicure
The lake is always worth checking in winter. Goldeneye occur in small flocks and Red-breasted Merganser are regular. Shorebirds take refuge around the lake and on the island in the northern half when disturbed, and as well as Turnstone, Knot and Dunlin, Purple Sandpiper is also possible. There used to be a large Mute Swan herd here, but now only small numbers occur. In storms there is always a possibility of wrecked seabirds on the water, but less so than on other marine lakes in the north-west. Scarce birds seen in the past have included Great Northern Diver, Scaup and Iceland Gull.

As well as the car park mentioned for Rossall Point it is possible to drive down to the lakeside at SD322480 and view the southern half of the lake from the car.

ROSSALL SCHOOL
SD312449, sorters.clashing.blizzard
The playing fields in the school grounds, and particularly the adjoining reed and scrub, can hold migrants. Wheatears and grounded Meadow Pipits are frequent whilst Yellow Wagtails and Ring Ouzels are annual. Short-eared Owls sometimes hunt the area in winter. Scarcities have included Lancashire's first accepted Woodchat Shrike, Red-backed Shrike and Bluethroat.

Although the school grounds are private they can be watched reasonably well from public paths. If driving, park at the north end of the car park that runs off Cleveleys Promenade (SD313443) and walk north to view the school grounds to the right. Beyond the school buildings a public footpath crosses the school premises coming out on Broadway. Scrub to the north, beyond the school sports centre, can be productive.

BROADWAY/FLEETWOOD FIELDS
SD317444, cute.tailwind.sheet
The fields inland of here between Rossall School and Rossall Beach tram stops often hold large numbers of Pink-footed Geese in winter. There are also several ponds and floods which though not obvious from the road hold Wigeon, Teal and often waders; these can be viewed with a telescope. If the geese are further east they can be viewed from the northbound and southbound lay-bys at SD325477 and SD326444 respectively, or from Fleetwood Road near Farmer Parr's Animal World.

FLEETWOOD CREMATORIUM
SD319478, dockers.rainfall.glider
The cemetery is bordered on three sides by trees and bushes and in the right conditions in spring and autumn can hold passerine migrants. Yellow-browed Warblers are seen most autumns, whilst in spring Wood Warbler, Redstart and Pied Flycatcher all occur. Keep to the perimeter path and show appropriate consideration of other cemetery users. Early mornings are best before birds are disturbed.

MOUNT PARK/MEMORIAL PARK
Mount Park: SD332481, amazed.commutes.withdrew
Memorial Park: SD328475, published.inhales.galaxy

Mount Park is another good site for grounded passerine migrants, with the best area being the sheltered trees inland of the 'mount' itself at SD333481. The mount also gives a good, elevated view over Morecambe Bay and can be a good place from which to observe visible migration and even to seawatch in calmer conditions. Rarities have included Pallid Swift, Pallas's Warbler has occurred and there have been multiple Yellow-browed Warblers in most autumns. Memorial Park is further inland but has more cover; it is more challenging to work than Mount Park but also gets regular migrants.

FLEETWOOD DOCK MOUTH
SD335475, crispy.birthing.memory
When the tide is out the mouth of the dock can be a good place to look for waders. Peregrine and, in winter, Twite also occur in this area. It can be accessed from the north at the end of the row of shops and cafes at SD340482 or from the south on the opposite side of the road to the ASDA store on Dock Street.

FLEETWOOD MARSH NATURE PARK
SD335464, busy.niece.attend
The nature park has several pools which hold waterfowl. Shoveler, Tufted Duck, Little Grebe and Great Crested Grebe are regular, and Pochard still occur but are infrequent. Cetti's Warblers have colonised and Reed and Sedge Warblers also breed. Regular ringing occurs in private parts of the site and in 2025 a Western Subalpine Warbler was caught here. The short-grassed area used for model plane flying can be productive for Wheatears and Whinchats on passage and Lancashire's only Short-toed Lark was seen here in 2011. There is also a reed-fringed pool over the railway bridge at SD329454 which can be worth checking en route to the nature park for ducks and grebes.

To access Fleetwood Nature Park from the south take the right turn at the 'Eros statue' roundabout on the southern approach to the town, turning left soon after between caravan parks at SD326454. Follow the road past the landfill site to the car park at the end. At present this road is in very poor condition beyond the tip and care is required.

'BURGLAR'S ALLEY'
SD331448, headlight.rollers.vast
The footpath east from Fleetwood Road, known locally as Burglar's Alley, can be productive for thrushes and other passerines, and Cetti's Warbler occur in the fenced off areas alongside. The disused railway line intersecting this path can be worth checking for migrants. Where the path meets the Wyre Estuary can be productive for wildfowl, and on high tides Snipe and Jack Snipe, and Rock and Water Pipits. To the south of here is a reservoir covered in the Wyre Estuary site text.

Park carefully off Fleetwood Road on or in the vicinity of Springfield Terrace and follow the path east between trees, then cross the former rail line passing between industrial areas.

ACCESS
All sites are close to the A585. From outside the Fylde take the M55 to junction 3 and then follow signs. A new bypass at Little Singleton has greatly reduced congestion at peak times.

The Fleetwood area is well served by public transport, with a frequent tram service to Blackpool and two bus services (one down the Promenade and one via Poulton passing Jameson Road and Burglar's Alley). It is possible to use both buses and trams to commence at one of the above sites and finish at another with a combined day ticket.

YOUR VISIT

Spring calm-weather seawatching is best in early morning, ideally two to three hours before high tide. Autumn seawatching is best during periods of sustained westerly winds, particularly south-westerly gales over a couple of days, irrespective of tide. During migration time passerine migrants quickly begin to move inland, and this is more pronounced at sites with disturbance. When deciding on an itinerary bear in mind that dog-walkers will turn up at Rossall Point and Rossall School from first light and earlier.

Nearby site: Knott End

Although it is only a few hundred metres across the Wyre Estuary, Knott End cannot easily be combined with Fleetwood on a birding trip unless the ferry across the river mouth is running (timings available in advance each month). The birding is generally similar to Fleetwood, though most of the many passerine highlights over the years have been in private gardens or sound-recorded as they passed overhead. The particular appeal of Knott End in winter is that it continues to attract a Twite flock, and these birds are best observed coming to supplementary feeding on the jetty/slipway at approximately SD346485.

In late summer the tern roost in the Wyre mouth and Morecambe Bay is generally best viewed from the Knott End side and regularly holds several hundred Sandwich Tern, whilst Black, Common and Arctic Terns all occur and an Elegant Tern was present in 2021. The golf course behind the seawall here can attract Wheatears and other migrants, and a Chough has been attracted to the short turf in the past. Black Redstarts are regular winter visitors to the area around the waterfront apartments next to the jetty.

There are bus services to Knott End from Blackpool and Lancaster, but they are relatively infrequent and Fleetwood is the better option for a birding trip by public transport.

42 BLACKPOOL PROMENADE AND COASTAL SITES

New South Promenade
Blackpool FY4 1RW
lend.reduce.lunch

North Pier
Promenade, Blackpool FY1 1NE
lend.reduce.lunch

Gynn Gardens
Warbreck Hill Road, Blackpool, FY1 2JR
SD308379
these.noises.shock

North Shore seawatch point
SD305376
shot.green.body

North Shore shorebird roost
SD306386
bags.ballot.grit

North Pier Starling roost
SD301363
lend.reduce.lunch

Pleasure Beach railway bushes
SD307322
tested.goat.keys

Starr Gate
Blackpool, FY4 1SU
SD303320
single.yoga.judge

Watson Road Park
Watson Road, Blackpool, FY4 2BP
SD312334
froth.bind.looks

Website: blackpool.gov.uk
Phone: 01253 477477
Email: enquiries@blackpool.gov.uk
Opening times: All the coastal and park sites listed are accessible at all times
Parking: On-street and pay-and-display car parks
OS Landranger Map 102, OS Explorer Map 286

In many ways the seafront of a tourist resort like Blackpool might seem an unlikely candidate for inclusion in a birdwatching site guide, but some excellent birding experiences can be had from or close to the Promenade of the 'Las Vegas Of The North'.

SPECIES

Winter: large flocks of Common Scoter with occasional Velvet Scoter and Long-tailed Duck; Red-throated Divers with occasional Great Northern Diver; Purple Sandpiper at North Shore roost or feeding in the vicinity at low tide.

Spring: seabirds including skuas and auks, with increasingly regular Puffin sightings in May; diver and wildfowl passage offshore; passerine migration includes Wheatear, pipits, wagtails and warblers; birds of prey passing north offshore including Ospreys; large Whooper Swan flocks.

Summer: Manx Shearwater, Gannet and Sandwich Tern; Eider often offshore from local breeding sites to the north and south; Rock Pipits at North Shore cliffs. There are also regular sightings of Harbour Porpoise, Grey Seal and even Bottlenose Dolphin.

Autumn: westerly and south-westerly gales can produce Fulmar, Leach's Petrel, Sabine's Gull, Grey Phalarope and Long-tailed Skua; passerines include regular Yellow-browed Warbler and in good conditions visible migration.

ACCESS

By car follow the M55 from the M6 to its end, then bear left at the 'helter skelter statue' roundabout onto Squires Gate for Starr Gate, or straight on along Yeadon Way for other sites.

Coastal Blackpool is ideal for green birding. There are four coastal rail stations on two different lines (considerably more trains run to Blackpool North than Blackpool South), and both a tram line and regular bus (service 1) along the Promenade, with the tram now connecting to Blackpool North station. Safe, wide-pathway cycling is also available on the seaward side of the Promenade throughout its entire length.

There are public toilets at the Solaris Centre near the Starr Gate seawatching viewpoint, and Gynn Square near the North Pier seawatching and shorebird vantage points.

YOUR VISIT

SEAWATCHING

The Fylde coastline runs virtually straight and more or less due north for 16km from Starr Gate to Rossall Point. This means in practice that reasonable to excellent seawatching can be enjoyed just about anywhere on this stretch, including Blackpool. The Blackpool stretch is in fact better for Common Scoter flocks, which tend to be scarce off Cleveleys and Fleetwood.

The majority of seawatching in Blackpool is done from the South Promenade/Starr Gate stretch, at present just to the south of the 'mirrorball' sculpture and Solaris Centre opposite Abercorn Place. Most days one or two regular birders are present here for several hours in the morning. Sightings from Starr Gate are generally posted on the Trektellen website (trektellen.nl). This stretch of coast is relatively exposed in stormy conditions.

Some seawatching also occurs from North Shore. Parking in the vicinity of the Grand Promenade Hotel, observers descend to the Middle Walk section of the Promenade. The Middle Walk includes covered areas and in stormy conditions more shelter is available than at South Shore, so it is often used in autumn 'Leach's Petrel' gales as the preferred Blackpool site. Some parts of this stretch are currently fenced off due to structural concerns, but seawatching can still be undertaken in storms from the Middle Walk at Gynn Square near the Hole In The Wall café.

SHOREBIRD ROOST

There is a shorebird roost on the external walls of the former boating lake, which is now the Formulakart go-kart track at Queens Promenade. Depending on the height of the tide and the wind strength the birds here are often at very close range on the nearest part of the walls. As well as Redshank, Oystercatcher and Turnstone the roost also attracts a small number of Purple Sandpipers in winter, normally one or two individuals. Rock Pipits, formerly only winter visitors, have recently started breeding on the rocky slope above the go-kart track and this is currently the only known nesting site on the Fylde coast.

STARLING MURMURATIONS
The piers, and particularly North Pier, hold significant Starling roosts in winter. Standing adjacent to North Pier near the Metropole Hotel can produce a spectacular display as the birds wheel and circle for several minutes before finally settling down to roost. Evenings are better than mornings, as the birds normally make a direct departure after sunrise. As with any sizeable murmuration birds of prey, mainly Peregrine and Sparrowhawk, can be attracted.

PASSERINE MIGRANTS
If you are already in the resort in spring or autumn several sites have been productive for migrant passerines over the years. These include the bushes at the rear of the Pleasure Beach and alongside the adjoining Blackpool South railway line, Watson Road Park and to the north Gynn Gardens off the western end of Warbreck Hill Road.

RARITIES
Rarities and scarcities have included Roller, Desert Wheatear, Pallas's Warbler, three Surf Scoter and several Balearic Shearwaters. The first three and many other scarce passerines were found by the late Ed Stirling and show what regular coverage of green pockets in urban areas at the right time of year can produce.

EAST LANCASHIRE

43 STOCKS RESERVOIR

Stocks Reservoir
School Lane, Slaidburn, Clitheroe BB7 4TS
SD732564
milky.dancer.descended

Opening times: Accessible at all times
Parking: Pay-and-display car park at the bottom of School Lane, SD732565; additional parking at Gisburn Forest Hub (SD745560)
OS Landranger Map 103, OS Explorer Map OL41

Stocks Reservoir is a large upland reservoir on the edge of the Forest of Bowland, formed in 1932. It has a combination of open water, islands and surrounding farmland, which makes it an important site for wildfowl, waders and passage migrants.

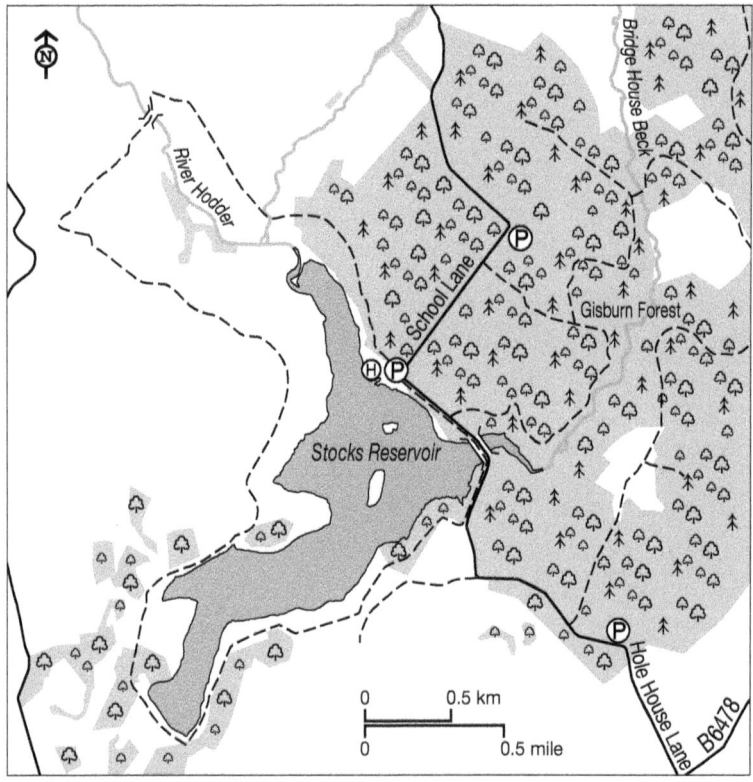

SPECIES

Stocks Reservoir was formed in 1932 by flooding the Dalehead valley and surrounding farmland. It was created by the Fylde Water Board to provide drinking water to the Blackpool area. Like many other such projects it saw the demise of a settlement, in this case the hamlet of Stocks-in-Bowland, which gave the water body its name. At around 3km long, and with variable water depth and islands, it has become an established birding site on the Lancashire side of the border with Yorkshire. At the right seasons It can be combined with visiting flashes at Hellifield (SD845573) just to the east, which are outside the area covered by this book.

Stocks Reservoir is known mostly for wildfowl and waders. In winter waterfowl include Great Crested Grebe, passage Whooper Swan, Mallard, Teal, Pintail, Wigeon, Goldeneye, Pochard, Tufted Duck, Red-breasted Merganser and Goosander. Small numbers of Red-breasted Merganser stay to breed, and large numbers of feral geese at this time include a flock of over 100 Barnacle Geese as well as Canadas and Greylags. Common Scoter drop in on overland migration.

Wader species can occur into double figures, particularly in spring. Regular species include Oystercatcher, Ringed, Little Ringed and Grey Plovers, Lapwing, Knot, Sanderling, Dunlin, Ruff, Common and Jack Snipe, Bar-tailed and Black-tailed Godwits, Whimbrel, Curlew, Redshank, Greenshank, Common, Green and Wood Sandpipers and occasional inland Turnstone.

Several species of raptor are regular, including Peregrine, Kestrel, Merlin, Buzzard, Hen and Marsh Harriers and Sparrowhawk. Osprey turn up in spring particularly, whilst Short-eared Owl can be seen on adjacent fells. A gull roost regularly forms and has attracted Little, Caspian, Glaucous, Iceland and Yellow-legged Gulls and Kittiwake. Tern passage includes Common, Arctic and Black Terns. Wood Warbler can sometimes be heard singing in the vicinity of the car park.

Rare and scarce birds recorded at Stocks Reservoir in recent years include Great Northern and Red-throated Divers, Black-necked and Red-necked Grebes, Cattle and Great White Egrets, Green-winged Teal, Bewick's Swan, Brent and Eurasian White-fronted Geese, Garganey, Ring-necked Duck, Scaup, Eider, Long-tailed Duck, Smew, Rough-legged Buzzard, Spotted, Pectoral and Curlew Sandpipers, Temminck's and Little Stints, Grey Phalarope, Ring-billed Gull, Great Skua and Great Grey Shrike. What remains the only Isabelline/Daurian Shrike recorded in the county was present here for a week in November 1996.

Footpaths and tracks around the reservoir generally do not get close to the banks. Viewing is therefore best from the causeway crossing the inlet from Bottoms Beck (SD737560) and from the two bird hides. The hides are accessed from the footpath running north-west from the car park, at SD730567 and SD728579 respectively. There is a log for recording sightings in one of the hides.

ACCESS

Approach from the unclassified road running from the B6478 Tosside–Slaidburn road, turning off north at the crossroads at SD748542. There is a pay-and-display car park at the bottom of School Lane at SD732565. There is no public access to the road at the south of the reservoir that leads to the fishing lodge.

Public transport is not a realistic option for this site given the need for a telescope and the length of the walk from the nearest bus stop in Slaidburn.

YOUR VISIT

Disturbance is not great at this site so timing is less critical than some other reservoirs. General guidance on good times to visit inland reservoirs applies though, with showers in migration time being good for dropping passing terns, Little Gulls and waders. Coverage of the site has diminished in recent years as some stalwart birders have left the area. If you are visiting, be sure to record your sightings in the logbook in the hide or on the East Lancashire Ornithologists Club website (eastlancsornithologists.org.uk).

NEARBY SITE: CHAMPION MOOR
When present, a flood pool on Champion Moor can be productive (SD749523). It can be viewed by parking carefully off the minor road to the north. Keep an eye on the sightings page of the East Lancashire Ornithologists Club ahead of a visit to help establish if the flood is there and attracting waders.

NEARBY SITE: GISBURN FOREST
Gisburn Forest, immediately to the east of the reservoir, has several paths including the 5km Dale Head Ramble trail, which can be joined from the Stocks Reservoir car park. Further parking at the Gisburn Forest Hub (SD745560) has toilets and a café. Species that can be seen in the forest include Redpoll, Siskin and Woodcock, and it is one of the best Lancashire sites for Crossbill.

44 DUNSOP AND LANGDEN VALLEYS

Dunsop Valley
Dunsop Bridge, Clitheroe BB7 3BB
SD659500
assist.headsets.gives
Opening times: Accessible at all times
Parking: Available along the road at SD631512
OS Landranger Map 103, OS Explorer Map OL41
SSSI

Langden Valley
Langden Intake, Trough Road, Dunsop Bridge BB73BH
SD632511
parsnip.surpassed.lingering

Both these sites in the scenic Trough of Bowland offer opportunities to walk through areas favoured by upland species. Dunsop Valley has the better track record for raptors, but this probably partly reflects coverage as it is an easier walk and rather more popular with birders.

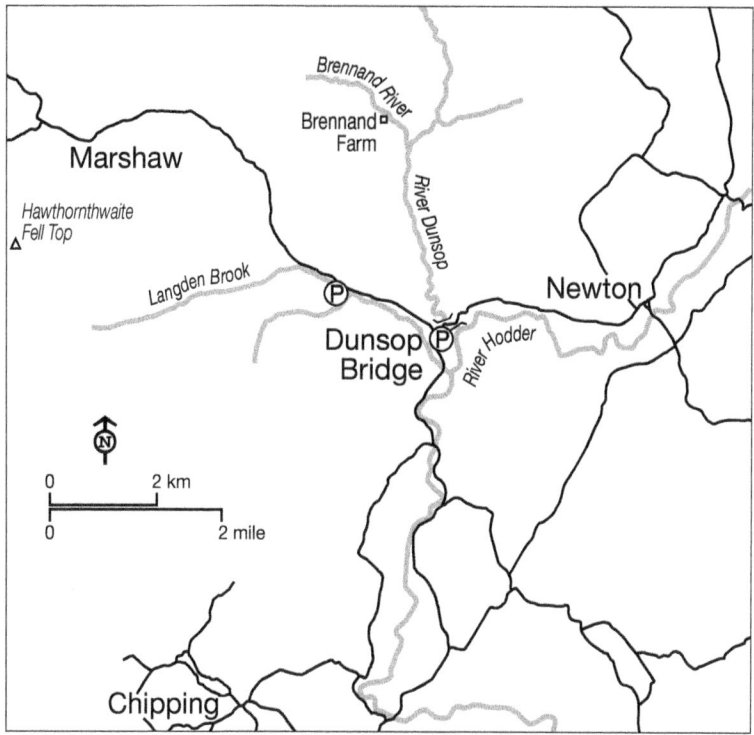

SPECIES (DUNSOP VALLEY)

The River Dunsop winds through a picturesque valley which gently gains height. There are extensive conifer plantations on the way up from the car parks at Dunsop Bridge, before the vista opens out as the path goes through heather moorland further north towards Whitendale.

The site is one of the best inland for raptor species in Lancashire. Hen Harrier breed in Bowland in variable numbers and can be seen here. Unfortunately persecution still occurs on some keepered moors, but the species does better on the United Utilities land in the vicinity. Osprey and Red Kite move through regularly, and Merlin also occurs, though it is less numerous than formerly. Both Golden and White-tailed Eagles have been recorded. In April and May 2017 a male Pallid Harrier held territory in the Whitendale Valley. Short-eared Owl can be seen on the heather moorland, and for several years Eagle Owls (almost certainly escaped from captivity) nested.

Other than raptors the typical range of upland breeding species that can be seen in summer includes Ring Ouzel, Pied and Spotted Flycatchers, Redstart, Whinchat, Wheatear, Dipper, Grey Wagtail, Cuckoo, Common Sandpiper and Tree Pipit. Raven and Red Grouse are resident. The conifer plantations often hold Siskin, and in irruption years Crossbills.

ACCESS (DUNSOP VALLEY)

There is an ample car park in Dunsop Bridge village at SD660501. Briefly pick up the road west, passing the special telephone box that celebrates the village's place (by one measurement) as at the geographic centre of Great Britain. Turn right before the bridge north and follow paths alongside the river. The tracks and paths are well surfaced and easy to follow until the track forks to Whitendale and Brennand Farm respectively (SD653535). This is pedestrian access only; vehicular access is not available to birders. From this point walking conditions are more challenging, and it is also advisable to be prepared for changes in weather as it is 3–5km back to the car park from here and relatively exposed.

There are toilets at the car park, and the tearoom (Puddleducks) does light refreshments. There is a bus (service 11) from Clitheroe which currently runs every two hours.

YOUR VISIT (DUNSOP VALLEY)

March and April are the best months for raptor watching. For other summer migrants April and May are generally most productive, but some species are declining and less easy to see here. If visiting in winter ensure that you are appropriately clothed for the elements.

SPECIES (LANGDEN VALLEY)

Larches beside the Trough of Bowland road and the vehicle track to the former water treatment works give way to a steep-sided valley with Langden Brook running through it, then heather moorland and a number of cloughs.

The range of species is similar to that of the Dunsop Valley. With more height gain than the initial walk up Dunsop Valley, and a number of steep-sided cloughs, good views of Ring Ouzel can often be had, whilst Cuckoo, Wheatear, Redstart, Tree Pipit and Common Sandpiper all occur. Grey Wagtail and Dipper can be seen on the beck. Further up the valley Red Grouse, Raven, Short-eared Owl, Oystercatcher and Stonechat all occur. In the past Langden Head (SD5851) had occasional trips of Dotterel but these are rare in the county as a whole nowadays.

Raptor watching can be productive in calm conditions. The range of species is also similar to the Dunsop Valley, including Hen Harrier, Merlin, Peregrine and Red Kite. At migration times other species pass through and these have included both Golden and White-tailed Eagles, several Honey Buzzards and annual Ospreys.

The larches by the road on the approach to the valley (SD631511) regularly attract Siskin, and Crossbill and Redpoll can also occur. Wood Warbler and Pied and Spotted Flycatchers have all been recorded and could be seen in spring.

ACCESS (LANGDEN VALLEY)

The approach road is by the Lancaster to Dunsop Bridge road (generally known as the 'Trough Road'). If arriving from the south it may be easier to leave the A6 at Scorton, picking up the Trough Road at Marshaw. There is abundant informal parking space along the road at SD631512. From here take the footpath over the bridge and follow the track south-west. The walk becomes strenuous so appropriate footwear, clothing for the conditions and plenty of fluids are recommended.

No public transport services use the Trough Road. If driving in from the west it can be worth stopping in the vicinity of Tower Lodge (SD604360), where the pine trees regularly attract Crossbills, and other woodland and riverine species occur close to the road.

YOUR VISIT (LANGDEN VALLEY)
April and to a lesser extent May are best for arriving summer visitors and overhead passage migration.

45 RISHTON RESERVOIR

Rishton Reservoir
Blackburn Road, Rishton BB1 4ET
SD716299, shot.drag.crown

Opening times: Accessible at all times
Parking: On Blackburn Road or adjacent streets
OS Landranger Map 103, OS Explorer Map 287

Rishton Reservoir is a 15-hectare concrete-banked reservoir located just west of Rishton town centre. Despite its urban fringe location, it supports a wide variety of wildfowl, waders and passage seabirds, making it one of the more interesting inland birding sites in East Lancashire.

SPECIES
The main expanse is the reservoir. The less disturbed western bank has a gently sloping field edge that can be attractive to wildfowl and waders. The site is bisected by the Blackburn to Accrington railway line, and the southern section has more tree cover round the edges.

As with many inland reservoirs in the area the site attracts passage terns, wildfowl and waders. One of the most notable recent examples was the first Purple Sandpiper recorded in Lancashire away from the coast, a juvenile in August 2017.

Other species noted in around 50 years of recording include Slavonian, Black-necked and Red-necked Grebes, Red-throated, Black-throated and Great Northern Divers, Leach's Petrel, Arctic and Great Skuas, Whooper and Bewick's Swans, Great White and Little Egrets, Brent Goose, Greenland and European White-fronted Geese, Ring-necked and Long-tailed Ducks, Smew, Common Scoter, Eider, Red-breasted Merganser, Pectoral Sandpiper, Osprey, Hobby, Marsh Harrier, Common, Arctic, Sandwich, Little, Black and White-winged Black Terns, Turtle Dove, Great Grey Shrike and Rock Pipit.

Although there is not a regular gull roost at the site, birds that have been feeding at Whinney Hill tip regularly drop in to bathe and drink. Glaucous, Iceland, Caspian and Yellow-legged Gulls can be present among hundreds of Herring and Lesser Black-backed Gulls, whilst Kittiwake and Mediterranean and Little Gulls occur on passage. Cut Wood is an area of trees between the reservoir, Blackburn Road and Cut Lane where Wood Warbler and Pied Flycatcher have been recorded. A Wryneck spent a few days on the dam wall of the reservoir in September 2024.

Lancashire

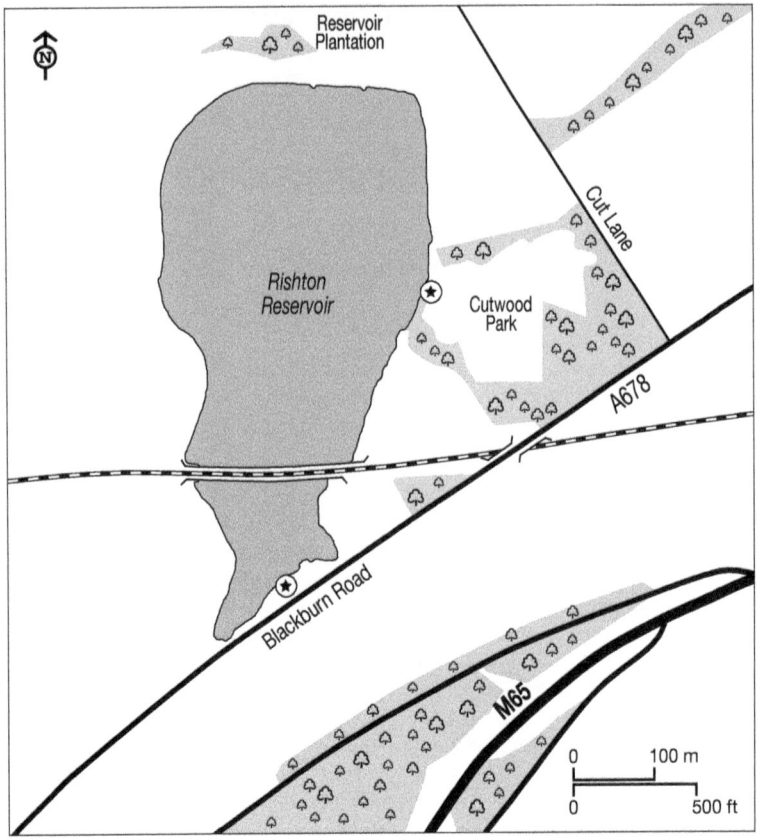

ACCESS
The reservoir is 1km west of Rishton town centre at SD715300. If driving and arriving from the motorway at the large roundabout adjacent to junction 6 of the M65, take the A678 for 3km. Park carefully on Blackburn Road or adjacent streets.

Buses to Blackburn and Preston (services 6 and 152 respectively) stop adjacent to the southern end of the reservoir. There is a regular rail service and Rishton train station is a less than 1km walk away.

From the junction of Blackburn Road and Cut Lane take the path through Cut Wood and then across the playing field/grassed area to reach the small viewing area with interpretive signage on the eastern shoreline. The southern section of the reservoir can be viewed from the A678 approximately 500m south-west of the above junction.

YOUR VISIT
East Lancashire Sailing Club is based on the reservoir, so given this and angling access (especially Friday to Sunday) an early morning visit is best before potential disturbance. As with other inland reservoirs during spring and autumn migration periods a visit after thunderstorms can often be productive before birds dropped by the weather have moved on.

Lancashire

46 FOULRIDGE RESERVOIR

Foulridge Reservoir
Skipton Road, Foulridge, Colne BB8 7LJ
SD891413
dance.doubts.serious

Opening times: Accessible at all times
Parking: On-site car park at Lower Resrvoir and roadside parking near Upper Reservoir
OS Landranger Map 103, OS Explorer Map OL21

The Foulridge Reservoirs lie either side of the southern edge of Foulridge village and form a compact but rewarding inland birding area. Foulridge Upper Reservoir is mainly bordered by fields, while Foulridge Lower Reservoir has a mix of willow fringe, gardens and fields. Both attract a wide variety of wildfowl and waders.

SPECIES
The twin Foulridge Reservoirs are either side of the road south of Foulridge village. Foulridge Upper Reservoir is surrounded by fields to the east of the road whilst Foulridge Lower Reservoir is bordered by a mixture of willow fringe, suburban gardens and fields to the west. Burwain Sailing Club operate on Foulridge

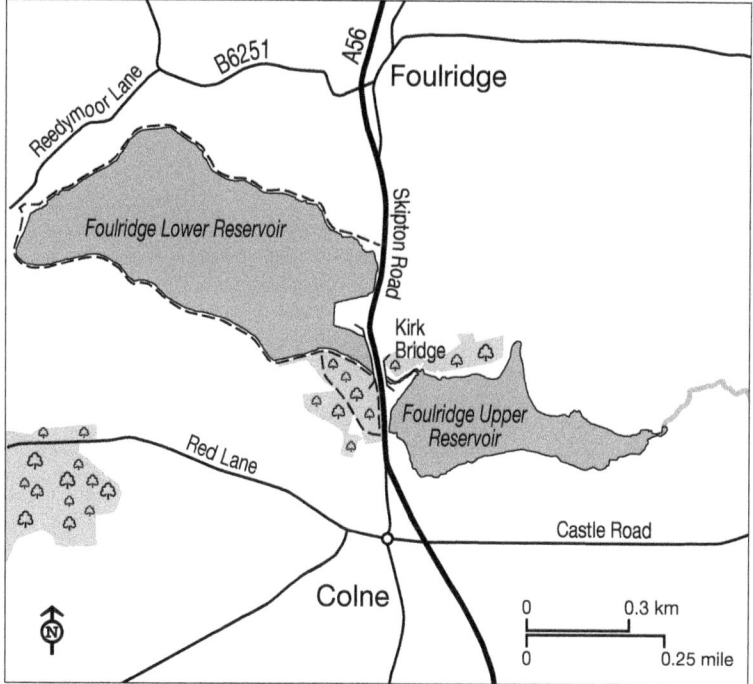

167

Lower Reservoir, which is sometimes referred to even among some birders by the alternative name of Lake Burwain.

Like many inland reservoirs interest is generally greatest in spring and autumn, though to some extent winter is not without rewards. Great Crested Grebe, Reed Bunting and sometimes Mandarin breed. Winter visitors include Little Grebe, Wigeon, Teal and diving ducks as well as good-sized flocks of Lapwing. A large gull roost attracts Glaucous and Iceland Gulls though in reduced numbers than previously. Regular passage waders include Little Ringed and Ringed Plovers, Dunlin, Redshank and Common Sandpiper. It is an increasingly popular site with naturalised Canada Geese, which can be worth checking for occasional other goose species.

Over the years the site has attracted a wide range of birds normally seen nearer the coast. These include Red-throated, Black-throated and Great Northern Divers, Shag, Arctic and Pomarine Skuas, Manx Shearwater, Gannet, Little Gull, Kittiwake, Black-necked and Red-Necked Grebes, Eider, Common Scoter, Scaup, Long-tailed Duck and in 2023, remarkably, two different Grey Phalaropes.

Other notable species recorded include Bewick's and Whooper Swans, Garganey, Ring-necked Duck, Mandarin, Ruddy Shelduck, Great White and Little Egrets, Sandwich, Arctic, Common, Black and Little Terns, Osprey, Hobby, Turtle Dove, Black Redstart, Rock Pipit and Yellow-browed Warbler. A very respectable passage wader list includes Spotted Sandpiper, Red-necked Phalarope, Little Stint, Wood, Green and Curlew Sandpipers, Black-tailed Godwit, Ruff and Greenshank. A popular Hoopoe was present for several days in the autumn of 2024.

ACCESS

From Junction 14 at the end of the M65 take the A6068 (left exit) and after about 1.6km go left on the A56 towards Skipton. After leaving the outskirts of Colne the reservoirs are either side of the road before reaching Foulridge.

As mentioned above, there are paths around the Lower reservoir. The Upper reservoir can be viewed to some extent from the A56, and also from footpaths to the north and particularly the causeway that separates it from Brownhill Reservoir at SD896412.

Two other smaller waters nearby attract good birds on occasion: Slipper Hill Reservoir (SD875420, chariots.shut.trailers) and Whitemoor Reservoir (SD877431, soonest.always.changed).

In terms of public transport access, the Foulridge Reservoirs are approximately a 1.6km walk from Colne station. There is an excellent bus service, with no fewer than nine routes currently stopping at St Michael's school, overlooking the Upper reservoir, and several of the buses go to Burnley as well as Colne.

YOUR VISIT

Early morning is best, particularly if sailing is taking place on the Upper reservoir. There are footpaths around virtually the whole of the Lower reservoir, so this is susceptible to general recreational disturbance also.

47 PENDLE HILL

Pendle Hill
The Avenue Car Park, Barley New Road, Barley BB12 9LD
SD823403
reception.plunge.stoops

Opening times: Accessible at all times
Parking: Pay-and-display in Barley village
OS Landranger Map 103, OS Explorer Map OL41
SPA

Pendle Hill is one of Lancashire's most iconic upland sites. Its extensive moorland, scree and grassy summit areas provide habitat for high-altitude species and passage migrants. The hill is particularly renowned for Dotterel in spring and Snow Bunting in winter.

SPECIES

At 557m, the summit of Pendle Hill can be seen for many miles within Lancashire and beyond. It is one of the best sites for two charismatic species, though sadly both are increasingly difficult to catch up with as numbers reduce.

Pendle Hill has been one of the most reliable staging posts for Dotterel in spring in the whole of England for as long as bird sightings have been recorded in Lancashire. Following increases in the Scottish population, numbers seen here rose accordingly, with highest May totals including 44 in 1982, 30 in 1988 and, this century, 14 in 2002. The last few years have seen brief ones and twos before a blank

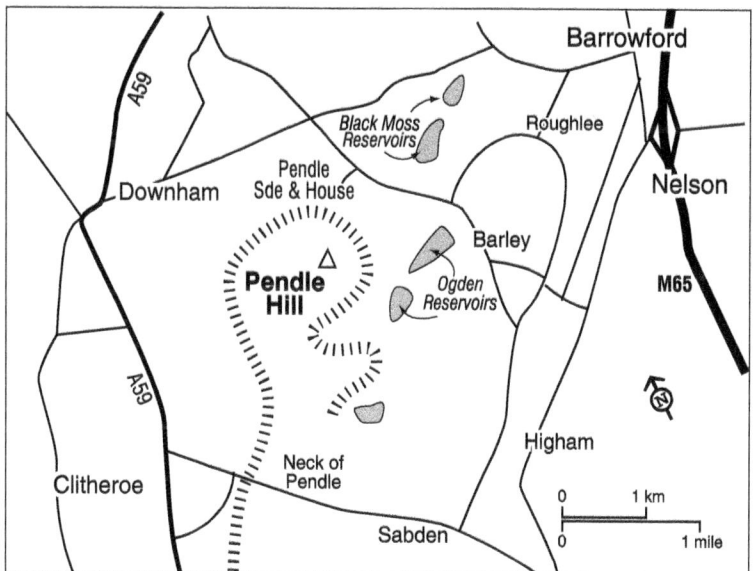

in 2024; it is hoped that this trend reverses. When Dotterel are present they show a preference for the short-cropped grassy areas of the summit. It is advisable to make the ascent early in the day in the first half of May if hoping for Dotterel before disturbance increases in the favoured areas.

Pendle Hill also holds one of the largest wintering flocks of Snow Buntings in Lancashire most years. Relatively recent counts include 25 in January 2018 and 12 in December 2021. As might be expected given the extensive suitable habitat, the flocks roam widely and can be difficult to locate. In recent times they have favoured the Downham slopes of the hill, but they can occur anywhere with extensive areas of Purple Moor-grass and even on scree areas over the ridge at the north end of the summit.

The slopes above Pendleside House and Farm regularly hold Ring Ouzels, staging on their northbound migration in spring. Numbers are very variable, but an exceptional minimum of 23 were noted on 14 April 2018. Late March through April is the best time for ouzels here. The quarry area and drystone walls in the vicinity of both tracks from Pendleside to the summit are a favoured area. Black Redstart has bred and also occurs on passage along with Wheatear, Whinchat and Stonechat. Two of the handful of East Lancashire records of Shorelark have occurred on Pendle Hill, once with Snow Buntings and once unaccompanied. Lapland Bunting has also occurred on a couple of occasions in recent times.

Red Grouse and Golden Plover may be seen on and near the summit and breed in the vicinity. Raptors can include Merlin and Hen Harrier, whilst in September 2019 a juvenile Montagu's Harrier was photographed on the lower slopes.

Within convenient walking distance of Barley village are the Black Moss Reservoirs (SD824411, brisk.changes.above, and SD828414, globe.feasting.radar) and Ogden Reservoirs (SD816399, notices.export.sprinting, and SD805396, myth.repayment.moss). All can be viewed from public footpaths. Passage waterfowl have included Common Scoter, Scaup and Smew.

ACCESS

From the north take the A59 and drive through the village of Downham then head south-east to Barley. From the south leave the M65 at junction 13, joining the A682 before picking up minor roads to Barley. A pay-and-display car park at The Avenue in the village holds over 100 vehicles. Walk north-east from here for approximately 1.6km before picking up the track to Pendleside Farm.

There are two well-marked routes to the summit. The left-hand route is longer but considerably less strenuous than the right-hand one, which is shorter and steeper. If taking the left-hand route bear right at the top to reach the trig point. A circular walk is an option, either by following the path south over Barley Moor from the summit or heading south-east to the Nick of Pendle over Pendleton Moor.

A bus service with up to seven departures daily runs between Clitheroe and Nelson and passes through Barley. It can be used to connect with train services from Blackburn and Preston to either of these towns.

YOUR VISIT

As an increasingly popular hike, the main paths are best used for birding walks on weekdays if an option, and ideally early in the day during migration time before any passage birds begin to get disturbed. Weather conditions can quickly change so particularly in winter be prepared, and wear stout footwear at all seasons.

Lancashire

48 FISHMOOR RESERVOIR

Fishmoor Reservoir
Delius Close, Blackburn BB2 3WS
OS SD700258
crowd.thus.scouts

Opening times: Accessible at all times
Parking: Roadside on or near Delius Close
OS Landranger Map 103, OS Explorer Map 287

Fishmoor Reservoir is the main holding reservoir for the Blackburn area catchment. It is essentially a significant birding site purely as a reservoir – and particularly as a gull roost – other habitat in the area having reduced over recent years.

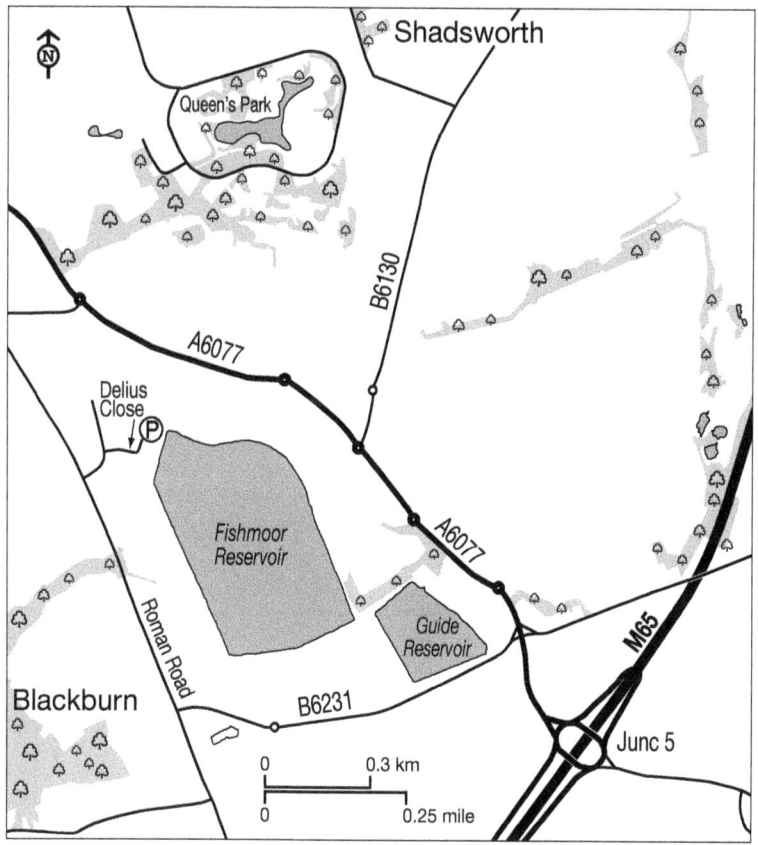

SPECIES

The site is renowned for its track record for scarce larger gulls. Glaucous and Iceland Gulls have both been regular over the years, and with advances in identification and changes in distribution Yellow-legged and Caspian Gulls are also increasingly being found. Alongside Whinney Hill tip at Accrington, where the birds feed during the day, it has become the pre-eminent site for Caspian Gulls in Lancashire. Other gull species seen include Ring-billed Gull, Kittiwake and Little Gull whilst Mediterranean Gull is regular, particularly now the species nests as close by as Belmont Reservoir in the West Pennine Moors.

Like most large inland reservoirs scarce or normally coastal waterbirds drop in on occasion. These have included several Common Scoter and Scaup as well as more notably Ring-necked Duck, Smew, Shag, and Red-necked and Slavonian Grebes. In the past waders have included Spotted Redshank, Little Stint, Sanderling, Ruff and Turnstone but low water levels are very unusual. The Golden Plover flock which attracted an American Golden Plover in 1995 no longer occurs.

ACCESS

From outside Blackburn take the M65 to junction 5. Head north-west on the A6077 and go left at the next roundabout joining Blackamoor Road. A track running alongside the smaller Guide Reservoir at SD701256 leads to Fishmoor Reservoir. It is possible to walk round the whole reservoir, but it can be muddy in winter.

Gull roost watching is an immobile activity and during cold weather birders often watch from their cars, but at the current time this is not permitted at the relevant business car parks. One option is to park sensibly on or near Delius Close (BB2 3WS). Walk through the close onto a path and cross a bridge over the drain then follow the bank of the reservoir near the 'outlet screws' to view the water. This location has the advantage of having the setting sun behind you when scanning the evening roost.

The reservoir is approximately 1.6km from Blackburn station. The area is served by several bus routes, being adjacent to Royal Blackburn Teaching Hospital.

YOUR VISIT

The gull roost forms towards dusk when birds have finished feeding. Arriving at least an hour before dusk is advisable as light can deteriorate quickly. Other than at the height of the breeding season in June and July any month can deliver, with Yellow-legged Gulls possible in late summer, but winter is best for Glaucous and Iceland Gulls. Caspian Gulls also turn up increasingly regularly including in winter at present, but this may reflect observer effort as well as the draw of the landfill site as a feeding resource.

NEARBY SITE: QUEENS PARK LAKE

A short distance north of the reservoir the lake in Queens Park (SD697272) holds loafing or bathing gulls during the day before they head to the roost, and recently this has included Caspian Gulls on occasion. The park is accessed from Queens Road, which encircles it.

49 ALSTON RESERVOIRS AND ALSTON WETLAND

Alston Reservoirs
Pinfold Lane, Longridge, Preston PR3 3BH
SD607363
twice.rejects.burden

Alston Wetland
SD605358
toggle.knocking.cadet

Opening times: Accessible at all times
Parking: Limited roadside parking
OS Landranger Map 103, OS Explorer Map 287

The reservoirs and wetland attract a wide range of passage wildfowl, grebes and occasional scarce ducks, terns and gulls, diligently recorded by keen local observers.

SPECIES

There are several reservoirs around the eastern side of Longridge, but Alston Reservoirs hold the main birding interest. Like many inland reservoirs they can be a draw for passing wildfowl, terns and gulls. High levels of coverage by dedicated local patchers means relatively little gets missed during passage seasons. In recent years the former Number 3 reservoir was decommissioned and to the benefit of both birds and birders has been repurposed as a 7.5ha wetland reserve. The site has developed a particular reputation for freshwater waders.

Alston Wetland is particularly known for the outstanding double of Killdeer in April 2013 and Wilson's Phalarope in 2017, a remarkable combination so far inland in the county. A Red-necked Phalarope was seen in June 2015. The roll call of other passage wader species is lengthy and includes multiple Pectoral Sandpiper, Temminck's Stint, Curlew Sandpiper, Little Stint, Wood Sandpiper, Ruff, Spotted Redshank, Greenshank, Bar-tailed and Black-tailed Godwits, Green Sandpiper, Grey Plover, Knot, Sanderling, Turnstone and a single Dotterel. In spring Whimbrel stop off in nearby fields and large flocks can drop in to roost. Little Ringed Plover frequently breed, as do Lapwing, Redshank and Oystercatcher. Jack Snipe as well as Snipe occur in winter and on passage.

As well as Great Crested and Little Grebes the reservoirs have held Slavonian and Red-necked Grebes, whilst Black-necked Grebe has occurred on several occasions. Divers are rare, with just a brief Great Northern in recent years. Common Scoter drop in on overland migration, and large flocks have also been seen passing through overhead. Several records of Scaup have involved up to three birds, there have been several Long-tailed Duck whilst Smew have also visited, though this is less likely these days. Feral Greylags have attracted Eurasian White-fronted Geese to drop in and there were two Tundra Bean Geese in March 2017. Whooper Swans occasionally appear on passage.

Passage Little Gulls and Arctic and Black Terns occur in favourable conditions,

Kittiwakes have also been recorded at these times and a White-winged Black Tern lingered on No. 2 reservoir in 2023. The site does not attract a particularly large roost or pre-roost of gulls but Iceland, Caspian and Ring-billed have all been seen as well as regular Mediterranean. Notably, both Pomarine and Long-tailed Skuas have been seen moving through in autumn. Ospreys are regular in spring and Hobby during the summer months.

The reservoirs are not particularly favoured by interesting passerine migrants, but coverage over the years has produced Richard's Pipit and Blue-headed Wagtail. Rock Pipit is probably an overlooked annual visitor to the banks of the reservoir. Whinchat, Redstart, White Wagtail and Twite move through, though the latter is disappearing as an English breeding bird. Garganey has been recorded on numerous occasions on the wetland. Other regular ducks of note are Goosander, Teal, Tufted Duck and Wigeon.

Other species, in some cases of unknown provenance, have included Great White and Little Egrets, Barn and Short-eared Owls, Red-Crested Pochard, Egyptian Goose, Ruddy Shelduck, Hooded Crow and White Stork.

ACCESS

Alston Wetland is accessed down Pinfold Lane but there is nowhere to park so leave your car considerately sited on the access road to the housing estate opposite (SD600354), then cross the busy Preston Road and walk down the lane. There are two viewing screens, the second one generally giving the best views. There is no general public access to the remaining reservoirs, but they can be viewed distantly from Preston Road, further to the north. Several footpaths between the two, while not giving views of the water because of the elevated banks, can afford views of birds around the reservoirs, including Whimbrel and Wigeon feeding in the spring and winter respectively. Dog-walkers will often be seen heading up to the banks of the reservoirs themselves but this is not permitted and any birder copying them would be doing so at their own risk.

The site is exceptionally well served by buses as the Preston to Longridge service stops just north of Pinfold Lane at Alston Meadow; it runs every 10 minutes or so Monday to Saturday and half hourly on a Sunday. This service also stops at Sunnybank, adjacent to the entrance to Grimsargh Wetlands, and an option is to alight at one, walk the couple of kilometres between the sites and return home from the other location.

YOUR VISIT

Migration times are best for both the current reservoirs and Alston Wetland, particularly after thunderstorms or other showers that might drop birds in. Water levels at Alston Wetlands will affect the mix of birds present, so it is worth checking news information services to see what has been reported recently.

NEARBY SITES

Dilworth Upper Reservoir between Forty Acre and Higher Road (SD616383, bronzes.imply.holdings) is small but has cover around the margins and is worth checking for diving ducks and grebes. Spade Mill Reservoirs (SD618375) are private but can be viewed distantly from the footpaths that run north-west off the B6243 at SD615372.

Lancashire

PRESTON/GARSTANG AREA

50 BROCKHOLES LWT RESERVE

Brockholes LWT Reserve
Just off junction 31 of the M6, Preston New Road,
Samlesbury, Preston PR5 0AG
SD588306
trails.bucket.modest

Website: lancswt.org.uk/nature-reserves/brockholes-nature-reserve
Phone: 01772 872000
Email: brockholes@lancswt.org.uk
Opening times: Reserve open 5am–7pm; visitor centre 10am–4pm
Parking: Main car park at visitor centre, £5 daily
OS Landranger Map 102, OS Explorer Map 286

Brockholes is a flagship Lancashire Wildlife Trust reserve that occupies the site of a former quarry and floodplain adjacent to the River Ribble. Its mix of lagoons, reedbeds, wet grassland and woodland creates a hotspot for both breeding and passage birds.

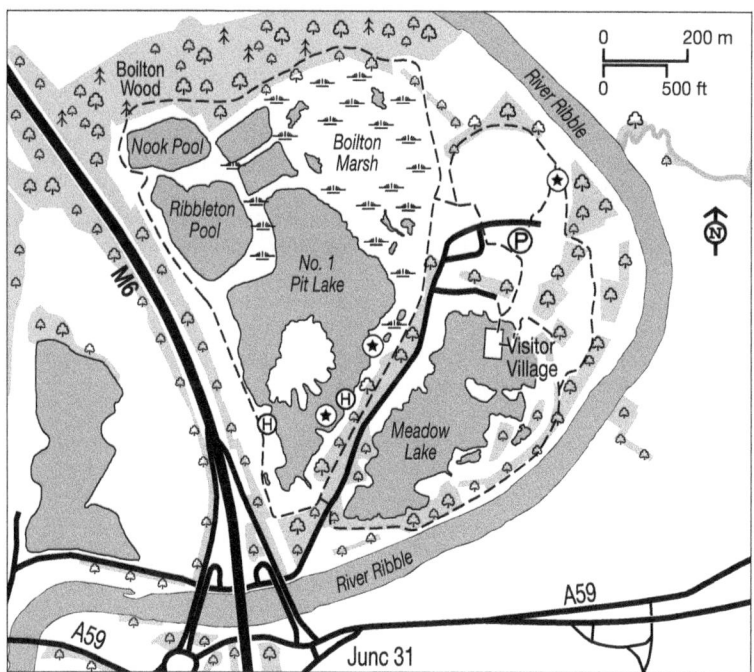

SPECIES

The potential of Brockholes Quarry as a birding site was clear before it was purchased by the Lancashire Wildlife Trust in 2007. Access granted to local birders to the lagoons not being actively quarried at the time had produced an enviable list of scarce and rare birds. Brockholes is now the Trust's flagship nature reserve, as well as being a conference and wedding venue to generate income for the charity's work. Whilst there are mixed views as to the relative merits of the current and previous incarnations of the site for birding, it remains a major destination for birdwatching in the Preston area and indeed within Lancashire.

The reserve includes all the water bodies to the east of the M6; there are other pits to the west. The main waters on the reserve are Meadow Lake, where the futuristic floating visiting centre is located, and Number One Pit lake. The former is a regular site for wintering Bittern, and the latter site has rafts utilised by Common Terns and a Sand Martin nesting bank. To the north of Number One Pit is Boilton Marsh, to which there is no access, but views can be obtained from the surrounding path. Further north again is Boilton Wood, which has woodland species and can attract migrants on passage.

The more interesting breeding birds include a mix of freshwater reedbed species. The Reed Warbler population is currently going from strength to strength. Sedge Warbler, Blackcap, Garden Warbler and Chiffchaff also nest whilst Cetti's and Grasshopper Warblers do so most years. Breeding wader populations have declined with *Crassula* encroachment but still include Redshank, Snipe, Common Sandpiper and Little Ringed Plover. There are about five pairs of Great Crested Grebe on site.

Several species of duck occur in good numbers outside the breeding season, including Teal, Wigeon, Gadwall, Goldeneye, Tufted Duck, Shoveler and Pochard; the last species is declining in the county. Over 100 Coot also winter. Up to 500 Lapwing linger outside the breeding season, whilst at different times in the spring Curlew, Oystercatcher and Whimbrel flocks gather on passage. At migration times other species regularly seen include Green Sandpiper, Black Tern, Little and Mediterranean Gulls and Common Scoter on overland migration. As well as Bittern there are increasing records of Great White Egret and particularly Little Egret.

Raptors include Hobby, which is a frequent visitor from local nest-sites every summer. Increasingly, Ospreys linger with birds in summer as well as passage periods, suggesting nesting may not be far off. Sparrowhawk and Peregrine are regular, as is Kestrel; a male of the last species, generally known as Kevin, has become habituated to people and has afforded exceptional photographic opportunities over many years.

Tree Sparrow occurs in small numbers. Marsh Tit used to be resident in nearby woodlands and still occurs very occasionally in Boilton Wood. Stonechat is regular in winter.

Although not part of the reserve the River Ribble is an intrinsic part of the birding experience at Brockholes. The river can often hold interesting wildfowl, including Goosander all year and Goldeneye in winter. Kingfisher is seen regularly, and in late autumn 2021 a Belted Kingfisher from America was on this stretch of river, though never seen on the reserve.

Rarities seen over the years include the county's first Semipalmated Sandpiper and Laughing Gull. Other notable species have been Pallid and Montagu's Harriers, Spotted Sandpiper, Lesser Yellowlegs, Night and Purple Herons, Savi's Warbler, Ashy-headed (Yellow) Wagtail, Red-necked and Black-necked Grebes, Red-necked

Phalarope, Crane, Smew, Pectoral Sandpiper, Temminck's Stint and remarkably for an inland Lancashire site Arctic, Great and Pomarine Skuas.

ACCESS AND FACILITIES

The reserve occupies 100ha adjacent to junction 31 of the M6 motorway. This proximity to the motorway makes it a very accessible venue if visiting by car, and it caters well for disabled visitors with 16 dedicated car spaces close to the visitor village. Kissing gates on the level trails can be navigated in smaller wheelchairs; for those using a large mobility vehicle a key can be obtained for a refundable deposit (currently £10).

The nature trails feature a number of hides and viewpoints. The Lookout Hide gives views over Meadow Lake, which can also be watched from the visitor village, and two hides overlook Number One Pit.

The visitor centre has toilets and eating establishments. The welcome centre serves drinks and snacks from 10am–4pm, seven days a week. Peckish café is open the same hours from Thursday to Sunday. The larger Kestrel Kitchen is open at weekends and during school holidays, also 10am-4pm. Indoor facilities are available every day for those preferring to bring their own packed lunch. For those visiting with younger children there is an adventure playground by the car park. The binocular and telescope retailer In Focus has a store overlooking Number One Pit lake where you can try out optics before potentially purchasing binoculars or a telescope.

In terms of public transport there are hourly services (route 59) from Preston bus station. Whilst it is some distance from Preston rail station an option for those with time and appropriate levels of fitness is to walk the Ribble Way alongside the river and birdwatch en route, including through Avenham and Miller Parks.

YOUR VISIT

The reserve car park is open from 6am-7pm each day and currently costs £5 for all users including LWT members. The visitor centre is open from 10am–4pm, and if planning to visit to try optics the In Focus shop is currently open 10am-4pm from Wednesday to Sunday. When the reserve car park is closed the reserve may be accessed by parking carefully on the entrance road and walking in.

51 PRESTON DOCK

Preston Dock
Riversway, Ashton-on-Ribble, Preston PR2 2YN
SD514297 (car park)
packet.critic.grants

Opening times: Car park open during daylight hours
Parking: The free car park is accessed from Mariners Way
OS Landranger Map 102, OS Explorer Map 286

Preston Dock is a former commercial port. The eastern end remains relatively undisturbed and is particularly notable for its Common Tern breeding colony, passage terns and a regular gull roost. The surrounding dockside and river areas also offer opportunities to see grebes, seaducks and occasional seabird visitors during stormy weather.

SPECIES

This former maritime dock feeding into the River Ribble closed as a port in the 1980s. It is now essentially a pleasure-craft marina at the western end and generally undisturbed at the eastern, Preston, end.

The site is best known currently in birding terms for its Common Tern breeding population. Numbers fluctuate, but thanks to management work by volunteers from the Fylde Bird Club to reduce predation by gulls, Coots and Moorhens the site continues to be favoured with young successfully fledged. In some years a small number of Arctic Terns have also joined the colony. The birds can be viewed from the free car park on the north side off Mariners Way. Tyres and nest boxes used by the terns are on the pontoons off here on both sides of the dock. The site also regularly attracts passage Black Terns. Roseate Terns have made fleeting visits on two occasions whilst Little Terns have also been seen. In the past overland tern migration was observed annually on the adjacent river and this presumably still occurs unrecorded.

The site has a decent track record of rare and scarce gulls. These include Ring-billed, Caspian, Glaucous and Iceland Gulls. Mediterranean and, to a lesser extent, Little Gulls are regular among the Black-headed Gulls. It is a good site for Darvic gull ring reading, as birds are often out of the water close to the car park on the pontoons. As well as large numbers of birds caught locally, Scandinavian-ringed individuals are regular. In winter very large numbers of Black-headed Gulls drop in to bathe before heading out to roost on the estuary, and a rarer small gull is arguably overdue.

Great Crested Grebes nest most years, and several birds are usually also present in winter. The site has also held Red-necked, Slavonian and Black-necked Grebes. Perhaps surprisingly there are very few records of divers here or on the adjacent river. Scaup have occurred on several occasions including small flocks, and Common Scoter turn up on overland migration.

During storms seabirds can take respite on the dock. Species seen over the years include Gannet, Shag, Manx Shearwater, Great Skua and Kittiwake. Kingfisher is regular on the main basin and also in the water in the locks between here and the river.

ACCESS

The free car park with views over the dock is accessed from Mariners Way; it is one way and entered from the eastern entrance. The car park is locked at night, with times displayed on site.

There are regular bus services from the town centre to the docks. It is also possible to walk from the town centre along the river, with woodland species in Avenham and Miller Parks and the potential for Grey Wagtail and Goosander plus Otters.

YOUR VISIT

If visiting to observe the nesting terns, June and July are best, and good background reading can be found at fyldebirdclub.org/preston-dock-common-terns/.

Spring and autumn visits are best after showers, which may drop passing terns and potentially other species. In winter the best plan is probably for a late afternoon visit, staying until dusk as gull numbers increase into the thousands.

52 GRIMSARGH RESERVOIRS

Grimsargh Reservoirs
Off Preston Road, Grimsargh, Preston PR2 5JR
SD592346
sits.paces.increases

Website: grimsarghwetlands.org
Phone: 07954 159074
Email: grimsarghwetlandstrust@gmail.com
Parking: Roadside parking
OS Landranger Map 102, OS Explorer Map 286
BHS

This site of three decommissioned United Utilities reservoirs is now managed as a nature reserve by the Grimsargh Wetlands Trust. The reserve is formed of three main areas: The Mere, Island Lake and The Fen. The 12ha site has a hide and viewing area, accessed off Preston Road opposite Oban Court (SD589345), with panoramas over both The Mere and Island Lake. A public footpath allows good views across the Island Lake to the west and the scrub of The Fen to the east.

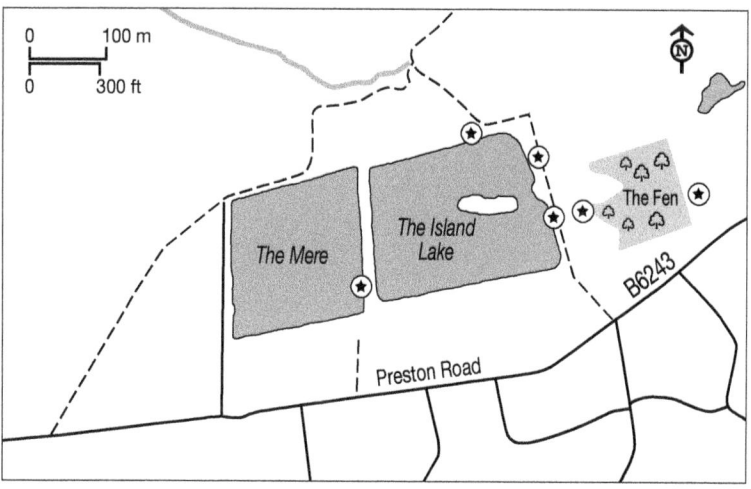

SPECIES

The site attracts a range of waterfowl and waders. Flocks of Lapwing are regular outside the breeding season, Curlew and Whimbrel flocks occur particularly on spring passage, and other wader species seen regularly include Snipe, Redshank, Oystercatcher, Green and Common Sandpipers and Little Ringed Plover. Wildfowl include Canada and Greylag Geese, Whooper Swan, Shelduck, Mallard, Teal, Shoveler, Wigeon, Tufted Duck, Coot, Cormorant, Grey Heron, Kingfisher and Little and Great Crested Grebes. Raptor species include Kestrel, Peregrine and Sparrowhawk whilst Osprey occur on passage. Mediterranean Gulls occur regularly among the Black-headed Gulls. Passerines include Reed Bunting and warbler species including Reed and Sedge.

Although no rare species have been recorded as yet, a range of scarce birds has occurred at the site, and as continuing habitat improvements are made further highlights can be expected. Scarcities seen to date include Slavonian Grebe, Great White Egret, Garganey, Common Scoter, Honey Buzzard, Avocet, Wood and Pectoral Sandpipers, Grey Plover, Little Gull and Black Tern.

ACCESS

By car, leave the M6 at junction 31A and follow the B6242 for Longridge. Take the B6243 towards Grimsargh, then continue through the village, passing St Michael's Primary School and The Plough pub. The wetlands are after the mini-roundabout on the left. Park sensibly on nearby streets rather than on Preston Road itself; do not park in the village hall car park.

Public transport is very good with the number 1 bus service from Preston bus station running at 10- to 20-minute intervals throughout the day and a stop at Elston Lane close to the site. Secure bike hoops are available beyond the gate to the viewing area for those cycling to the site.

There is disabled access to the main viewing screens from the Oban Court entrance via a mobility ramp.

YOUR VISIT

The public footpath means it is possible to visit the site at all times. Late April and early May is the best time for Whimbrel roosting on the lagoons or feeding during the day in nearby fields.

53 ABBEYSTEAD, BARNACRE AND GARSTANG LAKES AND RESERVOIRS

Cleveley Mere
Cleveley Bank Lane, Scorton, Preston PR3 1BY
Website: lancashire.gov.uk

Abbeystead Lake
SD556538
befitting.livid.boom

> **Cleveley Mere**
> SD501504
> stubborn.justifies.turkey
>
> **Grizedale Reservoir**
> SD523483
> ordeals.pictured.screeches
>
> **Scorton Picnic Site Lake**
> SD504507
> shuttered.shook.solving
>
> **Street Bridge**
> SD518521
> rejoin.milky.ghost
>
> Opening times: Scorton car park is open during daylight hours
> Parking: Car parks at Scorton and Stoops Bridge
> OS Landranger Map 102, OS Explorer Map OL41

The reservoirs, lakes and riverside sites around Abbeystead, Barnacre and Garstang offer a varied mix of birding opportunities, from inland reservoirs attracting large spring gatherings of Whimbrel to quieter stretches of the River Wyre holding Kingfisher, Dipper and Goosander.

SPECIES

There are several freshwater sites in this area that either hold a variety of species all year round or are worth a look at passage periods.

STREET BRIDGE ('WYRESIDE LAKES FISHERIES')

This complex of lakes is best known for a Pied-billed Grebe that was present for several days in 2021 and has also had Ring-necked Duck in the past. It holds diving duck, particularly Tufted Duck, every winter and also draws in Great Crested and Little Grebes, Gadwall, Shoveler, Goosander and a regular wintering flock of Greylag Geese. Whimbrel and Oystercatcher flock in the area on spring passage. Gulls drop in to bathe, and Grey Wagtail, Dipper and Kingfisher can be seen on the River Wyre, which runs through the site. Marsh Tit has been seen in private areas of the site in the last few years and may still be present.

To access the site park at Street Bridge (SD518522) and follow the Wyre Way footpath east, initially alongside the river viewing Banton's Lake and Fox's Lake. Alternatively walk towards the Fleece Inn, picking up the footpath at SD516524, to get a more elevated view. Unfortunately access to the remaining lakes has been closed to the public, but some can be viewed distantly from the Wyre Way to the south-west of the fisheries centre.

CLEVELEY MERE AND SCORTON PICNIC SITE

Cleveley Mere is private but can be viewed from the minor road to the west of Cleveley Bridge and the public footpath that runs south from the bridge towards Scorton. It regularly holds diving duck and grebes, whilst Night Heron has been reported on the River Wyre here and Dipper, Goosander, Grey Wagtail and Kingfisher are all regular. The adjacent Scorton Picnic Site (SD504503) has ample

Lancashire

car parking and a pleasant nature trail including views over a small lake. Siskin, Nuthatch, Great Spotted Woodpecker and Redpoll are all regular here.

BARNACRE/GRIZEDALE/GRIZEDALE LEA RESERVOIRS

The two Barnacre Reservoirs are the most productive in this complex. They attract outstanding numbers of Whimbrel on spring passage, including for example 883 on 1 May 2022. Black-necked Grebe has occurred more than once and Red-necked Grebe has also been recorded. Unfortunately there is no general access to the site. A public footpath along the north-west bank of Grizedale Reservoir (SD526484) can be accessed via a popular walk over the summit of Nicky Nook from Scorton village.

ABBEYSTEAD LAKE

This site is difficult to view well but attracts a range of waterbirds and is worth checking if already in the area. Good numbers of Teal can occur as well as other duck species, and waders rest on the margins of the island. The best approach is probably to park at Stoops Bridge car park (SD563543) and complete a circular walk crossing the River Wyre just downriver of the weir.

To the east of Abbeystead at Tarnbrook a large gull colony persists despite control measures to protect water quality, which is particularly important for its Lesser Black-backed Gull population.

ACCESS

Public transport is unfortunately not an option for any of the above sites, with no bus services running to any of these villages other than school services.

54 MYERSCOUGH QUARRY

Myerscough Quarry
Moss Lane, Bilsborrow, Preston PR3 0RU
SD504388
loops.sunblock.leathers

Opening times: Views from the canal bridge available at all times
Parking: Small parking/turning area on Moss Lane (SD502390); alternative parking in Bilsborrow with a short walk along the canal
OS Landranger Map 102, OS Explorer Map 286

Myerscough Quarry is a series of flooded gravel pits south-west of Bilsborrow and east of the A6.

SPECIES

Since extraction ceased it has become attractive to wildfowl in particular. Several hundred Wigeon winter, and Tufted Duck and dwindling numbers of Pochard also occur. Goosander commute between here and the adjoining Lancaster Canal, and Shoveler, Gadwall and Teal can be present in variable numbers.

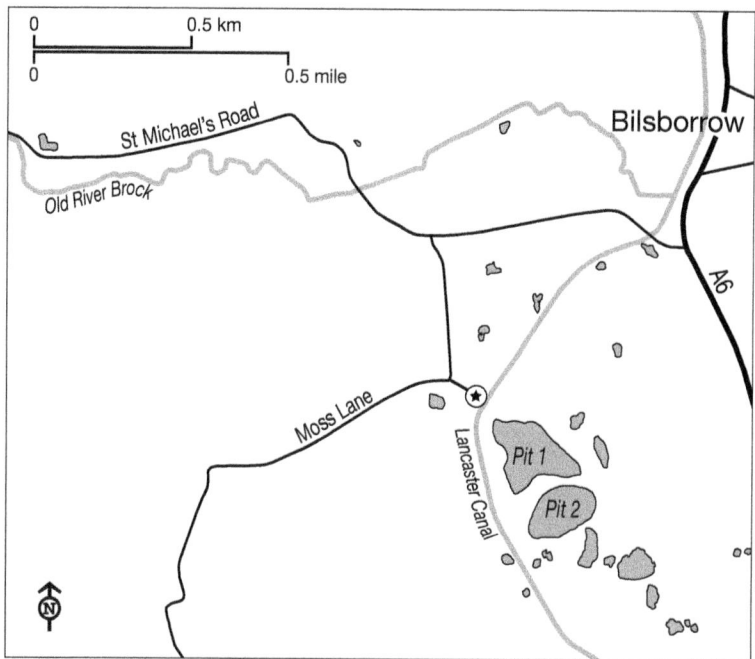

The site also attracts passage waders, though *Crassula* growth has probably reduced the attraction to some species recently. Green Sandpiper and Little Ringed Plover are regularly seen whilst flocks of both Bar-tailed and Black-tailed Godwits have occurred and Spotted Redshank and Wood Sandpiper have also been recorded. Breeding birds include Great Crested and Little Grebes, Reed and Sedge Warblers and Lesser Whitethroat. Sand Martins have nested in the bank behind Pit 1 on occasion.

Like other similar freshwater sites close to the A6 and M6 Myerscough Quarry has produced some rare and scarce birds. Particular highlights have been Lesser Scaup, Ring-billed Gull and several Black-necked Grebes, most recently one in 2022. There have also been several Scaup and Garganey, and Ospreys occasionally stop off to fish in spring.

ACCESS

The site is private, but much of the largest water body (Pit 1) can be viewed from an access bridge over the Lancaster Canal at SD503390. Most birds are on here or Pit 2 to the south and regularly move between the two when disturbed. There is a car parking and turning area on Moss Lane at SD502390. If visiting by public transport, bus services 40 and 41 stop in Bilsborrow, a 1.6km walk from the viewpoint. Walking via the canal gives a chance of Kingfisher and could also be combined with looking for Dipper on the River Brock to the north.

YOUR VISIT

The site's convenience to the A6 makes it a handy option to stop at when already travelling. For those accompanied by non-birding family or friends the Flower

Bowl complex is nearby with a garden centre, cinema, indoor golf and even a curling rink.

As with other inland waters, during or after storms can be productive in spring and autumn passage periods. Even though the site is private unauthorised visitors sometimes disturb the waterfowl, and for this reason early morning visits can be better.

SOUTH-WEST LANCASHIRE AND NORTH MERSEYSIDE

55 SEAFORTH NATURE RESERVE (PERMIT ONLY)

Seaforth Nature Reserve
Royal Seaforth Docks, Liverpool L21 1JD
SJ318971
move.rungs.trails

Website: lancswt.org.uk/nature-reserves/seaforth-nature-reserve
Phone: 07740 419187
Opening times: Accessible at all times
Parking: Reserve car park
Permit: You need to obtain a day pass from Lancashire Wildlife Trust (more information below).
OS Landranger Map 108, OS Explorer Map 275
Ramsar, SAC, SSSI

This site has been disaggregated from the combined treatment with Crosby Marina in the previous edition because of the different access arrangements for the two at present. The focus is on species using the reserve itself rather than flyovers that also pass over sites with more general access, and seabirds in the Mersey.

SPECIES

Seaforth is a Lancashire Wildlife Trust reserve that sits within the Port of Liverpool, managed by the Trust on land owned by Peel Holdings. The reserve comprises two lagoons, one saltwater and one freshwater, as well as the causeway between them, which can hold significant numbers of roosting waders and gulls. There is a small reedbed, and though the footprint has been diminished by development there is some grassland and scrub.

Seaforth initially came to prominence as a gull hotspot. Rarities seen over the years have included American Herring Gull, Laughing Gull and no fewer than three Ross's Gulls and five Bonaparte's Gulls. Little Gulls formerly gathered in spectacular large flocks of several hundred birds in spring; the species still occurs annually but the exceptional passage numbers no longer occur very often. Ring-billed, Iceland and Glaucous Gulls were more than annual but are all less frequent at present. Caspian Gull occurrences have increased, and the site is one of the best in Lancashire to see both this species and Yellow-legged Gull.

The proximity to transatlantic shipping berths led to a remarkable run of American passerines being seen on site. These have included Song and White-crowned Sparrows and Blackpoll Warbler. Passerines arriving presumably with less assistance, from the east and south, have included several rarities, among them Pied Wheatear, Red-throated Pipit, Red-rumped Swallow, Rosy Starling and Pallid

Swift. The site has a good track record for wagtails: as well as regular White and *flavissima* (British) Yellow Wagtail, passage Citrine, Blue-headed and Grey-headed Wagtail have all occurred. Remarkably, two singing Iberian Chiffchaffs have occurred, in 2017 and 2019. Scarce passage migrants have included Serin, Wryneck, Bluethroat, Yellow-browed Warbler and Little Bunting, whilst Black Redstart, Ring Ouzel, Whinchat and Wheatear are more frequent. Bearded Tit have occurred in the small reedbed.

A significant high tide roost of shorebirds forms on the causeway. This includes Redshank, Oystercatcher and Knot. Regular passage migrants include Spotted Redshank, Curlew Sandpiper, Little Stint, Common, Wood and Curlew Sandpipers. Rare wader species seen over the years include White-tailed Plover (2010), both Pacific and American Golden Plovers, Long-billed Dowitcher, Wilson's Phalarope (2008 and 2010), Buff-breasted, Marsh, Terek, White-rumped and Pectoral Sandpipers and Temminck's Stint.

Provision of tern rafts encourages Common Tern to breed, with perhaps 1 per cent of the national population using this highly urban site. There have been more than 200 pairs in recent years, with 208 pairs in 2021, for instance. As well as Arctic and Sandwich Terns, Roseate Terns are attracted to the throng of the colony annually; these often sport 'Roseate rings', having been marked at Irish colonies. Black Tern are double passage migrants and there have been half a dozen White-winged Black Terns as well as one each of Forster's and Gull-billed Terns. Regular wildfowl include Teal, Tufted Duck, Pochard, Scaup and Goldeneye, whilst scarce species have included Taiga Bean Goose, Green-winged Teal, Ring-necked Duck, American Wigeon, Long-tailed Duck, Garganey and Smew. The site has held several Black-necked Grebe over the years.

ACCESS

To access the reserve you need to obtain a day pass from Lancashire Wildlife Trust (seven days' notice is required). The application form is available at lancswt.org.uk/nature-reserves/seaforth-nature-reserve and requires the full names of everyone in the party and the make, model and registration of all vehicles accessing the site. LWT members may apply for an annual port pass (currently £35 per annum) from the ISPS office at the port (ISPS@peelports.co.uk).

Permit holders can find the freeport entrance at SJ327967 (the SD318971 reference on the LWT website is incorrect). If coming from outside Liverpool, exit the M57/M58 at junction 7 and take the A5036 south-west. At the junction with the A565 exit the dual carriageway and follow signs for the freeport. At the freeport entrance follow directions to the reserve car park. If there are scarce or rare birds on site, depending on species and where they are on the reserve, it may be possible to view them distantly from beyond the boundary fence at Crosby Marina.

There are three hides and a basic visitor centre. The site is not suitable for public transport birding due to health and safety requirements within the Freeport boundary prior to reaching the reserve. The paths on site are not particularly well suited to wheelchair access; Crosby Marina is a better option for birders with a physical disability.

Lancashire

56 CROSBY MARINA

Crosby Coastal Park
Cambridge Road, Liverpool L22 1RR
SD317977
award.bliss.upper

Website: https://www.sefton.gov.uk/around-sefton/parks-and-greenspaces/crosby-coastal-park/
Opening times: Car park open during daytime
Parking: Available off Cambridge Road, Blucher Street, Mariners Road and Hall Road West
OS Landranger Map 108, OS Explorer Map 275
Ramsar, SAC, SSSI

Immediately to the north of Seaforth Nature Reserve is Crosby Marine Park. Non-permit holders can view more conspicuous birds on Seaforth NR distantly through the port security fence. As might be guessed from its name, Crosby Marina has significant recreational use, particularly on the larger marine lake but also on the smaller boating pool. Nevertheless this large expanse of water close to the Mersey Estuary attracts diverse birdlife. The shore and dunes are also important bird habitats in their own right.

SPECIES

Crosby Marine Lake has a good track record for divers and grebes sheltering from adverse conditions at sea or recuperating. All three regular divers have occurred, Slavonian Grebe has been recorded and the site has a particularly good record for Black-necked Grebe, including three together in 2021. As well as regular Goldeneye, occasional Scaup, Red-breasted Merganser, Long-tailed Duck and Smew, both Common and Velvet Scoters have been recorded. Red-necked and Grey Phalaropes have been seen here, with the latter more frequent.

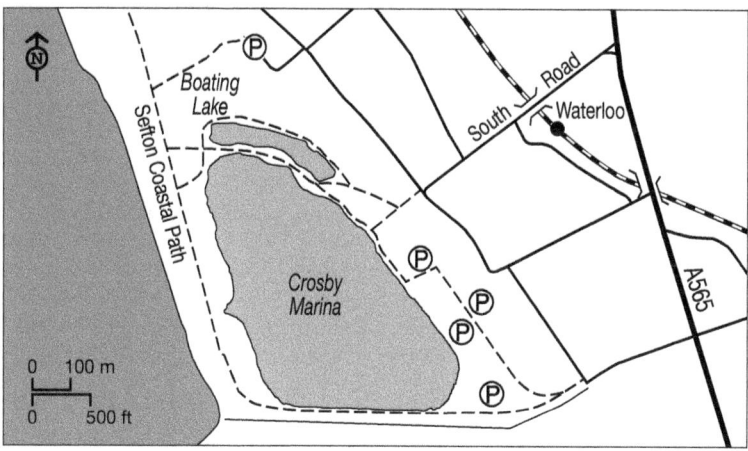

The dunes and beach can be very busy, not least with people visiting the internationally famous *Another Place* art installation, but they do hold birds. Snow Bunting is annual, whilst other species to have occurred include Lapland Bunting, Shorelark, Water Pipit, Black Redstart, Whinchat, Blue-headed Wagtail and Turtle Dove. There have been several Yellow-browed Warblers over the years in autumn. Rarities and scarcities have included Pallid Swift, Great Grey Shrike, Firecrest and Richard's Pipit.

As at Seaforth Little Gulls were formerly a significant spring spectacle on the marina, and although numbers are now much lower they remain regular. Terns are frequent: as well as Common Terns from the Seaforth colony these include regular Little, Arctic and Black, and White-winged Black has been seen on several occasions. The site is less favoured by rare and scarce large gulls than Seaforth but has attracted Caspian, Yellow-Legged and Ring-billed. A large Common Gull roost forms on the beach in winter; ring reading shows many of these birds come from Scandinavia.

The Sefton coast is a flightline favoured by raptors in spring and autumn. Osprey is often seen on calm spring days, whilst Hen and Marsh Harriers are annual and Peregrines from Liverpool nest-sites hunt the area. Egrets and Spoonbills pass through and will sometimes pause at the site.

Disturbance is an issue for freshwater waders but Common Sandpiper is regular on passage, and Wood Sandpiper, Little Stint, Curlew Sandpiper and Spotted Redshank have all occurred. Larger numbers of waders can be found on the shore, with flocks of Redshank and Curlew and on occasion Bar-tailed Godwit, Knot and Sanderling.

Crosby is one of the best places on the Lancashire side to watch seabirds battling out of the Mersey Estuary during strong north-westerly autumn gales. Species that can be seen in optimal conditions include Leach's and Storm Petrels, Great, Arctic, Pomarine and Long-tailed Skuas, Fulmar, Kittiwake, Sabine's Gull, Grey Phalarope and occasional early-season Little Auk. Wilson's Petrel has occurred, the only county record.

ACCESS

There are several car parks in the Marine Park. The most convenient for viewing the two lakes is at SD317977, reached by following the road in off Cambridge Road and heading past the sailing centre.

There are many public transport options to access the site, with several bus routes passing. Waterloo railway station is 300m from the site, accessed from South Road at SD317978. There are toilets and a café, the appropriately named Waterloo Sunset Café off Marine Crescent.

YOUR VISIT

Early mornings can be better for passerine migrants, with grounded birds more prominent before disturbance increases, and it is also the time when visible migration is most pronounced. Terns and Little Gulls are most likely in spring after showers and particularly with south-easterly winds. Seabirds are most likely to take refuge on the marina in the immediate aftermath of storm conditions. In winter the gull roost on Crosby Shore forms in the late afternoon before dusk.

57 MERE SANDS WOOD LWT RESERVE

Mere Sands Wood LWT Reserve
Holmeswood Road, Ormskirk, Lancashire, L40 1TG
SD448160
frail.eaten.acre

Website: lancswt.org.uk/nature-reserves/
mere-sands-wood-nature-reserve
Email: info@lancswt.org.uk
Opening times: Car park accessible at all times; visitor centre and cafe Wednesday–Sunday 10am–4pm
Parking: Reserve car park, £4 charge
OS Landranger Map 108, OS Explorer Map 285
SSSI

Close to Martin Mere, this charming Lancashire Wildlife Trust reserve can be combined with a visit to the Wildfowl and Wetlands Trust centre in the same day. The site name reflects the fact that the current mosaic of habitats was the result of sand extraction half a century ago; the sand was used in glass-making industries elsewhere in the north-west. Following cessation of quarrying in the early 1980s the LWT and the county council negotiated retention of the woodland areas and

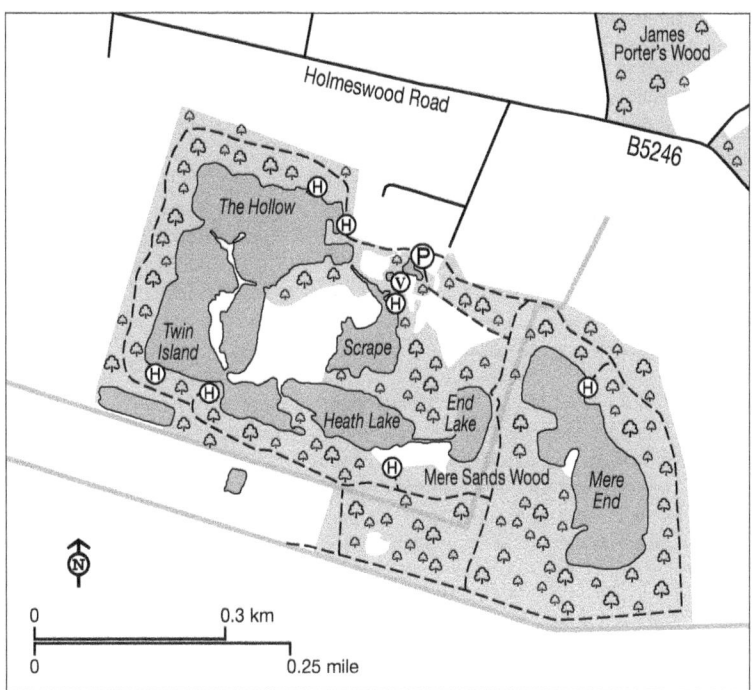

landscaping of the pits to create shallower edges and a mix of dry heath and marshland habitats. The site remains an LWT reserve to this day, with over 170 species of bird recorded, of which at least 60 have bred.

SPECIES

As might be expected, Martin Mere is favoured by wildfowl. In winter flocks of several hundred Teal are joined by lesser but still significant numbers of Shoveler, Wigeon and Gadwall. Great Crested and Little Grebes both nest and Kingfisher is a frequent visitor. Small numbers of Mandarin are regular. Water Rail are heard and occasionally seen. The site has become increasingly used by naturalised geese, with a total in excess of 3,000 Canada and Greylag Geese on occasion.

In terms of woodland birds Willow Tits still visit occasionally but are dwindling. Redpoll and several species of warbler can be seen including Reed and Sedge Warblers, Whitethroat and Lesser Whitethroat. Tree Sparrow is regular and Great Spotted Woodpecker breeds. Cetti's Warblers occur occasionally whilst several pairs of Reed Bunting nest.

Over the years a number of rare and scarce species have occurred at the site. Birders of a certain vintage will recall the stunning drake Surf Scoter that spent a day here in 2002, an exceptional freshwater occurrence. Since then other notable species have included Ruddy Shelduck, Green-winged Teal, Garganey, Ring-necked and Ferruginous Ducks, Scaup, Purple Heron, Great White Egret, Shag, Red-necked Grebe, Osprey, Wood Sandpiper, Black Tern, Waxwing, Firecrest and Wood Warbler.

ACCESS

From the A59 in Rufford join the B5246 heading for Holmeswood. Before reaching the village, after approximately 1.6km turn left at signs for Mere Sands Wood and kennels; the car park is at the end. There are six hides, all of which are wheelchair accessible, and there is a designated disabled parking space. Other facilities include a visitor centre, shop, café and toilets including accessible and baby-changing facilities.

Please note that, if arriving by car, parking charges apply to all vehicles, irrespective of whether all or some of the party are LWT members. Rates are currently £4 per day. There is the option to get a pass for three, six or 12 months, which can be purchased beforehand from the online shop.

In terms of public transport it is a walk of 1.6km or so from Rufford station to the reserve entrance road. There is a bus stop by the entrance road, and the Chorley–Southport service currently stops here several times a day.

YOUR VISIT

With the car park accessible at all times an early morning visit can be profitable before the levels of disturbance increase. A morning visit also fits well with going on to nearby Martin Mere WWT in winter where a visit in late afternoon can work well in terms of seeing Starling murmurations, raptors hunting and swans and other wildfowl gathering on the main mere for the public feed.

Lancashire

58 MARSHSIDE RSPB RESERVE

Marshside RSPB Reserve
Marshside, Southport, Sefton PR9 9PJ
SD352204
rear.urgent.tennis

Website: https://www.rspb.org.uk/days-out/reserves/marshside
Phone: 01704 211690
Email: Ribble.reserves@rspb.org.uk
Opening times: 8.30am–4pm (autumn/winter) / 5pm (spring/summer)
Parking: RSPB car park: £1.50 up to 2 hours, £3.00 over 2 hours; free for RSPB members
OS Landranger Map 108, OS Explorer Map 286
NNR, Ramsar, SSSI

The Marshside area is just north of Southport on the southern shore of the Ribble Estuary. The reclaimed freshwater marshes form the bulk of the RSPB reserve, with Sutton's Marsh north of Marshside Road and Rimmer's Marsh to the south. The saltmarshes of Crossens Marsh to the west of Marine Drive attract more marine and coastal species.

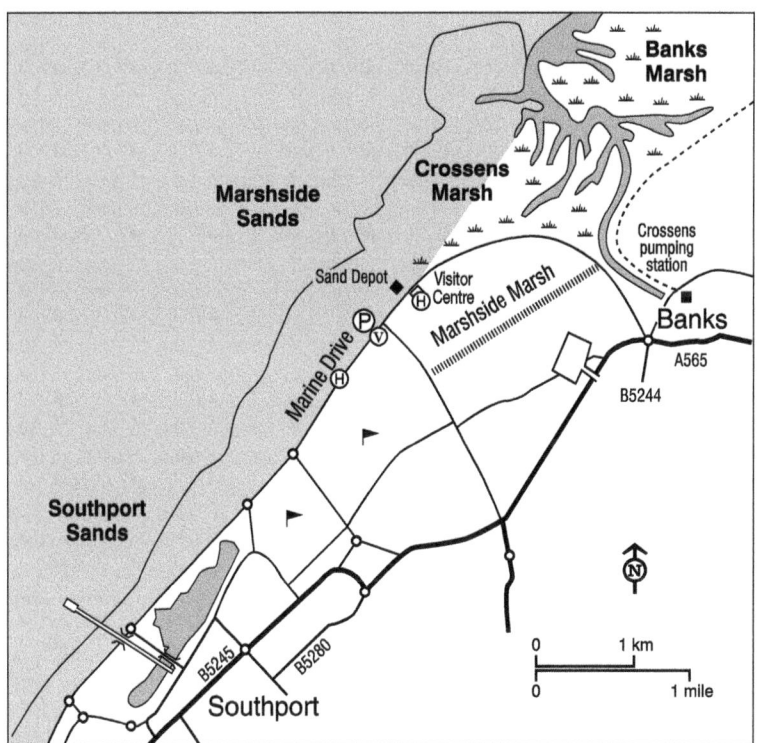

191

SPECIES

The reclaimed marshes flood extensively in winter. This attracts large numbers of ducks, particularly Wigeon, for which the Ribble Estuary is a stronghold, and also Teal in good numbers. As might be expected, several American Wigeon and Green-winged Teal have been seen over the years, whilst a Baikal Teal appeared in the 2013/14 winter.

Breeding populations of waders, including Lapwing, Redshank and Snipe, are augmented in winter by good numbers of Black-tailed Godwit and Golden Plover. Ruff have bred in the past and occur in good numbers in spring with a few wintering. Avocets have colonised and there are now around 50 pairs nesting. Regular passage waders include Spotted Redshank, Greenshank, Wood, Green and Curlew Sandpipers and Little Stint, whilst Pectoral Sandpiper is virtually annual.

The area has been attractive for Pink-footed Geese for many years, and detailed scrutiny and increased knowledge of identification features has led to annual records of rare and scarce species among them. These include more regularly Barnacle, Pale-bellied and Dark-bellied Brent and Eurasian White-fronted Geese. Tundra and Taiga Bean, Red-breasted, Snow, Ross's, Lesser White-fronted and North American Canada and Cackling Geese have also been recorded. A putative Grey-bellied Brent Goose was seen in February 2020. Bewick's Swans visit the area in small numbers most winters, but Whooper Swans are far more regular these days.

Marshside is perhaps the best site in Lancashire to view Garganey in spring and they sometimes stay to breed. Gadwall and Tufted Duck breed and are present all year round, the latter being joined by Scaup regularly and Lesser Scaup on a couple of occasions.

The saltmarsh and mudflats are important for Knot, Bar-tailed Godwit, Oystercatcher and Dunlin. The marshes north-east of the now-demolished sand-winning plant ('McCarthy's viewpoint') is the best area to view an impressive range of raptors in winter including Hen and Marsh Harriers, Peregrine, Merlin, Kestrel, Sparrowhawk, and Barn and Short-eared Owls. Little Egrets occur on both saltmarsh and freshwater marsh and are regularly also accompanied by Great White and Cattle Egrets. Spoonbills are increasingly regular both on the freshwater marshes and the estuary.

Skylarks still breed in numbers and relatively large flocks winter. Reed, Sedge and Cetti's Warblers all breed. Crossens Marsh is a regular wintering site for Water Pipit; high tides are best for viewing them as they get pushed closer to the embankment. Snow Bunting can still be seen most winters on the sandier beaches towards Southport. Twite flocks occur and generally favour the saltmarsh.

Close to the reserve the fairways and rough of Hesketh Golf Course can be productive for passerine migrants. As well as the expected Wheatear these can include Whinchat, Redstart, Pied and Spotted Flycatchers, Ring Ouzel in spring and Yellow-browed Warbler in autumn.

Whilst Marshside itself has a good track record for phalaropes (including Wilson's as well as Grey and Red-necked) and occasional seaduck, Southport Marine Lake to the south-west can still be worth checking for grebes, divers and other species taking refuge, particularly in storms.

An impressive list of rare and scarce birds has been seen in the Marshside area since the last edition of this book. These include Cackling and Todd's Canada Geese, Wilson's Phalarope, Black-winged Stilt, Broad-billed and Buff-breasted Sandpipers, Lesser Scaup, Blue-winged Teal, Temminck's Stint, American Golden

Plover, American Wigeon, Woodchat Shrike, Long-billed Dowitcher, Spotted Crake, Great Grey Shrike and Blue-headed Wagtail. Montagu's Harrier can now be added to this list, with a bird present in summer 2025.

Although published as long ago as 2001, the late Barry McCarthy's *Birds of Marshside* (Hobby Publications) remains a must-read for anyone interested in this site. Barry was a stalwart observer here and seawatching at Formby Point, and as a university lecturer and affable birder he effortlessly conveys the appeal of the area.

CALENDAR

March–August: Sand Martin, Wheatear and White Wagtail pass through; Lapwing and Redshank nest; wader passage includes Ruff, Little Ringed and Ringed Plovers, Greenshank and Dunlin; breeding Avocet, Gadwall and Shoveler; Cetti's, Reed and Sedge Warblers in song; Little Gull, Garganey and Spoonbill.

September–February: Pink-footed and Barnacle Geese with several other species among them; raptors hunt over the saltmarsh; Skylark flocks; Water Pipit on Crossens Marsh; regular autumn scarce waders including Little Stint, Curlew Sandpiper and often Pectoral Sandpiper.

ACCESS

Marshside is accessed north of Southport town centre. Coming from the M6 along the A565 take the last turning on the roundabout at Crossens to join the start of Marine Drive. The reserve car park is on the seaward side of the Marine Drive and Marshside Road junction at SD352204. Parking is free for RSPB members; non-members currently pay £1.50 for less than two hours and £3 for over two hours. The viewpoints, including that overlooking Junction Pool opposite the car park, are always open. The hides and the portaloo are open for varying times depending on the season (check the website before visiting). If also visiting the Marine Lake, walk down Marine Drive or alternatively there is a public car park at SD337185.

The two hides (Sandgrounders Hide, which doubles as the visitor centre, and Nel's Hide) are both wheelchair accessible.

If travelling by public transport the RSPB reserve is 3km from Southport railway station. Much of this distance can be a productive birding walk via the Marine Lake, checking the saltmarsh off Marine Drive. Several bus services run close to the site, including the 44 Southport to Crossens service with a stop at Elswick Road 200m from the reserve.

YOUR VISIT

Tidal conditions are an important factor if visiting for shorebird roosts and species flushed from the saltmarsh such as Short-eared Owls and Water Pipits. Spring tides are best and the timing of these can be found online. Showers in spring will sometimes drop passing hirundines and Black Terns. In a prolonged visit during winter it is generally advisable to look for harriers from the old sandplant towards dusk when they are more active.

Many of the scarcer spring passerine migrants such as Wood Warbler and Ring Ouzel are most frequently recorded on Hesketh Golf Course where there is more cover. There are no footpaths through the golf course so early morning viewing from its fringes before golfing gets under way is recommended.

59 HESKETH OUT MARSH RSPB RESERVE AND BANKS MARSH NNR

Hesketh Out Marsh RSPB
Dib Road, Hesketh Bank PR4 6XQ
SD421250
paler.televise.loudly
Website: rspb.org.uk/days-out/reserves/hesketh-out-marsh
Phone: 01704 211690
Email: Ribble.reserves@rspb.org.uk

Ribble Estuary NNR
Old Hollow Farm, Old Hollow Lane, Banks PR9 8DU
SD391226
flash.twin.random

Opening times: Car park open daily 8am–6pm
Parking: RPSB car park, no overnight parking
OS Landranger Map 102, OS Explorer Map 286
NRN, Ramsar

Hesketh Out Marsh and the adjacent Banks Marsh form a vital part of the South Ribble Marshes. They offer a rare expanse of managed wetland in southern Lancashire.

SPECIES

HESKETH OUT MARSH RSPB
The last edition of this book in 2007 predicted that bird recording on the South Ribble Marshes would be greatly improved by the imminent creation of new habitat at Hesketh Out Marsh. The initial restoration started that year, and the site has quickly established itself as a major Lancashire birding site. The expansion of the reserve in 2017 further enhanced the range of species and numbers of birds using the improved habitat. To date around 190 species have been recorded on the reserve, with others on the farmland inland of the embankment.

The reserve footprint occupies what had been reclaimed land for growing crops. Rising sea levels led to a partnership between the RSPB, Natural England and the Environment Agency to develop more natural flood protection by the creation of new habitat. Many new lagoons have been created, with several creeks being allowed to breach the old sea wall beyond.

In keeping with the wider South Ribble Marshes the area attracts large numbers of wildfowl. The adjoining estuary holds very large Wigeon flocks, which often use the reserve, and Teal also occur in large numbers. Unsurprisingly, careful scrutiny has led to several American Wigeon and Green-Winged Teal being found over the years.

Pink-footed Geese feed in the area regularly during the winter, and frequently strays can be found among them including Tundra Bean, Eurasian and Greenland White-fronted, and Barnacle Geese. Dedicated patch watching and increased knowledge of identification features has produced several vagrant Canada Geese

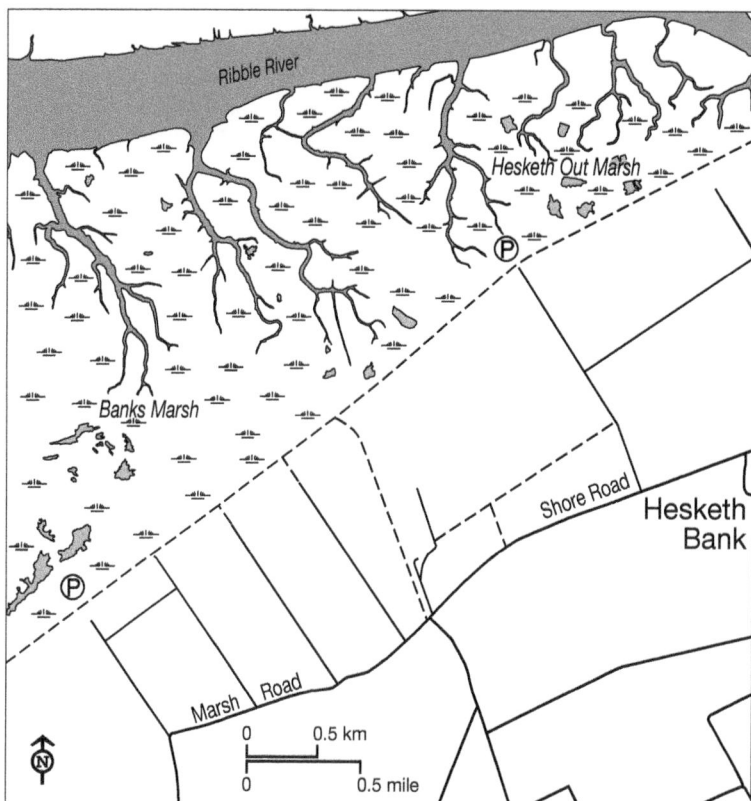

in recent years, with several Todd's Canada Geese and a Richardson's Cackling Goose in late 2021 (note feral Canada Geese use the marsh too and mingle with Pink-feet so care is advised). Other rarities include Snow Goose and Taiga Bean Goose, whilst a possible Grey-bellied Brent was present in the winter of 2020/21. The wild swans that use the site continue to include a small number of Bewick's as well as the more numerous Whoopers.

The lagoons are frequented by waders throughout the year. Breeding species include Avocet, Lapwing, Oystercatcher and Redshank. At passage times the site is one of the most reliable in Lancashire for Spotted Redshank, with several birds in autumn and usually at least one wintering. Other regular species include Ruff, Greenshank, Wood, Green and Curlew Sandpipers, Little Stint, Black-tailed and Bar-tailed Godwits and Dunlin. The growing list of scarce and rare waders includes several Pectoral Sandpipers, American Golden Plover, Buff-breasted Sandpiper, Temminck's Stint, Dotterel and Grey Phalarope. There is a historic record of Baird's Sandpiper.

Eider can be seen on the pools as the species nests on nearby parts of the estuary. Scaup and Long-tailed Duck have occurred in several winters, whilst Red-breasted Merganser and Goosander are both regular. Garganey linger on occasion, and Smew has been recorded a couple of times. Scarce waterbirds have included Black-necked Grebe and Ferruginous Duck. All three regular species of egret have occurred, as well as Spoonbills sometimes in small flocks. Glossy Ibis is an occasional visitor, and Crane has been seen in several years.

The large concentrations of waterbirds naturally attract a range of raptors. Peregrine and Merlin are regular. Marsh Harriers are present much of the year including in winter, when they are joined by Hen Harriers. Short-eared Owls winter and are easiest to see when high tides disturb them from roost sites. Ospreys stop off on passage and Hobby and Red Kite are seen on occasion.

Terns are regular in summer. As well as Common Tern, Arctic is seen most years and Little and Sandwich on occasion; there is one record of White-winged Black Tern. Little Gulls drop into the pools on passage, and Mediterranean Gulls breed in the vicinity.

The embankment affords views across neighbouring fields as well as the marsh and can itself be attractive to migrant passerines including Whinchat, Wheatear and White Wagtail. Blue-headed Wagtail is regular among a small local summering population of Yellow Wagtails, and Grey-headed Wagtail has occurred once. Twite still winter on the saltmarsh, and Snow and Lapland Buntings pass through. Grey Partridge, Corn Bunting and Tree Sparrow can be seen. Other records have included Richard's Pipit and Hawfinch, whilst Yellow-browed Warbler and Waxwing have both been seen around the car park. Chough has been recorded once and singing Quail have been heard a few times.

BANKS MARSH

South-east of Hesketh Out Marsh the national nature reserve at Banks Marsh is a similarly important habitat for nesting, passage and wintering wildfowl and particularly waders. The species seen in recent years overlaps significantly with Hesketh Out Marsh. Scarce and rare wader species recorded include Broad-billed, White-rumped, Buff-breasted and Pectoral Sandpipers, Lesser Yellowlegs, Collared Pratincole and Long-billed Dowitchers. One of only two Stilt Sandpipers seen in Lancashire was on the adjacent Hundred End Marsh. More or less all the scarce goose species that join with the Pink-feet flocks as referred to in the Marshside and Hesketh Out Marsh accounts have also been seen on Banks Marsh.

ACCESS

HESKETH OUT MARSH

By car, heading west from Hesketh Bank on Shore Road turn right at SD430235 onto the private Dib Road. Continue straight on, ignoring other tracks for 1km to reach the reserve car park at SD421250 (open 8am–6pm). Note that this track is not adopted and can have significant potholes so drive slowly and carefully. Pools can be viewed from the embankment in both directions.

It is a 1.6km walk from Hundred End junction on Shore Road. The hourly Preston to Southport bus route serves this stop.

BANKS MARSH

As with Hesketh Out Marsh the site is accessed from a track off Marsh Road, in this case at the junction with New Lane Pace heading north-west to Old Hollow Farm at SD391226. There is a bus stop by New Lane Pace Nursery, which is served hourly by the Preston to Southport service; Corn Bunting and Tree Sparrow are possible along Old Hollow Lane. The entire 3km of seawall between the two sites is a public footpath so both locations can be combined in a full day's public transport birding.

YOUR VISIT
Early morning visits are generally best for the light when viewing to the north from the seawall. Disturbance is also least at this time of day.

There are no facilities at either reserve. The nearest toilets (for customers only) are in Booths supermarket in Hesketh Bank, 4km from Hesketh Out Marsh. In winter in particular the embankment is very exposed, so if planning to patiently work through flocks of birds wear several layers of clothing.

60 SEFTON COAST

Ainsdale NNR
Southport, PR8 3QW
SD274082
suggested.flatter.swipes

Website: https://www.gov.uk/government/publications/merseysides-national-nature-reserves/merseysides-national-nature-reserves
Phone: 01704 578774
Email: enquiries@naturalengland.org.uk
Opening times: Accessible at all times
Parking: Public parking can be found at Ainsdale-on-Sea beach car park
OS Landranger Map 108, OS Explorer Map 285
NNR, Ramsar, SAC

The dunescape forming much of the coast between Liverpool and Southport has outstanding natural history interest. As well as Sand Lizards, Natterjack Toads and an endangered Red Squirrel population this includes diverse ornithological interest. The extensive dune system can be attractive to migrant passerines, and scarce and rare species stay to winter on occasion. Extensive sandflats offer feeding and roosting opportunities for large numbers of waders, and where the River Alt joins the sea near Hightown there are other habitats. Smaller areas of woodland and dune heathland can be productive. By virtue of its position protecting Liverpool Bay, Formby Point is one of the foremost seawatching venues in the county. The whole area, being between two large conurbations, is very popular for recreational activity, but there are several designated nature reserves which give birds and other nature some breathing space.

SPECIES
The site was made a national nature reserve in 1965, as arguably the best lime-rich sand dune system in north-west England. The main sand dune habitat of the reserve has breeding Skylark, Stonechat and Grasshopper Warbler. The wetter areas known as slacks can attract waders and wildfowl in winter, including Teal, Snipe and Jack Snipe. The conifer plantations at the eastern end of the reserve

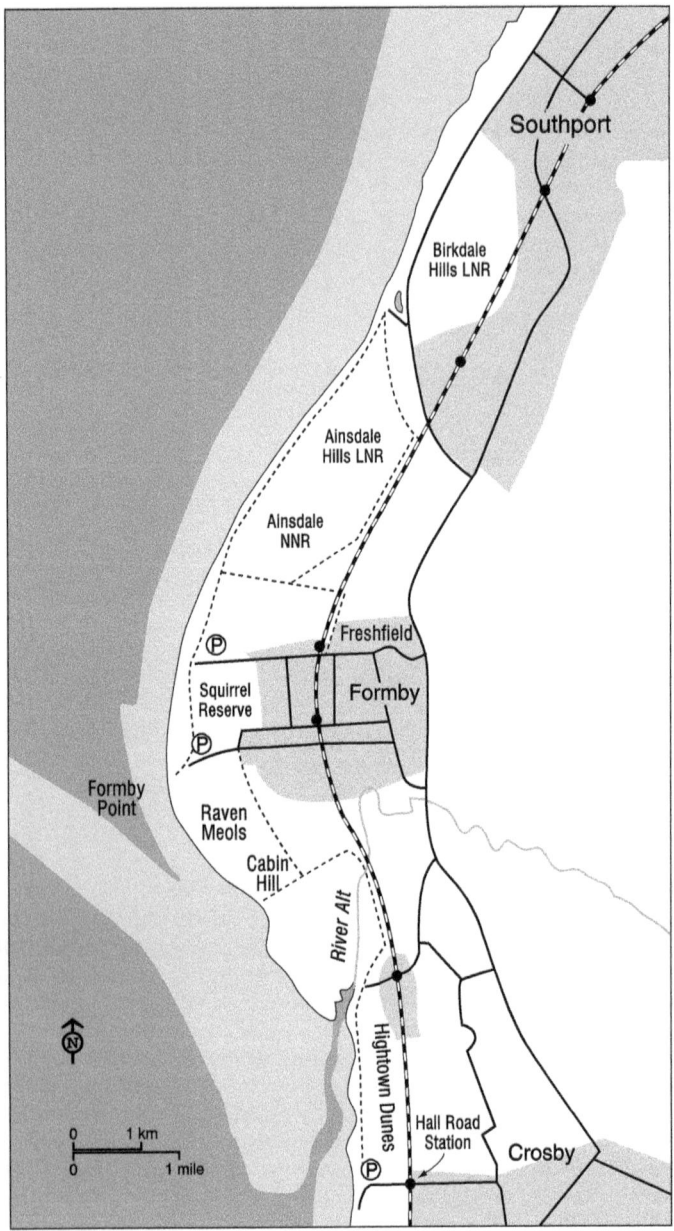

can be good for Crossbill, particularly in irruption years. In spring the reserve is one of the best coastal sites in the north-west to see passage Ring Ouzels, with several birds present on occasion. Wheatear, Redstart, Whinchat and to a lesser extent Wood Warbler and Pied Flycatcher are also regular migrants. Snow

Bunting feed on the strandline in winter, though these are dwindling in number; flocks of Twite can occur, and Black Redstart is seen with some regularity.

Among scarcity and rarity records at the site a remarkable series of Dusky Warbler records comprise one in autumn 2018 and two in the 2020/21 winter, one of which was a male singing before departure in the spring. Other scarce passerines in recent years have included Red-backed Shrike, Common Rosefinch, Wryneck, Bee-eater, Red-breasted Flycatcher, Great Grey Shrike, Lapland Bunting, Siberian Chiffchaff and Blue-headed Wagtail. Garganey have occurred several times on the slacks.

Seabirds are covered separately below but terns gather on the beach in autumn, and as well as occasional Roseate Terns an Elegant Tern was seen in 2021. Caspian Gulls have also been seen and are likely to be more frequent in the future.

There are 13km of footpaths, with colour-coded waymarks signposting the different trails, and several potential access points onto the reserve. There is a car park at the end of Victoria Road, Freshfield (SD274082), and organised pay-and-display car parking on Ainsdale Beach at SD298128; the latter is convenient for the Discovery Centre, with educational interpretive material about the reserve.

ACCESS

In terms of public transport, it is a 1km walk to the reserve from Ainsdale station. A half-hourly bus service (number 49) from Southport to Woodvale stops at the junction of Shore Road and Kenilworth Road, a short walk from the Discovery Centre.

SANDS LAKE

Shore Road, Ainsdale, Southport PR8 2PZ
SD301130, gloom.button.written

North of Shore Road this small lake often holds flocks of Tufted Duck as well as Little Grebes and is worth a look if in the area. Scaup, Long-tailed Duck and Little Auk have all occurred here. There is a boardwalk from the car park along the east shore, allowing coverage of the whole expanse.

ALT ESTUARY AND HIGHTOWN

Lower Alt Road, Hightown L38 0BB
SD296032, dentistry.gear.deed

SPECIES

Although much smaller than the estuaries of the Mersey and Ribble to the south and north, the Alt Estuary holds flocks of Grey Plover, Knot, Sanderling, Dunlin and Bar-tailed Godwit as well as Curlew, Lapwing and Redshank. Curlew Sandpiper, Little Stint, Spotted Redshank and Wood Sandpiper have been seen regularly over the years.

The regular breeding birds and migrant passerines are similar to those at Ainsdale, though Ring Ouzel is less regular. Scarce and rare species seen in recent years have included Woodchat Shrike, Richard's Pipit, Hoopoe and Siberian Chiffchaff.

Seaduck occasionally appear on the estuary, including Scaup, Common Scoter and more occasionally Velvet Scoter. The estuary does well for records of Ruddy Shelduck though their provenance is very uncertain. Terns roost here with a similar list of species to Ainsdale, including Roseate, and an Elegant Tern visited in 2021.

Fields around Hightown can attract Pink-footed Geese, with recent scarcities including Taiga Bean, Todd's Canada and Snow Geese.

ACCESS

The area is best explored on foot. Park either in Lower Alt Road, Hightown (SD296032), or at Crosby Coastguard Station (SD298004) at Hall Road on the northern edge of Crosby. The nearest train station is at Hightown, which is only 300m from the estuary and dunes.

BIRKDALE HILLS

SD308141, elated.changes.massaged

These dunes north of Ainsdale form part of the same chain of sand dune habitat. Species occurring in the dunes are similar to those listed for that site, and Garganey is seen most years on the slacks here. There is a healthy Grasshopper Warbler population and similar numbers of Cetti's Warbler have colonised in recent years. Wagtail flocks at Birkdale Cop often attract Yellow Wagtails. There is a remnant Corn Bunting population on Birkdale Moss.

Sanderling pass through in April and May, whilst good numbers of Jack Snipe and Snipe winter in the green beach area and can be flushed by large tides. A large gull roost can form at Birkdale as viewed from the Shore Road Car Park area, and careful scrutiny can produce scarce species including increasingly regular Caspian Gulls. As with other sites along this coast, seawatching is sometimes undertaken from Birkdale and this is covered below.

FORMBY POINT

Lifeboat Road, Formby, Liverpool L37 2EB
SD274082, cove.private.irony

SPECIES

As well as being a prime seawatching location (see below) Formby Point can be a productive area for observing passerine migrants. Several notable spring migrants are regular such as Whinchat, Redstart, Pied Flycatcher and Ring Ouzel. The extensive cover at a coastal headland means Yellow-browed Warbler is more or less annual, with up to four birds seen on occasion. Scarcities in recent years have included Woodchat Shrike and Hoopoe, whilst Snow Bunting appear on the beaches in many winters.

ACCESS

There is extensive car parking at the National Trust reserve at the end of Victoria Road, Formby. The pines here have also been a stronghold for Red Squirrels locally though numbers are currently lower due to squirrel pox. Other access is available at the end of Wicks Lane (SD283071) and Lifeboat Road where there is a car park (SD275064). There is a train station in Formby, which is most convenient for Lifeboat Road, whereas Freshfield station is the most suitable place to alight for the National Trust reserve.

To the south of Formby Point the nature reserves of Ravenmeols Hills and Cabin Hill are part of the chain of habitat stretching from Crosby to Southport. These can be accessed from Hightown to the south-east and Formby to the north.

SEAWATCHING

Formby Point is an obvious vantage point, and seawatching is frequently undertaken from the dunes south of the National Trust car park. Note, though, that it can be very exposed here with blowing sand an issue with the onshore winds that are often most productive for seabird passage. Alternative vantage points include Ainsdale, Birkdale and from the end of Lifeboat Road. Early morning is best on calm days as there is most movement at this time and visibility is generally good with the sun behind the observer. During gales, the periods either side of high tide are best as birds are pushed closer inshore; there is often something of a lull over high tide itself.

In winter, gales regularly bring Little Gulls inshore, though in reduced numbers these days. On calm days, Common Scoter rafts may be several thousand strong and can also hold Velvet Scoter and Long-tailed Duck. Red-throated Divers are regular, with Great Northern Diver seen most winters.

Spring passage includes large Arctic Tern movements as well as Arctic and Great Skuas and occasional Pomarine. Guillemots and Razorbills move through. This is generally the best time for Black-throated Diver, a very scarce species in Lancashire, and also for Puffin which is increasingly frequent.

During summer Manx Shearwaters and Gannets regularly feed offshore as well as Arctic, Common, Little and Sandwich Terns. These birds can attract several Arctic Skuas harrying for an easy meal. During strong gales, Storm Petrels can pass offshore. Roseate Terns are seen most years.

Autumn movements can include Guillemot and Razorbill, particularly the latter in some years. Black Guillemot is scarce but regular. During gales, Leach's Petrel is the main target, but after sustained westerlies Long-tailed Skua, Sabine's Gull and Grey Phalarope can all occur, with Little Auk later in the season.

Rarities include Lancashire only record of a Fea's type petrel, as well as Forster's Tern, Elegant Tern, and Balearic, Sooty and Cory's Shearwaters. In north-west England records of all shearwater species other than Manx are notable.

61 RIMROSE VALLEY

Rimrose Valley
Parklands Way, Liverpool L33 3YX
SD331980
discrepancy.liked.ozone

Website: rimrosevalleyfriends.org
Phone: 01515 581105
Email: enquiries@rimrosevalleyfriends.org
Opening times: Accessible at all times
Parking: Various car parks and roadside parking
OS Landranger Map 108, OS Explorer Map 275
LNR, LWS

Rimrose Valley is a large country park running between Crosby and Bootle. It provides an important corridor for wildlife in an otherwise urban setting. Its mixture of wetlands, grasslands and young woodland is important for both breeding and passage birds.

SPECIES

Rimrose Valley is a precious green space in a heavily populated area and has been reclaimed since the 1990s from being a domestic tip to provide a recreational and educational resource. It is bounded by the Leeds and Liverpool Canal and the Southport to Liverpool railway line. The wider site includes Brook Vale Local Nature Reserve. The mosaic of habitats includes reedbed swamp, damp meadow, willow-carr woodland and dry grassland. In 2017, the site was mentioned as potentially earmarked for a significant new road to Liverpool Docks but to date this has not been progressed.

Brook Vale LNR is at the very southern tip of Rimrose Valley Country Park. It uses land formerly occupied by allotments which were abandoned due to flooding. The area attracts some interesting wetland species including Reed Bunting, Reed Warbler, Common and Jack Snipe and Water Rail. Other warblers include Grasshopper Warbler and Common Whitethroat, and Cetti's Warbler has occurred. Barn and Short-eared Owls can be seen hunting grassland habitats and in summer Hobby hawk for insects. Wheatear, Whinchat and Ring Ouzel all occur on passage.

Scarcer species recorded have included Red-rumped Swallow, Honey Buzzard, Firecrest and Siberian Chiffchaff.

ACCESS

Brook Vale LNR is accessed from tracks off Beach Road, Litherland. The wider country park can be accessed from the Leeds and Liverpool Canal towpath, with car parks at Appleton Road (SJ331980) and off Kirstone Road West at Brindley Close (SJ334987). The site can be easily visited by public transport, with numerous bus services stopping near the wider site and Seaforth and Litherland station a short walk from Brook Vale LNR.

62 MARTIN MERE WWT

Martin Mere WWT
Fish Lane, Burscough L40 0TA
SD428144
repayment.archduke.moss

Website: www.wwt.org.uk/wetland-centres/martin-mere
Phone: 01704 895181
Email info@martinmere.org.uk
Opening times: The centre is open from 9.30am–4.30pm (winter) / 6pm (summer)
Parking: Free car park
OS Landranger Map 108, OS Explorer Map 285
Ramsar, SPA, SSSI

Lancashire

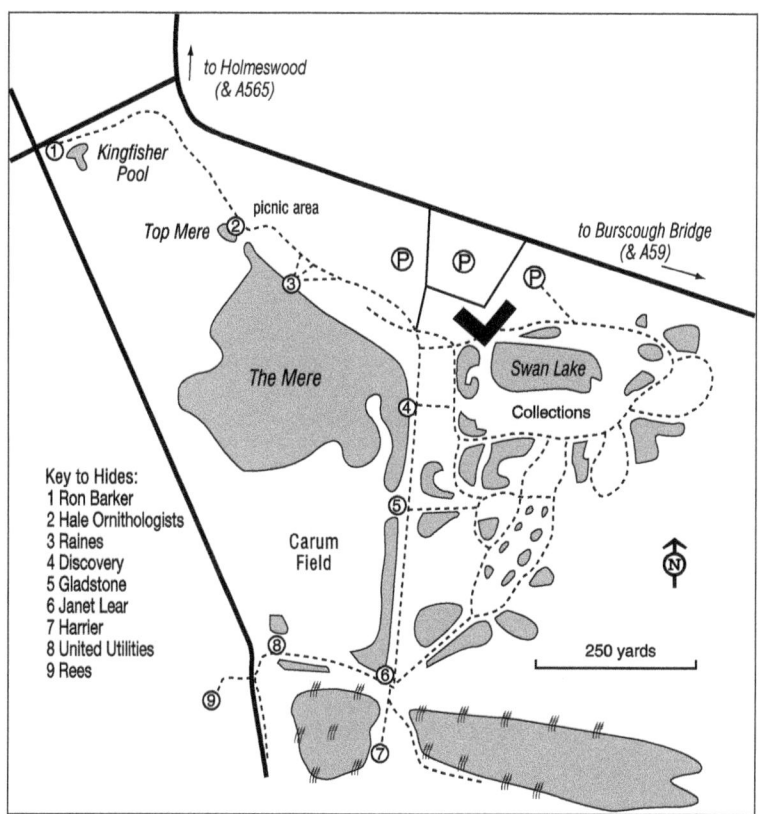

Martin Mere is the only Wildfowl and Wetland Trust reserve in north-west England. Historically it was the largest of several freshwater meres on the south-west Lancashire mosslands but had been drained. In the 1970s the WWT established a 150ha reserve restoring some of the previously lost habitat. More recently an extension of the reserve has seen a carrot field transformed to a reedbed wetland.

SPECIES

The reserve is perhaps best known for its wintering Whooper Swans. A supplementary feeding regime enables visitors to see hundreds of these birds at close quarters on the main mere at dusk. There are also internationally important numbers of Pink-footed Goose, particularly roosting but some feeding on the reserve during the day. Common Teal, Shelduck and Pintail occur in good numbers, and though numbers are declining Pochard still occurs, unlike at many other county sites. Bewick's Swan is occasional in very small groups.

Winter raptors include both Hen and Marsh Harriers, as well as Buzzard, Merlin, Peregrine and Sparrowhawk, and Barn Owl is also regular with Short-eared Owl occasional. Ruff is present among the wintering wader flocks, whilst Lapwing, Common Snipe and Redshank stay to breed in good numbers by current standards. Tree Sparrows have declined but there are still a couple of pairs.

The new reedbed has contributed to a burgeoning population of Cetti's Warbler, with 34 pairs in 2023. The Sedge Warbler population doubled to 93 pairs the same year, with a stable population of about 50 pairs of Reed Warbler. Bearded Tit has bred but is not recorded annually, and Bittern is seen most years with two pairs breeding successfully in 2024 as the reedbed became established. Autumn and winter warbler records include increasingly frequent Siberian Chiffchaff and Yellow-browed Warbler. In recent years a Starling murmuration of over 50,000 birds has been seen from the Harrier Hide before roosting in the reedbed.

As elsewhere, egrets have become an increasing feature. Little Egret is the most numerous but there are also occasional Great White Egrets, and Cattle Egret can be present in flocks of upwards of 20 birds. Spoonbill occurs with increasing regularity. Kingfisher is regular, often perching conspicuously in front of hides, including at the Millers' Bridge and perhaps unsurprisingly from Kingfisher Hide.

The Pink-footed and feral geese species attract all the expected scarce geese. Tundra Bean Goose is seen regularly in ones and twos, whilst both subspecies of White-fronted Goose are seen most years, including more than 30 Eurasian White-fronts on occasion.

Regular scarce species and semi-rarities include Little Gull, Black Tern, Garganey, Little Stint, Spotted Redshank, Wood and Curlew Sandpipers and Osprey. Pectoral Sandpiper occurs most years, and drake Green-winged Teal are regularly picked out in flocks of Teal. The reserve remains one of the prime sites in the north-west to see Spotted Crake though greater luck is required than in the past.

The list of rarities seen on the reserve is impressive. A pair of Black-winged Stilt attempted to breed in 2006, and another visited in 2023. Other species recorded include Lancashire's sole Franklin's Gull, White-spotted Bluethroat, Kentish Plover, Collared and Black-winged Pratincoles, White-winged Black Tern, Great Reed Warbler, Purple Heron, Wilson's Phalarope, Long-billed Dowitcher, American Wigeon, Gull-billed Tern, Montagu's Harrier, Red-necked Phalarope and Ring-billed Gull.

Close to the reserve the farmland by Curlew Lane, Rufford, has been one of the best sites in Lancashire for Yellow Wagtail in recent years and the species still occurs in summer in small numbers. Corn Bunting also still occurs in the area.

In line with several other older WWT reserves there is a captive 'ornamental' wildfowl collection. This includes a wide range of swans, geese and ducks from across the world grouped largely by geographical origin as well as other exhibits currently including Greater and Chilean Flamingos, White Stork and Asian Short-clawed Otter. Some wild waterfowl may be attracted to the collection area, and other birds including finches feed in the trees, where Waxwing has been noted.

ACCESS

Leave the A59 at Rufford and follow Mere Lane before taking Curlew Lane and finally Tarlscough Lane (becoming Fish Lane) to the reserve entrance at SD428144. The reserve is currently open from 9.30am–6pm in summer, 9.30am–4.30pm in winter. Hours may be extended if there is a rarity on the reserve.

Three rail stations are an extended walk from the reserve: Burscough Junction on the Liverpool to Preston line; and Burscough Bridge and New Lane on the Southport to Manchester line. The shortest walk is from New Lane, following Marsh Moss Lane, but until it picks up a footpath at the reserve perimeter there are no pavements, so care should be exercised given vehicles passing at speed.

Admission is currently £16.50 for adults, free for WWT members. Viewing opportunities include numerous hides, some of which are multi-level, offering different vantage points. There is also a reedbed walk. Additional hides are currently being added to the reedbed area. Facilities include a large visitor centre with gallery and gift shop, a café, toilets including disabled facilities and baby changing, a children's play area and the wildfowl collection.

YOUR VISIT

Pink-footed Goose influxes tend to have two distinct peaks, with an initial arrival in September and October and a second influx in February as birds return from feeding on sugar beet in Norfolk and Lincolnshire. Whooper Swans arrive from October, though a couple of pairs invariably summer, with peak numbers in December and January. Peak duck numbers are also in midwinter, when there is the greatest likelihood of Green-winged Teal.

Both spring and autumn can be productive for scarce and rare wader species. Given the importance of water levels on Sunley's and Vinson's Marshes it can be worth checking published sightings to decide if a visit is likely to be productive for passage waders.

In winter it is best to time a visit to include dusk where possible, to include the swan feed, increased chances of hunting raptors and the opportunity to observe the Starling murmuration.

63 LUNT MEADOWS LWT RESERVE

Lunt Meadows LWT Reserve
Lunt Road, Thornton, Liverpool L29 8YA
SD355021
feasted.tripled.price

Website: lancswt.org.uk/nature-reserves/lunt-meadows
Email: cashton@lancswt.org.uk
Opening times: The car park is open 9.30am–5pm; closed on Christmas Day and Boxing Day
Parking: Reserve car park
OS Landranger Map 108, OS Explorer Map 285

Lunt Meadows is one of Lancashire Wildlife Trust's flagship wetland reserves and an example of how flood management schemes can be harnessed for conservation. Created in 2014 as part of an Environment Agency project to store floodwater from the River Alt, it has developed into an important site for breeding, wintering and passage birds.

SPECIES

One of the most exciting and positive developments since the last edition of this book has been the creation of the Lancashire Wildlife Trust reserve at Lunt Meadows. Farmland adjoining the River Alt was earmarked by the Environment Agency in 2014 to be a flood storage reservoir, and a bonus from this has been the opportunity to manage the site for birds and other wildlife. Over the last

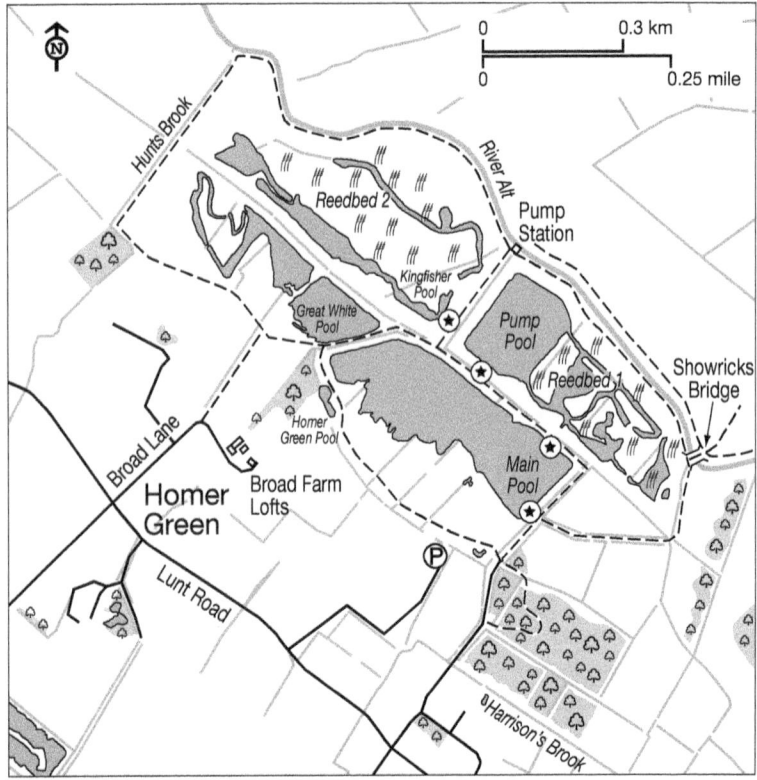

decade the site has established and attracted a range of wetland breeding species, including Bittern in 2021. It is also accumulating a growing list of rare and scarce visitors.

Bittern is an occasional breeder and is regular in winter. The Cetti's Warbler population is now in double figures, and Sedge and Reed Warblers and Reed Bunting use the site in good numbers. Water Rail is seen regularly in winter, and probably also nests in some years. Although Bearded Tit does not breed it has occurred several times and is a potential colonist as the site matures.

Several species of duck breed and are present all year, including Tufted Duck, Shoveler and Gadwall. Goosander also occur in double figures, and Garganey is seen most years. A flock of several hundred wintering Teal has held a drake Green-winged Teal. A range of scarce or normally marine ducks has also been recorded including Scaup, Ring-necked and Long-tailed Ducks, Common Scoter and a couple of Smew.

As the name 'Great White Pool' suggests, egrets and herons can be seen throughout the year. As well as the eponymous egret species and the expected Grey Heron and Little Egret there have been records of Cattle Egret, Purple Heron, Spoonbill and Glossy Ibis. Several pairs of both Great Crested and Little Grebes breed, and Black-necked Grebe is seen most years and has lingered on occasion.

It is perhaps waders for which the site has become best known. During a purple patch in 2019, Lancashire's second Stilt Sandpiper was present in May, with two

American Golden Plovers together in September. Regular visitors include Lapwing, Golden Plover, Black-tailed Godwit, Snipe and Jack Snipe whilst Little Ringed Plover breeds. Passage includes regular records of Wood and Green Sandpipers and Ruff and can also include Pectoral Sandpiper, Temminck's Stint, Little Stint, Bar-tailed Godwit, Sanderling and Curlew Sandpiper.

The site can attract passage marsh terns, with White-winged as well as Black noted, and Little Gull also pass through. The rough grassland around the reserve regularly hosts Short-eared Owls, particularly in 'vole years', whilst Hen Harriers are also fairly frequent. Osprey is regularly seen, particularly in spring.

Outside the reserve the arable fields nearby often hold Pink-footed Geese in winter. As with other sites that attract Pink-feet there is always a chance of other species, and at Lunt these have included Snow Geese and Greenland White-fronts. Grey Partridge still occurs locally.

YOUR VISIT

The 77ha reserve is accessed off Lunt Road. A footpath runs around the periphery of the site, and additional internal trails provide views over the Main Pool, Pump Pool and the smaller waters of Garganey Scrape, Kingfisher Pool and Great White Pool. There are hides or viewing screens at two points overlooking Main Pool and one each at Kingfisher and Pump Pools. Unfortunately, the paths are not currently suitable for disabled access, but the LWT have plans to address this and it will be worth checking their website for progress updates on this.

ACCESS

The car park is reached from the track heading north-east off Lunt Road at SD344021. Opening hours are 9.30am–5pm (closed Christmas Day and Boxing Day); vehicles left after this time may be locked in. Access outside these times involves walking from Lunt Road; if arriving by car, park in the lay-by opposite the entrance.

The 133 bus service from Waterloo Interchange stops near the reserve entrance and is an hourly service during the day from Monday to Saturday (no Sunday service).

CHORLEY AREA AND WEST PENNINE MOORS

64 WEST PENNINE MOORS

Belmont Reservoir
SD671169
remaining.spells.scoots

Entwistle Reservoir
SD719175
speedy.countries.family

Jumbles Reservoir
SD734142
september.blasted.bravest

Wayoh Reservoir
SD732168
redeemed.ashes.enjoys

Opening times: Reservoirs are accessible at all times
Parking: Available at all sites
OS Landranger Map 109, OS Explorer Map 287

This area combines unenclosed moorland, farmland, upland woodland, plantations and a chain of reservoirs, offering a range of habitats that support upland breeding birds, passage migrants and wintering waterfowl.

SPECIES

This upland region of some 230km² is bordered by Chorley in the west, Darwen in the north, Rossendale uplands to the east and the Greater Manchester towns of Bolton and Bury to the south. The area includes the unenclosed moorlands from Anglezarke Moor and Winter Hill in the west through to Holcombe Moor in the east as well as extensive areas of in-bye (arable and grassland) fields and upland woodland and softwood plantations, plus several chains of reservoirs. The core area of 76km² was notified as an SSSI in 2016 due mainly to its large expanses of priority-habitat blanket bog, its extensive mosaic of upland and upland-fringe habitats, its diverse assemblages of upland moorland and grassland, in-bye and woodland breeding birds, plus nationally significant numbers of two species of breeding gull.

The majority of the area is blanket bog moorland with a large plateau rising to 456m at Winter Hill, vegetated mainly with extensive tracts of cotton-grass and *Molinia* grasslands along with areas of heath. The area of heather moor is estimated to have more than halved. Land use is largely water-gathering grounds for the reservoirs with the vast majority of the area part of the United Utilities estate. Sheep farming predominates on the moors and is mixed with cattle on the moor-edge with some heather moors managed for grouse shooting.

There is a good range of breeding birds on the moorland including (latest estimated number of breeding pairs in the SSSI in brackets), Short-eared Owl (up to six), Golden Plover (30), Dunlin (eight), Curlew (100), Cuckoo, Stonechat (105), Linnet (90), Red Grouse (250), plus the ubiquitous Meadow Pipit and Skylark. Breeding species lost in recent years include Ring Ouzel and Yellowhammer, with Merlin, Twite and Whinchat now very scarce and Wheatear also declining.

Peregrine, Raven, Stock Dove, Kestrel, Barn Owl and colonies of Jackdaw breed in some of the abandoned quarries. Hen Harrier winters with some birds regularly lingering well into spring, and Marsh Harrier and Red Kite have become far more regular visitors, sometimes present for several weeks. In midwinter, the high moors are occupied by Red Grouse, parties of Reed Buntings and the occasional Snow Bunting whilst Golden Plover often start to return to the tops in January. The distribution of some breeding birds, particularly waders, is affected by avoidance of popular walking routes.

Upland farmsteads surrounded by in-bye pasture are in some cases now regressing with increasing rush cover. Field boundaries of mainly dry-stone walls are frequently in disrepair. Very few traditional meadows remain, the majority now under stewardship schemes. Improved pasture is cut for silage/haylage. In-bye fields retain typical breeding birds, often in good numbers, including (latest estimated number of breeding pairs in SSSI in brackets), Lapwing (115), Snipe (90) and Reed Bunting (160). Grey Partridge and Redshank are now very scarce but Grasshopper Warbler is relatively common in rush-dominated pastures. Farm buildings often hold Barn Owls and good colonies of Swallows, and House Sparrows have recently recolonised many upland farms.

The high water table with frequent wet flushes/springs drains into fast-flowing brooks, with most of the valleys now dammed to create numerous reservoirs, constructed to provide drinking and compensation water for the industrial towns downstream in Lancashire, Greater Manchester and Merseyside. Dipper and Grey Wagtail are common on the upland streams with Kingfisher in places. These upland reservoirs, where undisturbed, typically host breeding waders including Oystercatcher, Little Ringed Plover and Common Sandpiper together with breeding Great Crested Grebe, Tufted Duck, numerous Canada geese, the odd pair of Teal and increasing Greylag Geese.

Large gull roosts assemble on some reservoirs outside of the breeding season, particularly on Lower Rivington Reservoir, whilst Goosanders roost on several reservoirs, mainly Yarrow and Delph. Mallard numbers commonly exceed 100 on several waters but Teal are more localised. The once-regular herd of Whooper Swans is now a distant memory. The ornithological quality of many reservoirs is compromised by recreational pressures, with sailing and concessionary paths affecting distribution and numbers of winter wildfowl and breeding waders respectively.

Many steep-sided upland valleys host natural woodlands, typically Oak, Ash, Birch, Rowan and Holly, often with wet woodland of Alder and Willow. Many woods are now mixed with planted/self-seeded Beech, Sycamore and Rhododendron. Upland woodland, particularly in the Roddlesworth area, holds typical species including small numbers of Pied Flycatcher, Redstart, Tree Pipit, Spotted Flycatcher, Woodcock, Green Woodpecker and Wood Warbler. Buzzard have recently colonised the area and are now the most obvious diurnal raptor. Hobby, Goshawk and Long-eared Owl breed occasionally or in very small numbers with Nightjar also recorded. Lesser Spotted Woodpecker is now lost as a breeding species, but on the moorland-fringe woodlands above Bolton, Willow Tit still occurs and Tree Sparrow has recently recolonised.

Extensive coniferous and mixed plantations occur widely, mainly associated with immediate reservoir catchments. Plantations hold breeding Siskin and Redpoll, with Crossbill in most years, but the formerly large heronry at Entwistle is much declined. Large Carrion Crow roosts still form in mid-winter along with two huge Magpie roosts, one that held 730 birds in December 2016.

Large finch roosts also occur, largely in softwood plantations and Rhododendrons, with one roost hosting 1,882 Bramblings in November 2018. Woodcock winter in good numbers in many plantations and Rhododendron thickets.

Visible migration, particularly in the autumn, is often evident through valleys and along moor edges. On the Lancashire side of the border Darwen Moor and Belmont in particular reward regular watchers with some spectacular movements both in numbers and range of species. Numbers of migrating Meadow Pipit, Redwing, Fieldfare and Woodpigeon have run into many thousands on peak days. Passage birds often present in good numbers along the moorland edge with Wheatear numerous and Ring Ouzel regular and the Winter Hill mast complex a regular site for Black Redstart (and other grounded migrants). The moorland edges are worth checking for chats, flycatchers, redstarts and Wheatear with Great Grey Shrike recorded in this habitat too.

County rarities have included Lancashire's first Whiskered Tern, Dartford Warbler and Snowy Owl, with Golden Eagle photographed, Iberian Chiffchaff seen and sound-recorded and Dotterel and Firecrest encountered in the last couple of years.

SITES
Belmont Reservoir
This is an upland reservoir in the centre of the West Pennine Moors, surrounded by moorland, in-bye fields and some woodland and scrub. By far the most important and obvious feature of the site is the huge gullery. The Black-headed Gull colony here is the largest in the UK, hosting 13,528 pairs in 2021. Less obvious is the UK's largest inland colony of Mediterranean Gull, which held 119 pairs in 2023. The spectacle of the gullery is best enjoyed between April and June. Scarcer gulls, skuas and terns occasionally visit with Yellow-legged Gull, Kittiwake, Arctic Skua, and Sandwich, Black and Arctic Terns recorded.

Waders are well represented with Curlew, Lapwing, Oystercatcher, Snipe, Redshank, Common Sandpiper, Little Ringed Plover and the occasional Ringed Plover breeding. Large pre-breeding flocks of Lapwing and Curlew are obvious in spring. Scarcer waders occur with some regularity, particularly Dunlin and Ringed Plover, and also including Black-tailed and Bar-tailed Godwits, Greenshank, Whimbrel, Ruff, Wood and Green Sandpipers, Turnstone, Grey Plover, Knot, Sanderling and Avocet.

Besides the ubiquitous Greylag and Canada Geese, wildfowl variety is largely restricted to Mallard and Teal, albeit the latter can often number up to several hundred wintering birds. Scarcer waterfowl recorded include regular Whooper Swan and Common Scoter along with the rarer Smew, Scaup, Red-throated Diver, Bittern and Cattle Egret. Other local rarities in recent years have included Nightjar and Water Pipit with raptors such as Osprey, Hobby, and Marsh and Hen Harriers recorded annually.

Breeding passerines include Grasshopper Warbler, Reed Bunting, Sedge Warbler, Stonechat and Redpoll.

Viewing is particularly easy as the A675 bisects the site and, if you are able to stand the traffic along the busy main road, facilitates good birding from spring through to autumn, but the winter period can be quiet especially at times of hard weather. Whilst low water levels, especially in spring and late summer, are best for passage waders, if the water level drops too low the exposed reservoir bed becomes stoney rather than muddy, and hence less attractive. The only parking available is in a small lay-by on the west side of the A675 at the dam end (at SD672165); do not park on the A675 clearway or on farm tracks or in the bus turn-around (this is used by wagons as well as buses). Bus services from Bolton are infrequent with just six daily from Monday to Saturday only.

Entwistle, Wayoh and Jumbles Reservoirs
This chain of reservoirs stretches for 6.4km along the Bradshaw Brook valley with interconnecting woodland corridors and streams allowing for varied scenery and birding. Jumbles Reservoir is in the Greater Manchester recording area but included here with the neighbouring sites.

The plantations around Entwistle are best for coniferous woodland species such as Crossbill and Siskin with Spotted Flycatcher in clearings. The once-large heronry is now much diminished. Wayoh hosts the most varied and numerous selection of winter wildfowl with Tufted Duck, Goldeneye and Goosander regular along with the occasional Wigeon, Pochard and scarcer duck such as Scaup and Ring-necked Duck. The once-regular herd of Whooper Swans has now vacated. The woodland clough between Wayoh and Jumbles hosts a good selection of woodland species along with resident Kingfisher, Dipper and Grey Wagtail on the streams. Jumbles

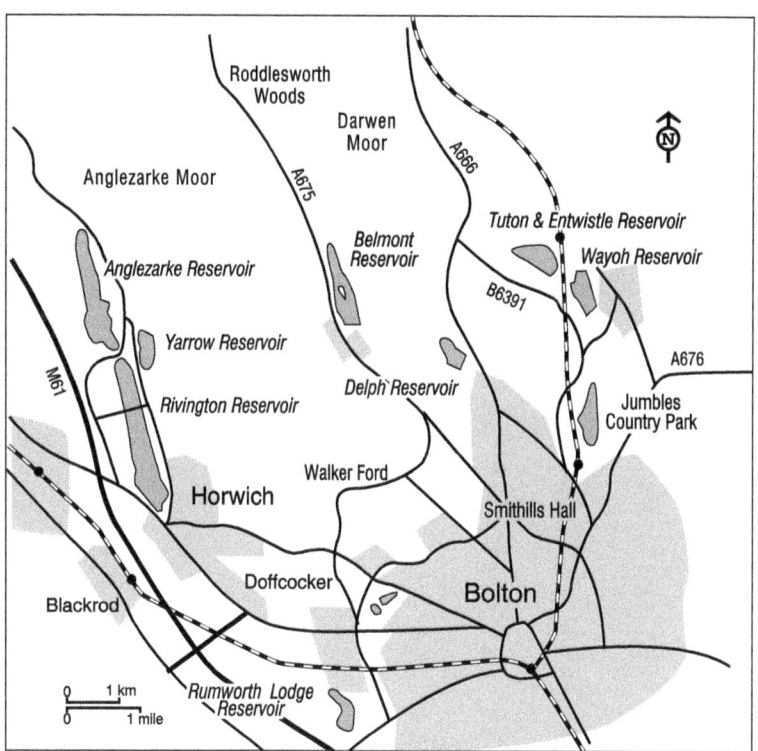

Reservoir has breeding Great Crested Grebe and regular Common Tern and has produced Bewick's Swan, Hawfinch and Black-headed Bunting (in 1995).

YOUR VISIT

The reservoirs and moorland are accessible via minor roads, footpaths and public rights of way. Belmont Reservoir has a small lay-by at the dam. Entwistle, Wayoh and Jumbles are reached via valley lanes and connecting paths. Paths can be steep, uneven or boggy, and there are no formal hides, so binoculars or a scope are recommended.

ACCESS

This extensive upland area is well traversed by footpaths giving plenty of scope to explore if visiting by car. The valley is accessible by train from either Bolton or Darwen with stations at Bromley Cross (for Jumbles) and Entwistle (for Wayoh/Entwistle). Bus services on the Bolton to Darwen and Bury to Darwen route are irregular (Monday to Saturday) and rural stops infrequent with the route via Edgworth, which stops at Bromley Cross, Chapeltown and Edgworth, the best. Check timetables carefully as these are subject to change or suspension of service.

CHORLEY AREA

65 ANGLEZARKE/RIVINGTON RESERVOIRS

Anglezarke Reservoir
SD616163
casino.lazy.lower

Lower Rivington Reservoir
SD625133
nanny.haunt.editor

Upper Rivington Reservoir
SD621146
rooting.regulator.filed

High Bullough Reservoir
SD618168
grub.diplomas.stylists

Yarrow Reservoir
SD625156
certified.glance.grandson

Opening times: Accessible at all times
Parking: Rivington Hall Barn (SD631145)
OS Landranger Map 109, OS Explorer Map 287

This group of five reservoirs near Rivington offers accessible walking paths and vantage points that provide opportunities to observe gull roosts, waterfowl and seasonal passerine migrants.

SPECIES

The five reservoirs near Rivington (Anglezarke, High Bullough, Yarrow and Upper and Lower Rivington) are accessible from causeways and footpaths. The area is perhaps best known in birding terms currently for the gull roost at Lower Rivington Reservoir. This has regularly produced Caspian, Glaucous, Iceland and Mediterranean Gulls in recent years with Kittiwake and Little Gull also occasional.

The reservoirs are also one of the best areas in the region to see naturalised Mandarin, with flocks in excess of 50 birds. Other regular waterfowl include Goosander and Great Crested Grebe, whilst Common Scoter occur occasionally on migration. Osprey also turns up on migration, and local rarities have included Black Stork, Crane, Ring-necked Duck, Slavonian Grebe and Wryneck.

A small population of Wood Warbler persists in the Chorley area and with good fortune birds may be seen in the area as they pass through en route to these sites. Other passerines that can be seen in the surrounding fields and woods include Redstart and Wheatear.

ACCESS

There is extensive car parking at Rivington Hall Barn (SD631145), which is a short walk from the causeway between the Rivington reservoirs and footpaths to Anglezarke. In terms of public transport Adlington station on the Preston to Manchester line is a 2km walk from the causeway. In summer a 'Rivington Rambler' bus operates between Horwich Parkway train station and Chorley, stopping at Rivington Village Hall.

YOUR VISIT

As with many reservoir sites, spring and autumn can be productive before any disturbance and particularly after showers. If visiting in winter, late afternoon viewing for wildfowl can then be combined with checking the gull roost. Vantage points for wildfowl and roosting gulls include from the A675 Bolton to Chorley Road just west of Horwich (SD626127) or from a minor road in the vicinity of the castle (SD628130). Other points to view from include the causeway roads between reservoirs, particularly between Upper and Lower Rivington Reservoir.

66 YARROW VALLEY PARK

Yarrow Valley Park
Off Birkacre Road, Chorley PR7 3QL
SD570152
couch.best.oath

Phone: 01257 515151
Email: contact@chorley.gov.uk
Opening times: Accessible at all times
Parking: Free parking off Birkacre Lane
OS Landranger Map 108, OS Explorer Map 285

This 283ha country park with a ranger service provided by Chorley Council has high recreational use but can still be rewarding. It has a good range of species that are relatively easy to see on the waters of Big Lodge and in the surrounding woodland and parkland.

SPECIES

Waterbirds and riparian landbirds present year-round include Water Rail, Great Crested Grebe, Grey Heron, Dipper, Kingfisher, Grey Wagtail and Little Egret. Resident woodland birds include Willow Tit and Great Spotted Woodpecker. Summer visitors include Grasshopper Warbler. More unusual visitors have included Red-throated Diver, Little Gull, Osprey, Yellow-browed Warbler and Hawfinch.

Lancashire

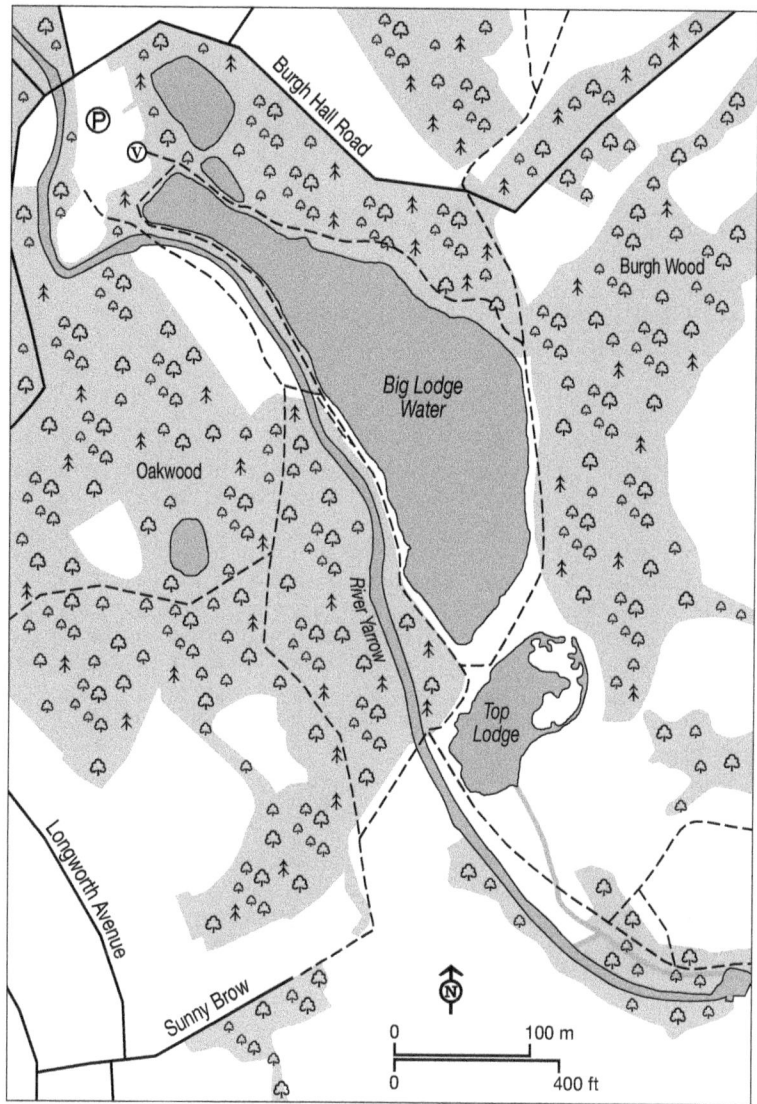

ACCESS

There is free parking at the information hub off Birkacre Lane. A frequent bus service (632) between Chorley and Wigan stops on Birkacre Road adjacent to the country park entrance.

The Treeface Cafe is open daily and toilets and accessible facilities are available to all park users. A play area called Yarrow Rocks encourages adventure play but younger visitors would require supervision by a non-birder.

Lancashire

67 CROSTON AND MAWDESLEY MOSS

Croston and Mawdesley Moss
Meadow Lane, Mawdesley, L40 2QA
SD466179
garage.divisions.juggles

Website: info@lancswt.org.uk
Phone: 01772 324129
Opening times: Accessible at all times
Parking: There is no car access on the mosses
OS Landranger Map 108, OS Explorer Map 285

Croston and Mawdesley Moss is an open lowland wetland offering excellent winter and migration birdwatching. Its network of footpaths provides access to a variety of waterfowl, raptors and passerines throughout the year.

SPECIES

The Croston and Mawdesley Moss area is particularly worth visiting in late autumn and winter. There is the possibility of good numbers of Whooper Swans and Pink-footed Geese, with the latter occasionally harbouring other goose species, as well as Lapwing, Stock Dove, Skylark and wintering thrushes, finches and buntings. Several species of raptor can be seen including Buzzard, Sparrowhawk, Peregrine, Kestrel and Merlin with occasional Hen and Marsh Harriers. Hunting Barn and Short-eared Owls are also possible. If floodwater is present other species may include Little Egret, Snipe, Jack Snipe, Green Sandpiper, Kingfisher, Dunlin, Ruff and Black-tailed Godwit. Stonechat is regular.

Summer generally requires more work for rewards but birds that can be seen include Whitethroat and scarce passerines, notably Tree Sparrow, Yellow Wagtail, Corn Bunting and Yellowhammer. Shelduck, Oystercatcher and Lapwing also nest in the area. Spring migration usually sees lots of Wheatear passing through and occasional Whinchat.

ACCESS

There is no car access on the mosses so you need to park where you can at access points such as at Great Hanging bridge, off Meadow Lane from Croston.

The site is accessible by train to Croston or Rufford train stations, with several footpaths criss-crossing the site. One option would be to alight at one station and walk across the moss footpaths to the other. There are regular bus services to Croston from Chorley and Leyland/Preston.

GREATER MANCHESTER

Despite being a relatively small (at just under 500 square miles), land-locked and highly urbanised county, Greater Manchester affords excellent year-round birdwatching and a wide variety of habitats, apart from a coastline of course. Its current 317 recorded species include such diverse rarities as Canvasback, Pied-billed Grebe, White-tailed Plover, White-billed Diver, River Warbler, Dusky Thrush, Desert Wheatear, Black-faced Bunting and Northern Parula, all proving what can be possible, but it is the broad range of commoner species and their abundance for which it is most reliable and productive.

In the west, the Flashes of Wigan and Leigh National Nature Reserve comprise a number of post-industrial coal-mining subsidence flashes which include the most diverse range of habitats along with the most comprehensive birdwatching. Though they are perhaps best known for their nationally important population of Willow Tits they also attract large numbers of wildfowl, along with passage waders and terns and winter gulls for those who desire.

To the north and east, the upland moors can rise to 540m above sea level, providing essential habitat for Red Grouse and breeding sites for upland waders, Wheatear and Ring Ousel, with some more geographically gifted locations providing the best opportunity for the impressive spectacle of visible migration watching in spring and particularly autumn. The east also includes some of the oldest and most extensive deciduous woodlands in the county, which still possess ever declining small populations of Pied Flycatcher and Redstart.

The south of the county is gifted with the lowland raised bogs of the 'mosses', important areas many of which were previously exploited for peat extraction but

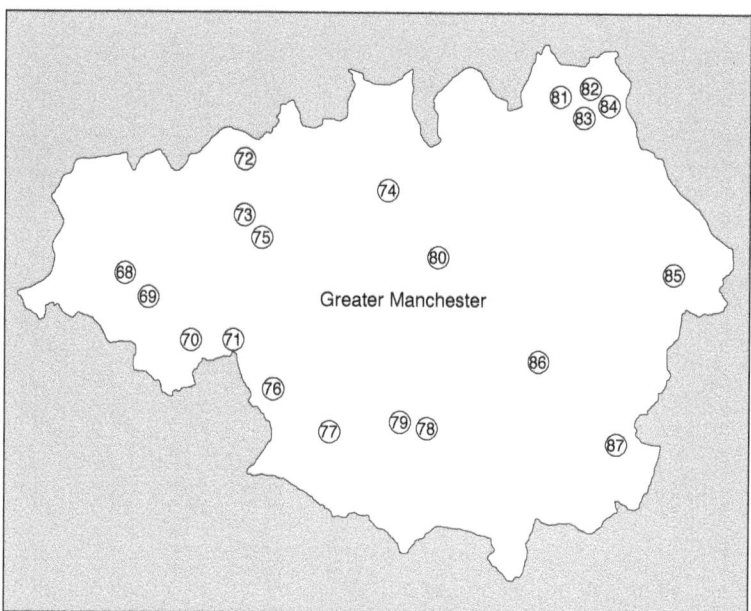

are now being restored and along with adjacent farmland provide important breeding habitats for species such as Yellow Wagtail, Corn Bunting and Yellowhammer, with a seemingly ever-increasing population of Hobbies.

68 WIGAN FLASHES

Wigan Flashes
Wellham Road, Wigan, WN3 5PA
SD579032
saves.upgrading.soap

Scotman's Flash
SD581037
boat.pets.prime

Pearson's Flash
SD581038,
friend.super.sake

Westwood Flash
SD580041
inch.loyal.figure

Ochre Flash
SD581029
nearly.maple.groups

Bryn Marsh
SD583026
keen.march.token

Horrock's Flash
SD590024
pines.porch.tries

Turner's Flash
SD586031
bits.slower.warm

Website: www.wigan.gov.uk/Resident/Leisure/Greenheart/Wigan-Flashes.aspx
Opening times: Open at all times
Parking: There's a small amount of roadside parking at Wellham Road at play.vets.pokers
OS Landranger Map 108, OS Explorer Map 276
NNR, SBI

This large area of eight colliery-subsidence flashes, covering about 260ha, lies 1km south of Wigan town centre and is connected by the Leigh branch of the Leeds to Liverpool Canal. Since the 1980s, these flashes have been developed for nature conservation and now form part of the Flashes of Wigan and Leigh National Nature Reserve. There is a wide variety of wildlife habitats, including open water, reedbeds including the largest in the county, woodlands, willow carr, scrub and rough grassland. Six viewing screens and 10km of paths around the site offer a range of short- to medium-length walks.

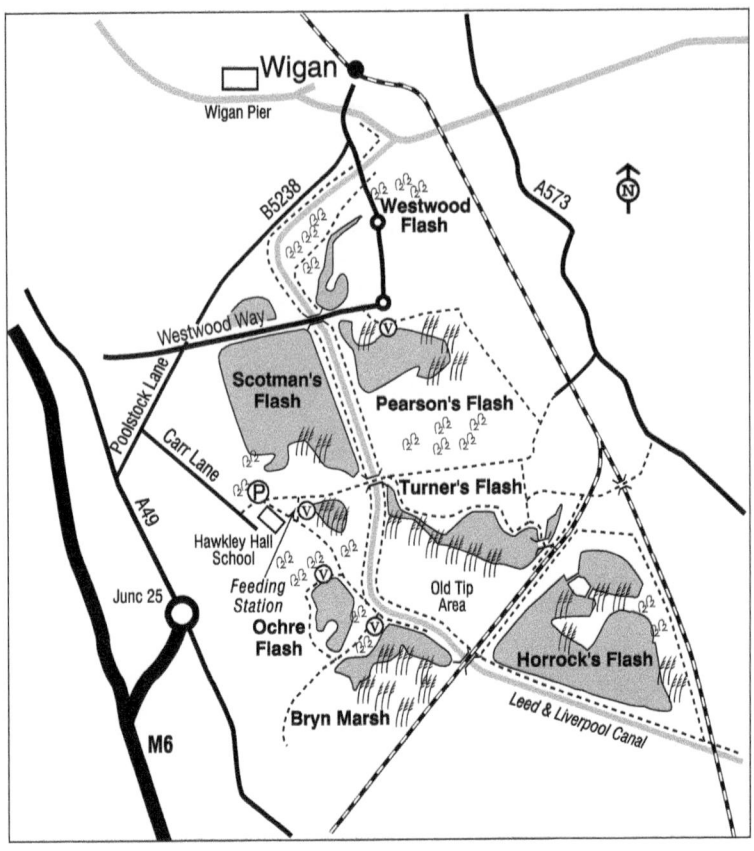

Since the area is very large it can be difficult to cover in a single day so can be conveniently divided into two sections:

SCOTMAN'S FLASH, PEARSON'S FLASH, WESTWOOD FLASH, OCHRE FLASH AND BRYN MARSH

Scotman's Flash and Pearson's Flash are separated only by the canal and when sailing disrupts birds on the former, they relocate to the latter, which is undisturbed and has a reedbed at the eastern end. Good views of both flashes can be had from the canal banks and there are crossing points over the canal at the lock at Westwood Flash and at Moss Bridge at the southern end of Scotman's Flash. There is a viewing screen and feeding station near Hawkley School and footpaths connecting Scotman's Flash to Ochre Flash and from there to Bryn Marsh, which has the largest reedbed in the county. On the east side of the canal the former landfill site is now covered with trees and there is some rough grassland between Pearson's Flash and Turner's Flash.

HORROCK'S FLASH AND TURNER'S FLASH

Horrock's Flash is bounded by railway lines on two sides and the canal on the third side, but there is a circular path which gives good views of the various pools and reedbeds that make up this site. A track across the level crossing on the loop line leads through some fishing ponds and under a railway bridge; turn immediately left for Turner's Flash, negotiate the barrier that prevents vehicles, and a good view of this water can be obtained from its east end. Alternatively, Turner's Flash can be accessed from Moss Bridge on the canal and a large reedbed runs the whole length of the site. Both flashes have small peaty islets and Horrock's Flash also has areas of reedbed and often mud, especially on its southern and eastern edges.

SPECIES

Winter: although Bitterns are now practically resident, this season occasionally sees their numbers swell and they can occur in any reedbed, though Hawkley reedbed, the south end of Scotman's Flash and Bryn Marsh are preferred. Typically, they are rarely easy to see. Wildfowl may congregate, particularly on Scotman's and/or Pearson's Flash with Wigeon, Shoveler, Gadwall and Teal, along with occasional Whooper Swans and Smew. The large numbers of resident Water Rails are supplemented by migrants from continental Europe, and the wooded areas, particularly that between Pearson's and Turner's Flash, can hold Lesser Redpoll and Siskin flocks.

Breeding: the entire site is important for breeding birds, with a pair of Marsh Harriers in recent years, the highest population of Water Rails in the county, a small number of Common Terns and Black-headed Gulls on Horrock's and Turner's Flashes, several hundred pairs of Reed Warblers across the site, Sedge and Grasshopper Warblers, Redshank, Common Snipe and Redpoll, not to mention various wildfowl. The Willow Tit population across the site is the highest in the county, particularly around Pearson's and Westwood Flashes, and although they can be found at any time of year, the feeding station near Hawkley reedbed in winter almost guarantees sightings. Similarly, Cetti's Warblers are resident and are found in the highest concentrations in the county.

Passage: in late summer and early autumn the site can be nationally important for Gadwall and Shoveler, and flocks of Wigeon pass through. In autumn, Common Snipe numbers swell and Horrocks Flash is usually best for the scarcer passage waders, with Black-tailed Godwit, Dunlin, Greenshank, Curlew and Ruff all annual. Pearson's Flash is usually best for duck congregations in late summer and autumn, and if water levels are low there can be exposed mud.

Rarities have included Red-necked Phalarope, White-winged Black Tern, River Warbler, Ring-necked Duck, Green-winged Teal, Ring-billed Gull, Collared Pratincole, Lapland Bunting and Richard's Pipit.

ACCESS

SCOTMAN'S FLASH, PEARSON'S FLASH, WESTWOOD FLASH, OCHRE FLASH AND BRYN MARSH

There is a car park for the Wigan Flashes in Worsley Mesnes, off Poolstock Lane (B5238) just north of Scotman's Flash at SD575040, middle.push.baked. Alternatively, there is a small amount of roadside parking near Hawkley School, off Welham Road, which is just south of Scotman's Flash at SD579032, public.sheep.gross. The main footpaths are generally well maintained, and the canal towpath is concreted so suitable for wheelchair users.

HORROCK'S FLASH AND TURNER'S FLASH

Parking near Horrock's Flash is not possible and access requires a walk either south down the towpath from Pearson's Flash (parking as per under Scotman's Flash) or north from Platt Bridge. To access via the latter, from the A573 Warrington Road, turn into Miller's Lane at Platt Bridge and then into Victoria Road almost immediately which becomes Stratton Drive. After the playing field on the right, the road bends sharp left and there is a stile on the right. Park here and take the track from the stile to the canal at SD599023, firm.nests.snows. Turn right on the towpath and under the railway bridge to access Horrocks' Flash. After about 100m, it is possible to either turn right into woodland to begin an anticlockwise circular walk round the flash or continue along the towpath to do the same walk clockwise. On both routes, turn off along the level crossing to access Turner's Flash. Alternatively, having viewed part of Horrock's Flash from the towpath, continue north under the Liverpool to Wigan railway bridge for about 1km to Moss Bridge and the northern section of the Wigan Flashes (as discussed under Scotman's Flash).

YOUR VISIT

Early morning visits reduce the risk of water-sports activity disturbing the waterfowl on Scotman's Flash. When sailing or other water sports are taking place on Scotman's Flash, focus on Pearson's Flash as birds flushed from Scotman's generally take refuge there. Canal towpaths are shared spaces and can be busy; they may also be used on occasion by unauthorised trail bike activity, so caution is advised.

69 ABRAM FLASHES

Abram Flashes
Warrington Road, Abram, WN2 5XY
SD607006
cloak.hello.monday

Website: www.wigan.gov.uk/Resident/Leisure/Greenheart/SSSI/Abram-Flashes-SSSI.aspx
Parking: Very small free car park available at cloak.hello.monday but can fill quickly. Alternative parking is available along Warrington Road, Abram.
OS Landranger Map 109, OS Explorer Map 276
NNR, SSSI, SBI

Dover Basin
SD613002
verse.asset.yards

Lightshaw Hall Flash
SD615997

Western viewing screen located at flattens.frost.sideburns; eastern viewing screen located at restriction.mavericks.voucher

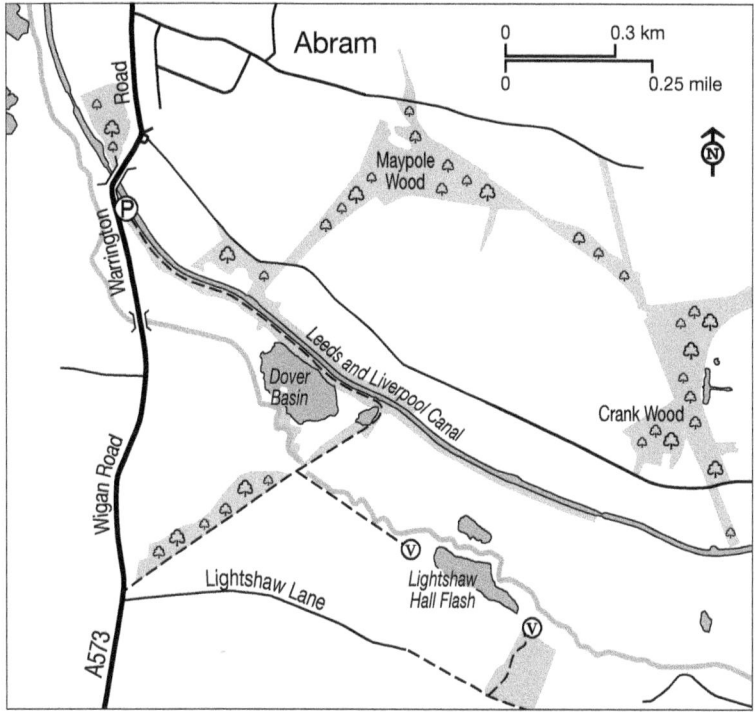

The Abram Flashes lie adjacent to the Leeds and Liverpool canal along the Hey Brook and form part of a series of subsidence wetlands and flashes stretching for some 10km between Wigan and Leigh, now designated the Flashes of Wigan and Leigh NNR.

SPECIES

Just south-east of the Dover Lock Bridge car park along the canal towpath lies Dover Basin, which is viewable from the towpath and can attract waders including Black-tailed Godwits in spring and wildfowl year-round. The reedbed at its southeastern end has breeding Reed Warblers and the occasional wintering Bittern, and there are resident Cetti's Warblers round the fringes of the flash. A little further south along the towpath is Lightshaw Meadows SSSI, now under the stewardship of Lancashire Wildlife Trust. There are two viewing screens at either end of Lightshaw Hall Flash. The western screen is accessed off the footpath between Lightshaw Meadows and Dover Basin, but the eastern screen is only accessible off Lightshaw Lane. The site can be heavily impacted by flooding, but it can afford excellent birding and if water levels are low can be very good for waders with Avocet, Black-tailed Godwit, Greenshank and Wood Sandpiper all annual. The area also has breeding Shelduck, Shoveler, Redshank, Sedge and Grasshopper Warblers, plus Yellow Wagtail, Skylark and Yellowhammer in the surrounding farmland and Lesser Whitethroat and Willow Tit in the hedgerows and small woods.

Rarities have included Green-winged Teal, Purple Heron, Broad-billed Sandpiper, Red-necked Phalarope, Temminck's Stint and Wryneck.

ACCESS

There is a small car park by the Dover Lock Bridge along Warrington Road in Abram (SD607006, begun.theme.curl), which lies just north-west of Dover Basin. If this is busy, park in the car park by the Plank Lane swing bridge (on the western border of Pennington Flash at SJ630997, talkative.quarrel.sums) and walk 1.6km west along the canal towpath to reach the footpath to Lightshaw Meadows.

YOUR VISIT

The comments about birding from busy canal towpaths on other Wigan area sites also apply here. Given the importance of water levels at Lightshaw Flash it is advisable to check recent sightings (e.g. on the Manchester Birding Forum) to get a feel for whether conditions are enticing waders to land.

70 PENNINGTON FLASH

Pennington Flash
St. Helens Road, Leigh, WN7 3PA
SJ642991
credited.movements.spicy

Website: www.bewellwigan.org/location/attractions/pennington-flash/
Telephone: 01942 489848
Parking: The main pay and display car park is open at all times. Hides are usually unlocked early to mid-morning and locked prior to dusk. This site can be very busy and car parking full on weekends and particularly bank holidays. It is also worth checking their website for any planned events prior to your visit as these increase visitor attraction.
OS Landranger Map 109, OS Explorer Map 276
NNR, SBI

A large colliery subsidence flash 1.6km long, south-west of Leigh, Pennington Flash is the most southerly of eight local flashes which make up the Flashes of Wigan and Leigh NNR. The site was developed as a 200ha nature reserve by Wigan Council in the early 1980s and in October 2022 was designated as a National Nature Reserve by Natural England. There is an excellent range of habitats on site with the large area of open water of the flash itself, along with various scrapes, pools, marshy areas and smaller reedbeds around the nature reserve area, more wooded areas especially around the southern side of the flash and open grassland on the now transformed old spoil heap of Ramsdale's Ruck in the north-west of the site. A full circuit of the flash can take just 90 minutes but realistically could take as long as three hours when birding and enjoying the full range of habitats. A shorter visit could take in just the area of the nature reserve and all its hides in the north-east of the site.

Greater Manchester

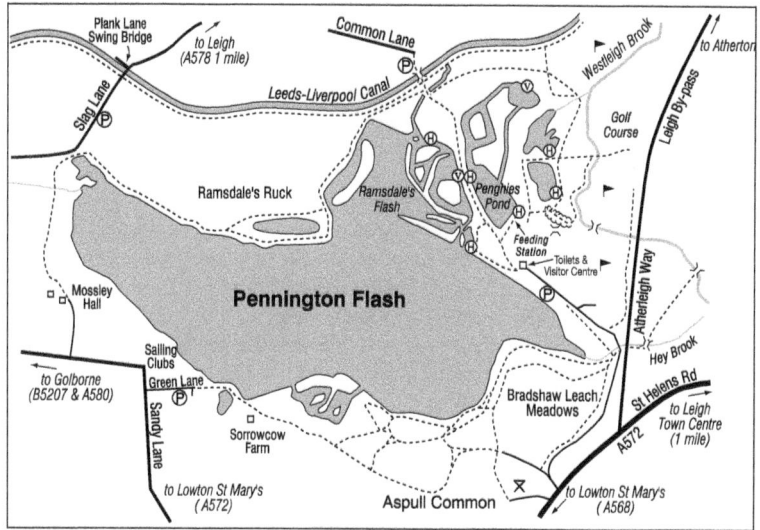

SPECIES

Up to July 2023, 250 species had been recorded, the highest number for any single site in the county and a testament to the site's diversity of species. It remains perhaps best known for the discovery of Britain's first-ever record of Black-faced Bunting in 1994.

Winter: commoner waterbirds can usually be found in good numbers, including Gadwall, Shoveler, Tufted Duck and Great Crested Grebe, with both Goldeneye and Goosander increasing in recent years. Lapwings roost in large numbers on the spit in front of Horrocks Hide during the daytime, as do Cormorants. The nightly gull roost can often exceed 10,000 and is a great spectacle, best viewed from the southern shore of the flash an hour prior to dusk (just east of the sailing club by Sorrow Cow Pond, accessed via Green Lane) and includes regular Mediterranean Gull, scarce but annual small numbers of Caspian and Yellow-legged Gulls and occasionally Iceland and Glaucous Gulls. Bittern is annual but difficult to see, generally in and around Ramsdale's Reedbed, and there are occasional visits from Whooper Swans, Wigeon and Scaup, with scarcer records of divers, rarer grebes and Smew. The feeding station at Bunting Hide provides reliable and very close views of Willow Tit, Bullfinch and often Water Rail and is usually a bustling area with many species, unless one of the regular Sparrowhawks has been through! Siskins can be found, especially around the patches of alders in the area between Teal Scrape and the Leeds to Liverpool Canal and along the very southern edge of the flash around Aspull Common. Little Egrets are now resident and Great White Egret is increasingly seen. Pennington Flash has also become one of the most reliable sites to see Kingfisher, with most of the hides being adorned with perching posts outside them, which regularly attract very close views and photographic opportunities. Elsewhere, they can regularly be found along the southern shore of Ramsdale's Ruck and the shore around Sorrow Cow Pond.

Spring passage: wader passage is generally focused on the spit in front of Horrocks Hide or the shore of the sailing club, though the latter usually necessitates a very early morning visit, and Dunlin, Black-tailed Godwit, Greenshank, Common

Sandpiper, Turnstone and Sanderling are all annual. Resplendent breeding-plumaged Black-necked Grebes are occasional visitors and tern passage can be excellent in late April and early May with Common and Arctic Terns often in good numbers, the latter especially in conditions encouraging their overland movement. Little Tern is annual but very scarce and Black Terns are generally dependent on prevailing winds and weather conditions, but Little Gulls are often seen regardless, occasionally in small flocks. Hirundines and Swifts amass over the water at this time of year, often in huge numbers, and Grasshopper Warblers often reel from the rough areas along the northern fringe of the site just south of the canal, while Willow Warbler and Chiffchaff are common around all areas. Both Redstart and Spotted Flycatcher are annually recorded in very small numbers, most reliably from Ramsdale's Ruck or the area around Ramsdale's Hide.

Breeding: whilst the breeding season is the quietest period, there is always something to see. There is much activity around the small Black-headed Gull colony off the spit and the noisy heronry in the nature reserve area, whilst the reedbeds hold Reed Warbler, Reed Bunting and small numbers of Sedge Warbler. Gadwall and Tufted Duck breed, with occasional Ringed and Little Ringed Plover as well as Redshank. Kingfishers remain active and often highly visible at this time of year, as are the several pairs of Willow Tits around the site, whilst Skylarks can still be found on Ramsdale's Ruck.

Late summer and autumn passage: Tufted Duck begin to gather in late summer with their numbers swelling into the hundreds and are worth scanning for rarer species, whilst Common Scoter are most regularly seen in this period, though can occur at any time. Garganey are annual in small numbers, and the spit once again attracts return passage waders, often including Whimbrel and Greenshank. Black Terns occur most reliably in September, and the gull roost begins to form in late autumn, often containing the first of the Caspian and Yellow-legged Gulls. Visible migration watching has recently been realised from the highest point of Ramsdale's Ruck and can be very productive, with Tree Pipits now annually recorded in late August/early September and large movements of flocks of Woodpigeons and winter thrushes in early October to mid-November along with noisy skeins of Pink-footed Geese and occasional Whooper Swans crossing the skies above. The trees and bushes around Ramsdale's Ruck can hold commoner migrants, and the open grassland there has small flocks of Meadow Pipits with the occasional Stonechat.

Rarities have included Blue-winged Teal, Green-winged Teal, American Wigeon, Canvasback, Lesser Scaup, Ring-necked Duck, Glossy Ibis, Kentish Plover, Black-winged Stilt, Pectoral Sandpiper, Red-necked Phalarope, Ring-billed and Sabine's Gulls, Whiskered and White-winged Black Terns, Little Auk, Alpine Swift, Penduline Tit, Little Bunting, Richard's Pipit, Tawny Pipit, Red-footed Falcon, Great Reed Warbler, Marsh Warbler, Serin and Black-faced Bunting.

ACCESS

Signposted from the A580 East Lancs Road. Follow the signs and turn onto the A579 Atherleigh Way then left onto the A572 St Helens Road. The main entrance is situated on St Helens Road opposite Leigh Fire Station, located at WN7 3PA, SJ644990, resolved.molars.binds. There is also a small car park at the western end off Plank Lane (SJ630996, quantity.providing.column), which is useful for accessing Ramsdale's Ruck and the western end of the flash. The Sorrow Cow viewpoint is accessed off Sandy Lane by turning into Green Lane and continuing just past the sailing club entrance, parking close to the field side of the road (but not the sailing club side of the road as large goods vehicles require access, and not in front of the cottages) at SJ632987, rural.volume.savings.

YOUR VISIT

Facilities on site include a large pay-and-display car park, six bird hides including one with a bird feeding station and a wildlife-themed coffee shop and café with toilets and information centre. All bird hides and paths around the nature reserve area and the paths on the southern side of the flash are wheelchair accessible. The site can be very popular and extremely busy with the public and water sports at weekends and bank holidays and is often best avoided at these times. Much attention at this site is focused on the spit, particularly for waders and resting terns, where the view over it from Horrock's Hide also provides an excellent panorama over the flash itself and affords some protection during the worst of the weather. There are also three other main viewpoints over the flash, as follows:

The Point on Ramsdale's Ruck, opposite the spit, which offers a view over the end of the latter as close as it does from Horrocks Hide but without the obscuring vegetation which grows profusely during summer and autumn.

The southern shore of the flash by Sorrow Cow Pond, just east of the sailing club, can also be worthwhile, especially in spring where the panoramic view of practically the entire water is excellent for viewing terns and hirundines and permits the easiest view of the very western end of the flash. It is also consistently the best location for viewing the late autumn and winter's evening gull roost.

The leaning posts by Ramsdale's Flash, which is the arm of the main flash just north of the spit below the canal. This is where wildfowl disturbed by sailing seek refuge, and the leaning posts above the reedbed provide an excellent view of the flash and reedbed itself.

Weekday visits avoiding bank holidays are best as disturbance is lower. If visiting at weekends, early mornings are generally best before disturbance on the water increases. There is currently a park run on Saturday mornings at 9am which attracts many runners and walkers to a 5km course on some of the paths around the flash. The six hides face in different directions and depending on the position of the sun can be selected to optimise visibility.

71 HOPE CARR

Hope Carr
Hope Carr Lane, Leigh, WN7 3XB
SD664988
blocks.brink.vented

Opening times: Accessible at all times
Parking: A small free car park and roadside parking are available at overdrive.crunching.bleach
OS Landranger Map 109, OS Explorer Map 276
SBI

Hope Carr is a small but diverse wetland complex near Leigh, with old settling beds and lakes providing year-round opportunities to see waterfowl and passerines.

SPECIES

Consisting of old treated-effluent settling beds in the north half of the site and two purpose-built lakes in the southern half, it is unfortunately now a mere shadow of the former wader magnet that it used to be. The settling beds are now disused, and most are heavily overgrown whilst those that do still contain water have high water levels and no exposed mud. The lakes are now surrounded by woodland and viewing them is difficult due to a complete lack of management, but a good variety of birds can still be found and the well-watched January 2019 Blyth's Reed Warbler proved that it would be a mistake to ignore the site.

Whilst the exposed mud is gone and with it the passage waders that made this site the premier location in the county for them in the 1990s and early 2000s, Green Sandpipers are still present other than in mid-summer and can often be seen on the rotating filter beds within the sewage works itself, viewed through the perimeter fence. The rotating filter bed also provides warmth and food in winter for hundreds of Pied Wagtails, up to double figures of Grey Wagtails and, especially in harsh weather, Meadow Pipits. Whilst Water Pipits, once a site speciality, are no longer recorded they should still be looked for, just in case. The old settling beds are now very overgrown with long grass and some trees, but they provide excellent habitat for Reed and Sedge Warblers and Whitethroat in summer with Reed Bunting, Cetti's Warbler, Bullfinch and Willow Tit are all resident. The main lake has Teal, Gadwall, Tufted Duck, occasional Shoveler, Little Grebe and, in winter, Goosander. The woodland around the lakes has breeding Treecreeper and also Siskin and Redpoll in winter, both of which can also often be seen well in the alder trees around the United Utility car park. Along the western edge of the site, running alongside Pennington Brook, there is a garden on the opposite bank to the footpath which has well-stocked bird feeders. The feeders support a small population of Tree Sparrows, which are present all year but best seen during the winter months. The birds can often be found in the bushes running along the footpath just upstream of the green metal footbridge, flying back and forth across the brook to the feeders.

Rarities have included Temminck's Stint, multiple Pectoral Sandpipers, Grey Phalarope and Blyth's Reed Warbler.

ACCESS

From the Greyhound Roundabout on the A580 East Lancs Road turn north towards Leigh on the A574 Warrington Road, turning left at the first set of lights on to Greenfold Way. At the mini-roundabout turn left, following the road slightly to the right to park either by the concrete barriers or along the space on the left of the road (at SJ664988, pegged.precluded.footpath). From here footpaths run between the settling beds and south through the United Utilities car park.

YOUR VISIT

Check the filter beds for Green Sandpiper. Filter beds can be particularly attractive in hard weather for passerines unable to feed in other locations so periods of snow or frozen temperatures may be particularly productive.

72 THE HORWICH MOORS

The Horwich Moors
Walker Fold Road, Horwich, BL1 7PS
SD676121
clincher.cookbooks.trendy

Opening times: The entire moors are always accessible.
OS Landranger Map 109, OS Explorer Map 287
SBI

North George's Lane
SD651124
trailer.recruiter.sweeper

Higher Meadows
SD649130
drumbeat.avoiding.wildfires

Wildersmoor
SD652128
clincher.cookbooks.trendy

Parking: Parking for North George's Lane, Higher Meadows and Wildersmoor is at the roadside pull off along George's Lane at intruders. chest.fixture.

Walker Fold and Burnt Edge Loop
SD668125
unionists.clinic.grin

Parking: A large free car park is located on Walker Fold Road at rocks. rigid.boats and is always open but can be busy at weekends and bank holidays.

Scout Road Watchpoint
SD695134
bucket.marker.loses

Parking: There is very limited roadside parking at backup.ozone.slot and sage.pens.ruins. Paths around the quarry MUST NOT be used when red flags are flying.

High Rid Reservoir
High Rid Lane, Horwich, BL6 4DR
SD669102
pokers.intervene.outnumber

Parking: Limited roadside parking (on a very poorly surfaced road) is available at judges.looked.chips.
OS Landranger Map 109, OS Explorer Map 276
SBI

Greater Manchester

The Horwich Moors comprises many small plantations, woods, fields and moors, criss-crossed with a plethora of paths and tracks. The whole area affords some of the best birding in the county at the right time of year and in the right conditions. At other times it can be typically bleak and appear birdless, but even on these days there is usually something to see and the views during clear weather cannot be bettered anywhere in the county, particularly high on Smithills Moor, where from a single point it is possible to see the North Wales mountains, Snowdon and Anglesey to the south, the whole of Morecambe Bay and the peaks of the Lake District to the north and views across Lancashire to the north-east and Yorkshire beyond.

SPECIES

There is no bad time to visit the moors and even the poorest of weather has produced some great birding, but it is the spring and autumn migration periods that produce the finest moments and visible migration is more prominent and rewarding here than anywhere else in the county, rivalling any other inland UK site. Spring sees Ring Ouzels move through, often in small flocks, and Wheatears can sometimes be found in three-figure numbers across the whole site, but early autumn delivers perhaps the best birding. Site specialities such as Tree Pipit, Whinchat and Spotted and Pied Flycatchers can be found feeding in and around the plantations and, as an example, just one single early autumn day compiled 37 Spotted Flycatchers (with a maximum single flock of 11), eight Whinchats, 51 Willow Warbler, 15 Tree Pipits (including a flock of 10), 1,690 Meadow Pipits and 83 Wheatears. Twite is scarce but pass through in small numbers with other commoner finches, including Crossbill, amassing much larger daily counts. Migrant wildfowl often fly over in large flocks, including Pink-footed Geese and Whooper Swans, and raptors can feature heavily with double-figure numbers of hunting Kestrels and regular records of Osprey, Hen Harrier, Marsh Harrier and Merlin. Later in the autumn, winter thrushes and hirundines move through in

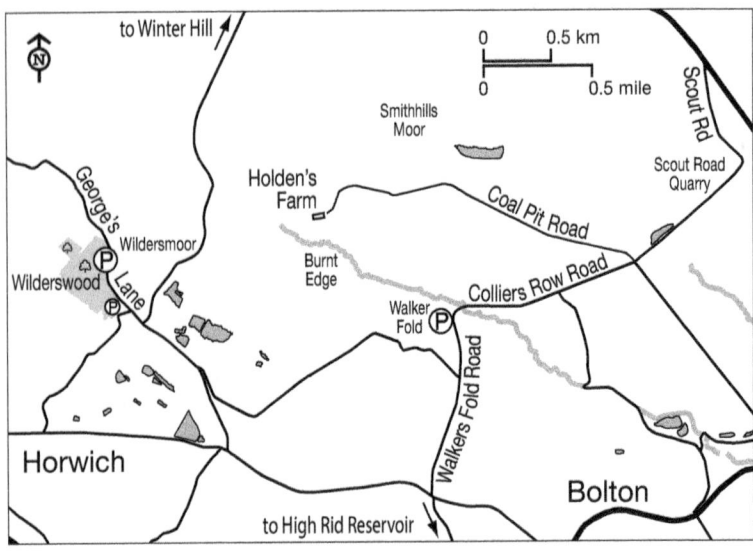

flocks well into the thousands. Red Grouse are resident breeders in small numbers as are Raven and Stonechats, with post-breeding numbers of the latter well into double figures. Winter can see Bramblings in the Beech woodlands, and although the moors tend to be bleak and rather birdless, Snow Buntings have been found on the tops on a few occasions.

Rarities and scarcities have included Glossy Ibis, Black Stork, Red-footed Falcon, Dotterel, Arctic Skua, Honey Buzzard, Wryneck, Richard's Pipit and Lapland Bunting.

ACCESS

As the whole area is so large it is best split into three main birding sites and a further location for visible migration watching alone. There are many more areas not covered though, as this guide only attempts to highlight the most regularly visited (and hence productive), so exploring other areas and paths could pay dividends.

NORTH GEORGE'S LANE, HIGHER MEADOWS AND WILDERSMOOR

Located on the western edge of the moors, just above the town of Horwich and accessed by turning north onto George's Lane off the B6226 Chorley Old Road. Continue for approximately 1.6km until you reach the Wilderswood car park (at SD651124, mush.clinic.unafraid). Parking is available here or a little further up the slightly rough track past Wilderswood itself on the broad bend of the road (at SD650126, finds.afterglow.dissolve). Wilderswood can hold migrants, particularly in the clearings on its northern side. Just north of this is Higher Meadows, lying between Wilderswood and the county border with Lancashire, just beyond the Rivington Pike Snack Shack. This rather small area of woodland, short grass sheep fields and hedges to the west of the road is highly productive for downed migrants like Redstart, Spotted Flycatcher and Tree Pipit and is also excellent for visible migration, with best viewing from around the bend in the road (at SD649129, imported.estimates.communal). Wildersmoor, to the east of the road, attracts Wheatears, Whinchats, good numbers of Meadow Pipits and often raptors. As many birds can be seen from the road itself, access for wheelchair users is possible, though the road is not well surfaced and parts do have a slight gradient.

WALKER FOLD AND BURNT EDGE LOOP

Arguably the most productive area, certainly for grounded birds, in the entire Horwich Moors site, this large valley on the southern edge of the moors has sheltered plantations, mature woodland, rough grassland and moorland. Access by turning off Walker Fold Road into the Woodlands Trust Walker Fold (free) car park (at SD676121, brands.chill.picked) where paths lead north into Walker Fold and Burnt Edge Valley beyond. At the junction of the northern end of Walker Fold and the southern end of Burnt Edge valley are some isolated trees and a small copse, which can be excellent for migrants and produce the county's most reliable migrant Pied Flycatchers in late summer/early autumn. From here there is a circular walk with a path leading off to the right up to Holden's Farm where short grass fields and a plantation (which lies below the farm and is of the same name) also give cover for migrants. Finches feed in the weedier fields and Willow Warblers can form sizeable flocks in the plantations during late summer/early autumn. At the farm, follow the path west and then eventually turn south along the bottom of the steep side of the valley back to the northern end of Walker Fold again.

Note that the path here is very rocky but during early autumn the moorland either side of it can hold chats, warblers and pipits, including Tree Pipit. There is also a much more stable path along the top of the ridge, which affords excellent views across Burnt Edge Valley and eventually leads directly back to the car park. Given the general roughness of the paths and the steep gradients of many, this area is generally unsuitable for wheelchair users.

Lying just to the east of Burnt Edge, Coal Pit Lane runs for almost 3.2km, eventually coming to Holden's Farm, so a circular walk from Walker Fold and Burnt Edge is possible and can be productive. It is best not to drive up Coal Pit Lane as it is narrow and has few parking or passing places. Instead, park at the side of Colliers Row Road near the junction with Coal Pit Lane (around SD689126, food.debit. deals) and from here follow Coal Pit Lane as far as Holden's Farm, checking the many small plantations, rough fields and well-vegetated gardens for migrants. The Beech woodlands that cross the lane can hold Bramblings in late autumn and winter and it is worth keeping an eye open across the valley for raptors, which have included two separate Red-footed Falcons. The road is well surfaced so suitable for wheelchair users though some of the gradients in and out of the small Beech-filled valleys can be slightly steep. Above Coal Pit Lane is Smithills Moor, running all the way up to the top of Winter Hill. The moor has breeding Red Grouse, Meadow Pipit, Skylark, Snipe, small numbers of Golden Plover and rarely Dunlin. There are many paths up onto the moor off Coal Pit Lane and a network of paths across the moor itself. Both Lapland and Snow Buntings have been found feeding along them so they can be worth a walk, and raptors hunt over the moors and occasionally include Hen Harrier. On the very top of Winter Hill there are a couple of long stone walls and mast cable compounds which during migration periods regularly harbour Black Redstarts, large numbers of Meadow Pipits, Wheatears, Whinchats and occasional warblers.

SCOUT ROAD VISIBLE MIGRATION WATCHPOINT

The quality and quantity of visible migration over the Horwich Moors was fully realised many years ago now and the watchpoint above Scout Road Quarry has not only provided the most consistent variety of species and actual numbers of birds but, just as importantly, affords the best views of them as they pass by, often at head height.

Early spring produces annual Osprey sightings, and the fields to the east of the watchpoint, immediately east of Belmont Road, attract large flocks of resting Curlews from late February to early April. The site really comes into its own for visible migration watching in autumn, and any time between August and mid-November can be productive with impressive numbers of birds moving south down the valley and past the watchpoint. August and September, in particular, see hundreds and often thousands of Meadow Pipits moving through daily, along with hundreds of finches and huge movements of hirundines, well into the thousands on peak days. Raptors move overhead with kettles of Buzzards, occasional Marsh Harriers, Ospreys and Merlins whilst the bracken around the watchpoint can hold passerines, especially warblers and chats. Mid-September sees the first of the Pink-footed Geese skeins overhead and a small number of Whooper Swans too, and these increase in number and frequency into October. Meadow Pipits continue to move through into October and finch flocks increase with good numbers of Siskins, Bramblings and Crossbills, the last of which can often drop into the conifers in Gale Brook Plantation immediately to the north-east of the watchpoint. The start of

October usually heralds the arrival of the winter thrushes and as the month progresses the flocks of Redwings and Fieldfares, often also including Ring Ouzels and rarely Hawfinches, sweep across the moors. Daily numbers are often well into the thousands in advantageous weather conditions. Similarly, Woodpigeon movements can include impressive daily five-figure counts, and their flocks, along with thrushes and geese, continue well into mid-November.

The watchpoint is located above Scout Road Quarry at the very eastern end of Scout Road, above the sharp bend. It is essential not to use the paths above or around the quarry when there are red flags flying as the quarry is occasionally used as a live firing range. There are two roadside parking locations, one outside the quarry itself (at SD697134, backup.ozone.slot) and another just further north up the road before the start of Gale Brook Plantation (at SD695136, sage.pens.ruins). There are paths up the sides and above the quarry leading to the viewpoint and affording excellent views to the south and east.

YOUR VISIT

If visiting for visible migration, calm conditions will generally be more productive, both in terms of the numbers of birds passing over and the ability to identify them on sight or by call. The Trektellen website gives extensive data on visible migration from sites across the north-west, the wider UK and beyond. It can be accessed to understand when species are moving in large numbers and a 'vis-mig' session could be productive.

NEARBY SITE: HIGH RID RESERVOIR

This small, stone-lined upland reservoir lies 4.8km west of Bolton, at the foot of the Horwich Moors. It is easily accessed (though is not wheelchair accessible) and can be efficiently covered, producing good, easy birding.

SPECIES

In winter Little Grebe, Tufted Duck and Goldeneye are present, with occasional Scaup and Whooper Swans. Gulls, sometimes including Yellow-legged, Caspian and Mediterranean, rest on the grassy areas or wash on the reservoir itself during the day. Spring brings passage terns, including Arctic, and Common Sandpipers on the reservoir edges, occasional Little Gulls and Common Scoters drop in and large flocks of hirundines feed over the water. Early autumn sees annual records of dispersing Black-necked Grebe and Mediterranean Gull from breeding sites in neighbouring counties and Spotted Flycatcher and Whinchat in the surrounding hedgerows and fields, along with good flocks of feeding Meadow Pipits. The proximity to the Horwich Moors and the open vista of the areas surrounding the reservoir also lends itself to the potential for good visible migration.

Rarities and scarcities have included Ring-necked Duck, Long-tailed Duck, Velvet Scoter, Glossy Ibis and Grey Phalarope.

ACCESS

High Rid Reservoir is accessed off High Rid Lane, which itself can be accessed either off the A673 Chorley New Road onto Fall Birch Road which then becomes High Rid Lane, or alternatively by turning off the B6402 Old Kiln Lane onto Old Hall Lane, which also then becomes High Rid Lane. Either way, the lane itself is poorly surfaced and there is parking for only a few cars around SD670100, judges.looked.chips. However, there is plenty of roadside parking along either of the two main roads and then a relatively short walk to the reservoir itself. There is a footpath beyond the parking place which leads through the fields and up to the pump house of the reservoir with a path then leading around the perimeter.

YOUR VISIT

High Rid Reservoir hosts a small water activity centre. This means that some disturbance on the water is possible at times, and visiting earlier in the day may be better to reduce the chances of this.

73 RUMWORTH LODGE RESERVOIR

Rumworth Lodge Reservoir
Beaumont Road, Ladybridge, Bolton, BL6 4JJ
SD681077
rally.hedge.scales

Opening times: Accessible at all times
Parking: parking available along Armadale Road around cross.modest.await. Parking must not be made along Beaumont Road itself or in pull-offs along it as access to farmers' fields is always necessary.
OS Landranger Map 109, OS Explorer Map 276
SBI

This large, shallow reservoir is located to the west of the A58 Beaumont Road, just west of Bolton and south-east of Lostock railway station. It provides compensation water so that the Middlebrook, which downstream becomes the River Croal, never runs dry. It has a gentle shore profile apart from the northern arm which has a dam and steep, wooded sides. The southern shore has a willow thicket and there is a marsh and large reedbed in the south-east corner. Around the reservoir are some damp meadows, arable fields and mature hedgerows. The reservoir itself has two small islands in its southern half and often has exposed mud around its complete shore.

SPECIES

Rumworth's main attraction is for waders, especially when the water levels are low enough to expose mud, with Ringed and Little Ringed Plovers, Dunlin, Redshank, Greenshank, Black-tailed Godwit, Curlew and Green and Common Sandpipers all annual. Wildfowl are most often found in the southern half of the reservoir and sometimes include Whooper Swan in winter, while the hedgerows and bushes around the marsh attract migrant warblers, Redstart and Spotted Flycatcher on passage. The marsh and reedbed have breeding Reed and Sedge Warblers and Reed Bunting and have contained Bittern in winter.

Rarities and scarcities have included Ring-necked Duck, Great White Egret, Pectoral Sandpiper, Temminck's Stint, Great Grey Shrike, Richard's Pipit and Marsh Warbler.

ACCESS

There is no parking along Beaumont Road. Do not park in any of the pull-ins off the road to permit access for farm vehicles. Instead, park along Armadale Road (which runs parallel with Beaumont Road) and access the site via the stile on Beaumont Road at SD681077, rally.hedge.scales. From here walk down towards the marsh following the hedge line and then along the marsh edge towards the shore by the two islands. It is possible to view the whole reservoir (other than the arm by the dam) from there. The site is not suitable for wheelchair users.

YOUR VISIT

Viewing the area is only possible from one footpath and viewing spot; access around the shoreline and the fields beyond is not possible. It may be worth considering the position of the sun at the time of a planned summer visit to avoid birding with glare or silhouette issues.

74 ELTON RESERVOIR

Elton Reservoir
Buller Street, Bury, BL2 2BR
SD790098
foster.crab.dots

Opening times: Accessible at all times
Parking: A small free car park is located just before the Elton Sailing Club off Buller Street at prime.feels.beyond.
OS Landranger Map 109, OS Explorer Map 277
SBI

Withins Reservoir
SD860087
lively.long.socket

This 1km-long reservoir in the Irwell Valley was originally built in 1803 to feed the now-disused Manchester, Bury and Bolton Canal and consists of a stone retaining wall on its east side and gently shelving banks on its west margin. Holding a wide variety of habitats and at a relatively high elevation, the site is one of the most productive in the county, especially during spring and autumn migration, and in more recent years has proven very fruitful for visible migration.

SPECIES

Winter: good numbers of Goldeneye and particularly Goosander (which can roost nightly in good numbers) can be seen, along with regular Wigeon and occasional Whooper Swans amongst the commoner wildfowl. Little Egrets are practically resident but can be more obvious in winter, and a small feeding station by the car park attracts Willow Tit, Bullfinch and occasional Brambling at this time of year. Water Rails inhabit the marshy areas and can be particularly conspicuous along the canal. The electricity pylons, especially those near the north end,

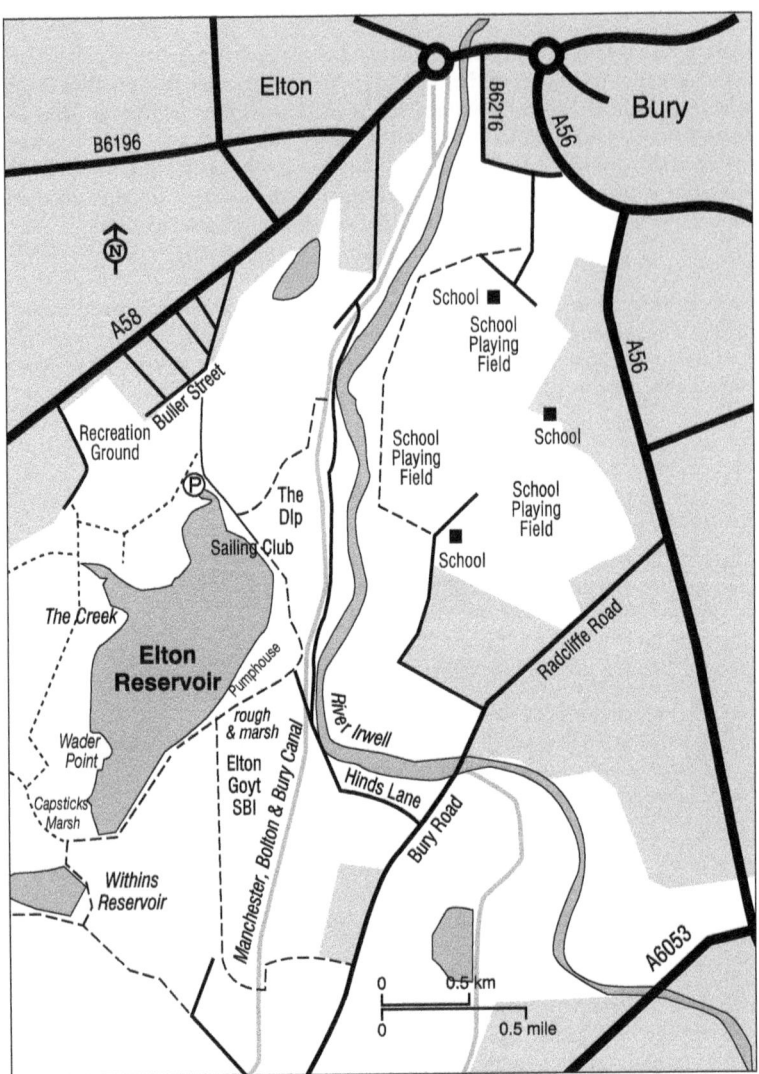

should be checked for Peregrines, which often hunt from them. On days when there is disturbance from the sailing club the wildfowl can relocate to Withins Reservoir, just to the south of Elton.

Spring: terns can include Black, Sandwich and Arctic which, along with Little Gulls, are more frequently seen during anticyclonic weather with associated easterly winds. Small numbers of Common Scoter pass through with scarce but annual Scaup and Garganey. Hirundine numbers build up over the reservoir and are joined by feeding Swifts. The surrounding hedgerows can hide Redstarts and Lesser Whitethroats along with more conspicuous Whitethroats, Willow Warblers and Blackcaps in good numbers. Wheatears and Yellow Wagtails particularly favour the dam wall and grass around it. Spring is probably the best time for waders with commoner Little Ringed Plover, Ringed Plover and Common Sandpiper joined by fairly regular Turnstone, Sanderling, Curlew, Whimbrel, Greenshank and sometimes Wood Sandpiper. The shoreline between The Creek and the aptly named Wader Point in the southern half of the reservoir and on the opposite bank to the dam wall are the most productive sites for any such waders. Rock Pipits are scarce but annual migrants usually arrive during March, and Elton is second only to Audenshaw Reservoirs in the county for their regularity.

Summer: Lesser and Common Whitethroats, Blackcap, Willow Warbler and Chiffchaff all breed, along with Reed, Sedge and Grasshopper Warblers in lesser numbers. Willow Tits remain and can still be found more easily around the feeders, as can increasingly Ring-necked Parakeets. Common Terns do not breed but can still often be on the reservoir at this time of year. Due to the site's often heavy disturbance, breeding wildfowl and waders are unfortunately poorly represented but Kingfishers are resident and can be located with luck anywhere around the shore.

Autumn: Elton has one of the best records in the county for autumnal migrant Redstart, Spotted Flycatcher and Lesser Whitethroat, with birds most often favouring the area around Crow Trees Farm just to the south of Withins Reservoir, or The Dip, which is immediately north-east of the sailing club. Stonechat and Whinchat prefer the area of Elton Goyt, just to the east of the reservoir, and returning waders often include Greenshank and Green Sandpiper. Visible migration can be watched from the pumphouse along the dam wall or just south of Withins Reservoir and has proven very productive, with only the Horwich Moors currently managing more impressive totals. Tree Pipits begin moving over in late July and through to early September with the last month seeing increasing numbers of Meadow Pipits and hirundines moving through. October and early November begins the movements of 'winter thrushes', which can pass over in daily totals of thousands rather than hundreds. During these months, finch flocks and often huge daily totals of Woodpigeons migrate overhead, accompanied by the sound of skeins of Pink-footed Geese and often rarer birds such as Hawfinch, which has become something of an annual event at this site. October also usually records a few Rock Pipits moving back through.

Rarities have included Black-eared Wheatear, Bee-eater, Little Bittern, Lesser Scaup, White-winged Black Tern, Sabine's Gull, Spotted Sandpiper, Common Crane, Alpine Swift and Rose-coloured Starling.

ACCESS

The main car park is signposted off the A58 Bury and Bolton Road. Opposite the Wellington pub, turn down Kitchener Street, turn right at the end and then left across a bridge, then turn right and follow the road to the car park by the sailing club at SD790098, losses.glitz.aware.

YOUR VISIT

A footpath runs around the entire periphery of the reservoir. Other footpaths run off it and can be explored. Though the footpaths around the western side of the reservoir are not wheelchair accessible, viewing most of the reservoir and the feeders is possible from the area immediately around the car park itself, and the road past the sailing club and path south along the dam wall are also suitable for wheelchair users.

Greater Manchester

75 CUTACRE COUNTRY PARK

Cutacre Country Park
Salford Road, Middle Hulton, Bolton, BL5 1BR
SD697054
rising.fears.magma

Opening times: Accessible at all times
Parking: Car parking available at all times. The free car park is located off Salford Road at super.draw.toxic.
OS Landranger Map 109, OS Explorer Map 276

This former open-cast coal mining site lies just off the A6 at Over Hulton, Bolton, and with mining ceasing in 2011 the site has since been developed with large industrial units on the northern section but with a 215ha country park to its south. The country park has a free car park off the A6 and a network of paths around it, including a 6.4km circular route. The habitat is mainly rough grassland with some young woodland, several small pools and marshy areas and a large lake on its southernmost border.

SPECIES

Winter: Grey Partridge, Stonechat and Willow Tit are resident but can be more visible at this time of year and the plantations hold Siskin and Redpoll, whilst Water Rail can often be heard from various small pools, if not seen. Swan Lake (the large lake on the southernmost border of the site, close to the railway line) has Gadwall and often Wigeon and Goldeneye and also attracts bathing gulls, regularly in their hundreds, and including Yellow-legged, Caspian, Iceland and Glaucous Gulls. The proximity of the lake to the path is such that there can be disturbance, but it also means views are usually very good.

Passage: spring sees Skylarks back on territory and the country park resounds to their song such is their abundance, but it is with migrants that the site often comes alive, with Wheatears moving through the grasslands and rough areas in good numbers, Sedge and Reed Warblers around almost every pond and Grasshopper Warblers in the rough grassland areas. Common and Green Sandpipers can be found, along with Little Ringed and Ringed Plovers, plus occasional Black-tailed Godwit, Dunlin and Greenshank. Late spring and early autumn bring more Wheatears to the site, along with Whinchats, Redstarts and Tree Pipits, though the last are usually overflying birds and perhaps hint at the site's potential for visible migration watching. As a relatively new location it remains under-watched and has huge potential given more coverage.

Scarcities and rarities have included Garganey, Ring-necked Duck, Black-necked Grebe, Bittern, Caspian Gull, Glaucous Gull and Shorelark.

ACCESS

Cutacre is located near junction 4 of the M61. Take the A6 towards Over Hulton and the car park is immediately off the A6 near the large warehouses on the left (if travelling west) at SD697054, rising.fears.magma. From the car park follow the footpath south past the large warehouses on the left where the country park opens to the east and there are many footpaths to explore. The western side of the park has rough grassland and small plantations whilst the eastern side of the park has more small pools and marshy areas.

YOUR VISIT
Other than the car park there are no other facilities on site, and whilst the paths are well surfaced and wheelchair accessible, some sections of path have fairly mild gradients, though these could be easily avoided.

76 LITTLE WOOLDEN MOSS LWT RESERVE

Little Woolden Moss LWT Reserve
Off Astley Road, Irlam, M44 5LR
SJ698953
riskiest.clap.casually

Website: www.lancswt.org.uk/nature-reserves/cadishead-little-woolden-moss
Phone: 01204 663754
Email: jlawson@lancswt.org.uk
Opening times: All the mosses are accessible at all times
Parking: Parking is available via very limited spaces for Little Woolden Moss LWTR, Little Woolden Moss and Cadishead Moss at the end of the track off the junction of Twelve Yards Road and Astley Road at ballots.pits.myths. For all other general sites south of the railway line or if the latter car park is full, the best parking is at Moss Farm Fisheries at Cutnook Lane, Irlam, Manchester M44 5NB (telephone 07540 898979), rumbles.spindles.trample, which is generally open 9am to 5pm. Contact the fishery prior to your visit on the telephone number to check parking availability and if a fee applies (usually not if purchase is made from the on-site café).
OS Landranger Map 109, OS Explorer Map 276
SBI

Cadishead Moss
Off Astley Road, Irlam
SD699952
recliner.rinsed.results

Chat Moss
Twelve Yards Road, Irlam, M44 5LY
SJ715965
merely.accompany.remarried

Irlam Moss
Astley Road, Irlam, M44 5LG
SJ708950
ranged.precautions.hacksaw

Greater Manchester

> **Astley Moss SSSI**
> Off Rindle Road, Astley, M29 7LG
> SJ698975
> sonic.eggplants.laminate
>
> Parking: Parking for Astley Moss, including the SSSI, is via very limited roadside parking only along Rindle Road at permanent.custard.bucket
> SSSI, SBI

This 115ha site showcases the 'restoration reversal' of peat extraction, which ceased here in 2012. Purchased by Lancashire Wildlife Trust, the reserve is now becoming a true peatland and has proven to be a haven and real attraction for birds. With many shallow pools and marshes it has become one of the best locations in the county for waders and the grassland around them is favoured by raptors.

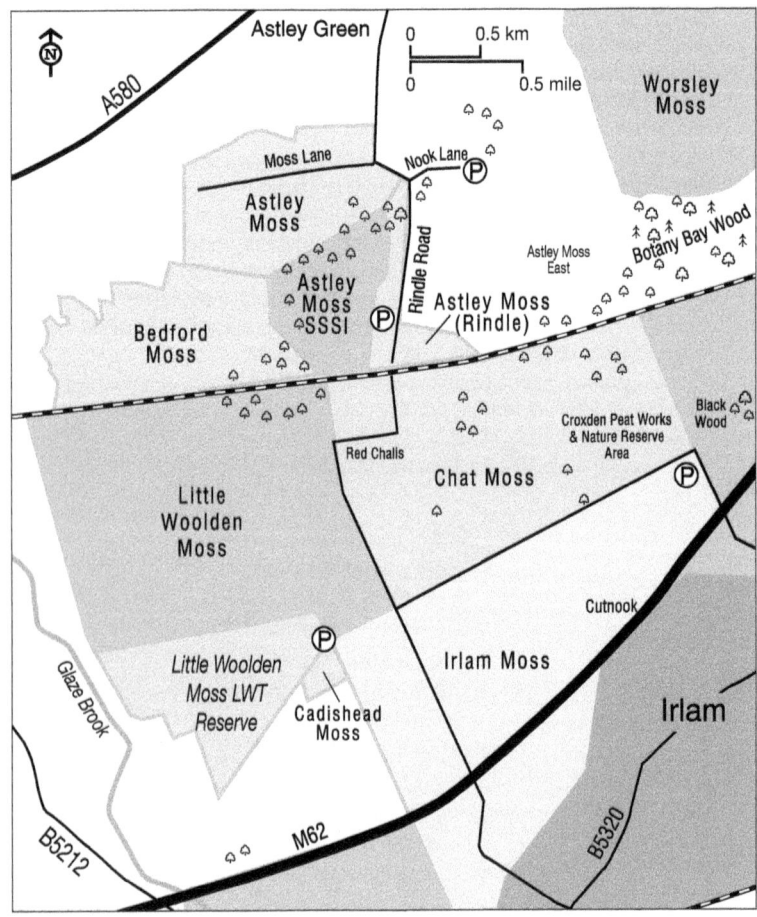

SPECIES

Winter: perhaps the quietest period for the reserve itself (unlike the fields around it) but raptors using the site include Merlin, Peregrine, seemingly ever-present Buzzards and occasional Short-eared Owl. There remains a good population of Reed Buntings, and Linnets and Stonechat are regularly seen whilst Pink-footed Geese use the reserve after feeding in the surrounding fields and Whooper Swans occasionally drop onto the pools.

Passage: either passage period brings waders, and if water levels are favourable Ringed and Little Ringed Plover, Greenshank, Wood and Green Sandpiper, Black-tailed Godwit and Dunlin are all regular. Wheatears and Whinchats can be seen out on the moss, the latter especially in autumn, and the tree-lined southern edge of the reserve has a healthy population of Willow Warblers and occasionally migrant Redstart.

Summer: breeding Curlew, Redshank, Snipe, Meadow Pipit, Yellowhammer, Reed Bunting and Yellow Wagtail all make the reserve a busy place during this period. Large numbers of feeding Swifts and an abundance of dragonflies provide one of the site's summer highlights as attendant Hobbies, often multiple birds in the sky at the same time, put on an impressive display which cannot be witnessed to this extent and regularity anywhere else in the county. Marsh Harriers are seen in every month and Ravens are regular over the reserve. Wildfowl such as Teal and Shoveler inhabit the pools but are often secretive and difficult to see.

Rarities and scarcities have included Greenland White-fronted Goose, Stone Curlew, Black-winged Stilt, Great White Egret, Common Crane and Yellow-legged Gull.

ACCESS

There is a small car park for the reserve at the end of Lavender Lane (at SJ699953, blanket.firework.balance) off Astley Road which can be accessed from the south by turning north off the B5320 Liverpool Road in Irlam or from the north by crossing the railway level crossing at the southern end of Rindle Road in Astley. Either way, both Astley Road and Lavender Lane are rough surfaced in parts and if the car park is full then park sensibly elsewhere along Astley Road, making sure to leave plenty of room for the farm machinery and large goods vehicles that also use the road. The site itself has a well-surfaced, wide path, purposely designed for wheelchair users, which runs around the perimeter of the reserve but is not completely circular.

NEARBY SITE: LITTLE WOOLDEN MOSS

Just north of the Lancashire Wildlife Trust Reserve of the same name, between it and the railway line to the north is an area of arable fields called Little Woolden Moss. It can be accessed by a gate on the northern side of the perimeter path of the reserve and these fields often have large flocks of Skylarks in winter along with flocks of finches, regularly hunted by Merlins and Sparrowhawks. In spring they can contain Wheatears and White Wagtails, particularly if they have been freshy ploughed. Corn Buntings, whilst now very scarce, can still be found on the tree-lined edges of the fields and the area has previously recorded Great Grey Shrike and Common Crane.

NEARBY SITE: CHAT AND IRLAM MOSSES

These are remnants of a huge, impassable peat bog that helped to isolate parts of north-west England in medieval times. Today, the area effectively separates Salford and Wigan and acts as a huge green lung for Manchester. The whole area is a large expanse of arable, vegetable and rough fields (with some grass turf fields) interspersed with small woods, hedgerows and some marshy areas. Chat Moss lies just south of the railway (which forms its northern border) and is bordered by Astley Road to its west, Twelve Yards Road to its south and Fiddlers Lane to the east. It has a large expanse of remnant peat extraction known as Croxden's Peat Works, which has pools along its southern edge close to the junction of Twelve Yards Road and Cutnook Lane, which can be attractive to waders and annually produces Wood Sandpiper. In summer, the fields just off Astley Road hold practically annual singing Quail and have a very healthy population of breeding Yellow Wagtail and Yellowhammer, with Wheatear and Whinchat in spring and autumn. Rarities have included Bean Goose, Red-footed Falcon, Great White Egret, Common Crane, Great Grey Shrike and Richard's Pipit. Barton Moss lies south of Twelve Yards Road and is bordered by Irlam town to its south and Fiddlers Lane to the east. It has similar breeding birds to Chat Moss, including Yellow Wagtail, but in winter the fields just to the west of Astley Road can hold feeding Pink-footed Geese and, like elsewhere on the other mosses, the field edges can hold good flocks of finches, including Brambling. Rarities here have included Great White Egret and Desert Wheatear. As most birding is carried out from the roads across the moss, the area is generally suitable for wheelchair users.

NEARBY SITE: ASTLEY MOSS

This site can be accessed just south of Astley village by turning south off the A580 East Lancashire Road onto Lower Green Lane and then Rindle Road. The southern border of the moss is formed by the Liverpool to Manchester railway line, where there is a level crossing that permits vehicular access onto Astley Road and out onto Chat Moss. To the east of Rindle Road is Astley Moss SSSI, a large area of restored peatland that harbours Meadow Pipits, Reed Buntings, Willow Tits and occasionally Cuckoo in summer. The hedge- and tree-lined fields either side of the southern end of Rindle Road can contain large finch flocks in winter (including Brambling) and winter thrushes, Yellowhammer, Tree Sparrow and in summer Yellow Wagtail. Spring used to see the annual gathering of small flocks of Whimbrel particularly in the fields immediately west of Rindle Wood, and whilst those numbers are no more, a few individuals still occur. Multiple Ring Ouzels have also been seen on these fields in spring, along with regular Wheatears, and rarities have included Great Grey Shrike.

77 CARRINGTON MOSS

Carrington Moss
Birch Road, Carrington, M31 4AP
SJ748915
apartment.birthdays.nibbles

Greater Manchester

Parking: There is limited roadside parking along Isherwood Road, off Carrington Lane around mops.demotion.cool
OS Landranger Map 209, OS Explorer Map 276/277

Carrington Moss is an extensive lowland farmland and wetland area near Manchester. Its arable fields, pastures and woodland support a variety of breeding and wintering birds.

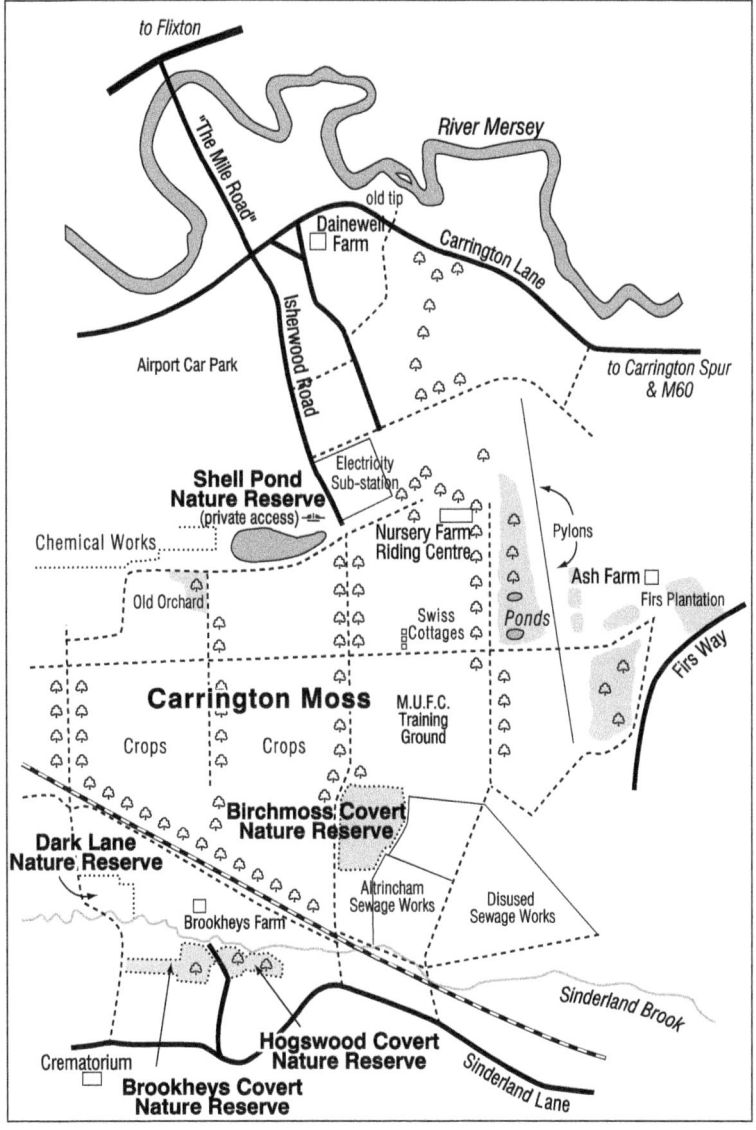

243

The area is bounded by Partington in the west, the A6144 to the north, Sale and Altrincham in the east and the Bridgewater Canal in the south. The habitat is arable and pastoral farmland with the private Shell Pool Reserve on the north edge within the chemical works complex.

Many small roads and footpaths cross the very large area of the moss and these are always worth investigating, but it is the area just south of the chemical works off Manchester Road in Carrington itself that receives most focus and produces the best results. The large arable fields hold Skylark and finch flocks in winter and breeding birds include Lapwing, Yellowhammer, Yellow Wagtail, Buzzard and Grey Partridge, whilst Hobby are regular in summer and Quail are practically annual. The fields here also attract Wheatears in spring along with occasional Redstarts in the hedgerows and practically annual but still scarce Ring Ouzels. Just south of these fields is Altrincham Sewage Works where the old, disused settling beds can hold some wildfowl and, rather rarely these days, waders such as Green and Wood Sandpipers. In summer there are Sedge and Grasshopper Warblers along with Whitethroats. Slightly west of here, on the north side of Sinderland Road, is the mature oak woodland of Brookheys Covert Nature Reserve and Sinderland Brook by Brookheys Farm, which often holds Little Egret with, immediately west again, the small scrub woodland of Dark Lane Nature Reserve.

ACCESS

From Sinderland Lane in the south or from Isherwood Road, off the A6144 Carrington Lane, in the north.

THE MERSEY VALLEY

The Mersey Valley contains 216ha of urban countryside from the boundaries of Manchester and Stockport in the east to the Manchester Ship Canal in the west.
There are two key sites:

78 CHORLTON WATER PARK

Chorlton Water Park
Maitland Avenue, Manchester, M21 7NH
SJ823919
hush.happen.anyway

Website: https://www.visitmanchester.com/listing/sale-water-park/14411101/
Opening times: Accessible at all times
Parking: Free car park at boss.desk.guess
OS Landranger Map 109, OS Explorer Map 277
SBI

Chorlton Water Park was formed during the construction of the M60 motorway in the 1970s when gravel was excavated from the site. The gravel pit was subsequently flooded, creating the lake that is central to the 69ha Local Nature Reserve that the site is today. Around the lake are paths leading into woodlands and meadows.

SPECIES

The main interest of this site used to be its wintering flock of *Aythya* ducks, which gathered in large flocks during the day, moving to feed on the Salford Quays by night. Unfortunately (in part due to the cleaning and aerating of the water on the Quays), such flocks, which used to also attract Scaup and Ferruginous Duck, are no more, but smaller numbers of Tufted Duck and the occasional Pochard still occur. The woodlands have Siskins in winter and Goosander and Goldeneye can be found on the lake during this period too.

Scarcities and rarities have included Ring-necked and Long-tailed Ducks, Fulmar, Shag, Great White Egret and Spotted Crake

ACCESS

A small car park is located off Maitland Avenue, Chorlton, which can be found at SJ822920, catch.leap.resort. If arriving by public transport the 86 and 23 buses both serve stops close to the park and run from the city centre. These buses can be boarded from Manchester city centre. The nearest tram stop to Chorlton Water Park is Barlow Moor Road on the Manchester Airport line. From the tram stop it's approximately a 15-minute walk to the park.

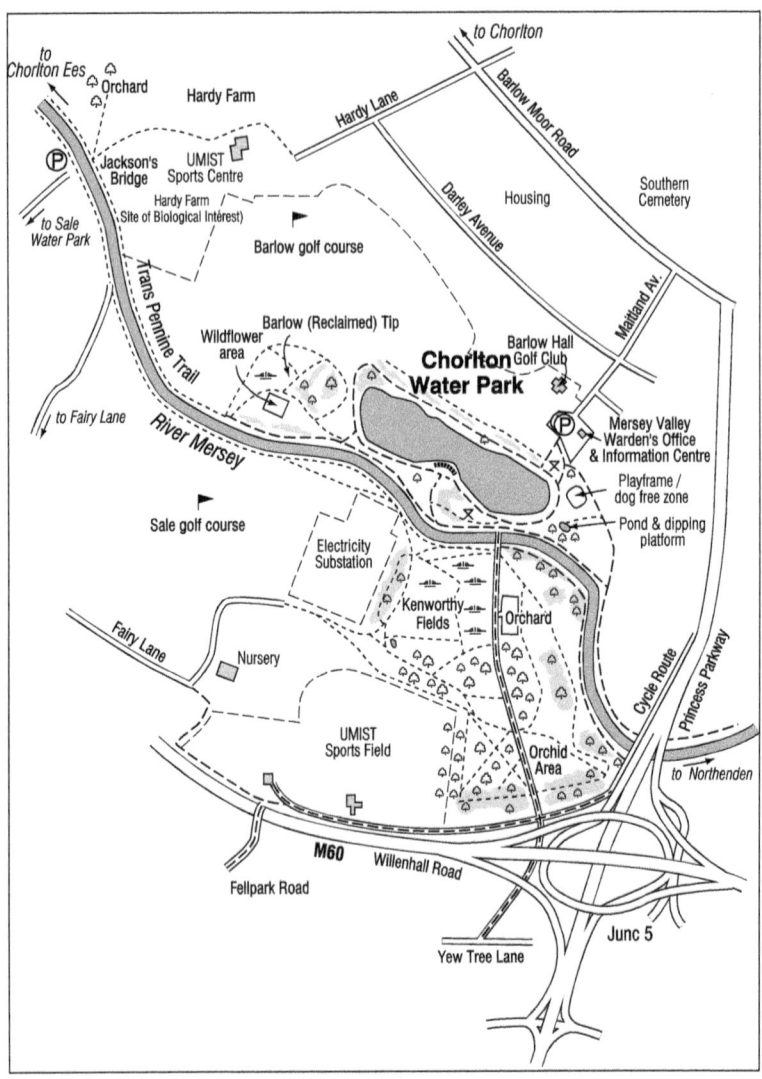

There are toilets here and usually some refreshment vendors (coffee, ice cream depending on season) between here and the lake.

YOUR VISIT

If arriving by car the small car park can get very busy at weekends. The site is very popular with dog-walkers and there can be some disturbance, particularly in areas where access to the banks is more straightforward. Early mornings can therefore be better, before the number of visitors increases.

Greater Manchester

79 SALE WATER PARK

Sale Water Park
Off M60, Sale Moor, Sale, M33 2LX
SJ801927
varieties.total.curving

Opening times: Accessible at all times
Parking: Free car park situated at stared.flips.sober
SBI

As with Chorlton Water Park, the lake at Sale was a gravel pit that was flooded after the extraction of gravel for the M60. A network of paths leading from the visitor centre and free car park includes a circular walk around the lake and there are well-wooded areas with some patches of grassland. At the nearby Broad Ees Dole, a hide overlooks a marsh and swamp with shallow water.

SPECIES

The lake itself has Gadwall, Goosander (in winter), Great Crested Grebe and Kingfisher, with daytime-roosting Cormorants on the pylons and Common Terns in spring and late summer. It annually attracts Common Scoter, particularly in early spring, and Wigeon in winter. The wooded and scrub areas around the lake have Willow Tit and Cetti's Warbler with commoner species such as Nuthatch, Treecreeper and Bullfinch along with Siskin and Redpoll in winter. Broad Ees

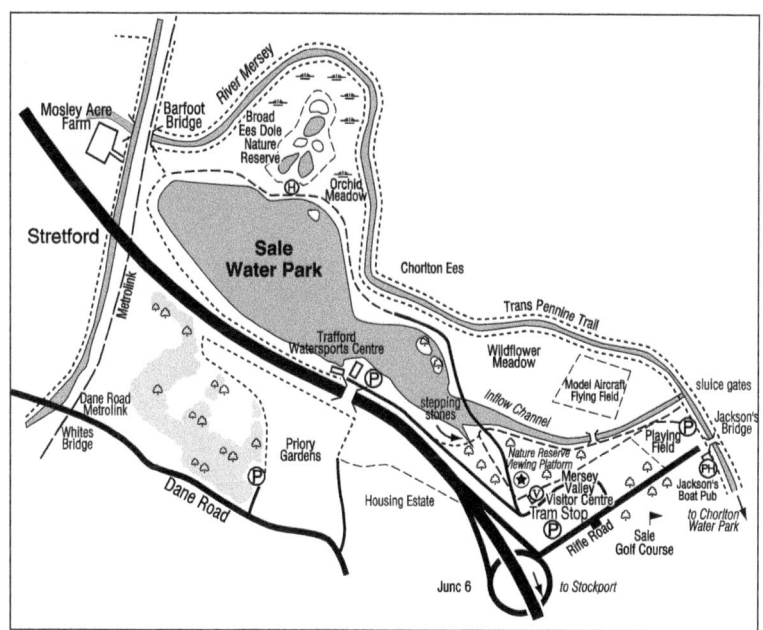

247

Dole, with its exposed mud, has Shoveler and Teal, often Little Egret, and can attract waders, with Snipe and Common Sandpiper regular.

Rarities and scarcities over the years have included Caspian Tern, Great White Egret, Fulmar, Gannet, Guillemot, Great Grey Shrike, Temminck's Stint and Shag.

ACCESS

Sale Water Park is located 0.8km from junction 6 of the M60 just north of Sale, off Rifle Road (M33 2LX, SJ806924, linen.hidden.joined). There is a large free car park and toilets, including wheelchair accessible, on site. There is also a Metrolink tram stop at Sale Water Park.

YOUR VISIT

The water park is very popular with walkers, and the lake is used for water sports so the area can be very busy and quite disturbed at weekends in particular but the paths are generally well maintained and are suitable for wheelchair users.

80 HEATON PARK RESERVOIR (PRIVATE SITE)

Heaton Park Reservoir
St. Margaret's Road, Besses o' th' Barn, Simister, Bury, M25 2GT
SD830047
drift.laptop.points

Opening times: The majority of the reservoir is viewable at all times from outside the perimeter fence at begin.mess.loved.
Parking: Pay and display car parking is available at St. Margaret's Car Park, St. Margaret's Road, Besses o' th' Barn, Simister, Bury, M25 2GT, swung.work.bronze
OS Landranger Map 109, OS Explorer Map 277
SBI

Heaton Park Reservoir is a large stone-lined balancing reservoir on the north side of Manchester, adjacent to but separate from Heaton Park, a large parkland area. The site is private and while access permits are no longer available, some of the reservoir can be viewed from outside the perimeter fence close to the BT Tower by the south-east corner of the reservoir. Because this is a considerable distance from the reservoir, a telescope is needed.

SPECIES

The reservoir used to host a large nightly roost for gulls that fed by day at Pilsworth Tip, less than 3km to the north. In its heyday, around 1,000 Great Black-backed Gulls roosted in early January, along with up to several thousand Herring Gulls with regular Yellow-legged, Caspian, Iceland and Glaucous Gulls among them. Nowadays, the night-time roost is severely reduced and is often not present at all, but a daytime roost remains and scarcer gulls are still picked out among the commoner species. Goldeneye and Goosander can be present in good numbers, with the latter increasing in the late afternoon as they come to roost. If water levels are low the exposed shore can attract passage waders such as Dunlin, Black-tailed Godwit, Redshank, Greenshank, Turnstone, Sanderling, Common Sandpiper, Curlew and Whimbrel. Wildfowl include Tufted Duck (with an annual late summer influx), and occasional Scaup, Wigeon and Common Scoter.

Scarcities and rarities have included Lesser Scaup, Ring-necked Duck, Purple Sandpiper, Grey Phalarope, Bonaparte's Gull, Kumlien's Gull, White-winged Black Tern and Red-rumped Swallow.

ACCESS

As stated above there is no access to the reservoir itself. There is a car park in Heaton Park at SD830046, which is a short walk from the BT Tower at SD831048.

81 WATERGROVE RESERVOIR

Watergrove Reservoir
Ramsden Road, Great Howarth, Wardle, Rochdale, OL12 9NJ
Website: https://www.visitrochdale.com/things-to-do/watergrove-reservoir-p12901
SD913178
drive.degree.begin

Opening times: Accessible at all times
Parking: Free car parking is available at wake.stray.global and there is a separate disabled car park, accessed with a key, which is available by contacting 01706 881049. The hide is usually locked but access may be available by contacting 01706 881049.
OS Landranger Map 109, OS Explorer Map OL12
SBI

Constructed between 1930 and 1938, this water supply reservoir lies in a south-facing valley 240m above sea level a little over 1km north of Wardle village. The southern edge consists of a dam wall, but the northern margin has exposed muddy islands and bays, and an arm reaching north into Higher Slack Brook Clough. Behind the reservoir to the north is a curving ridge of moors rising to 400m and the northern side of the reservoir has been planted extensively with native trees. The scenery and views are spectacular and there are many paths, including a circular walk around the reservoir and leading up onto the moorland.

SPECIES

Winter: small numbers of wildfowl, including Wigeon, Goldeneye and Goosander, with regular visits from Whooper Swans and occasional Peregrine and Merlin; Woodcock roost in the woodlands. The feeding station has Bullfinch and often Brambling, Redpoll and Siskin among commoner species, while Green

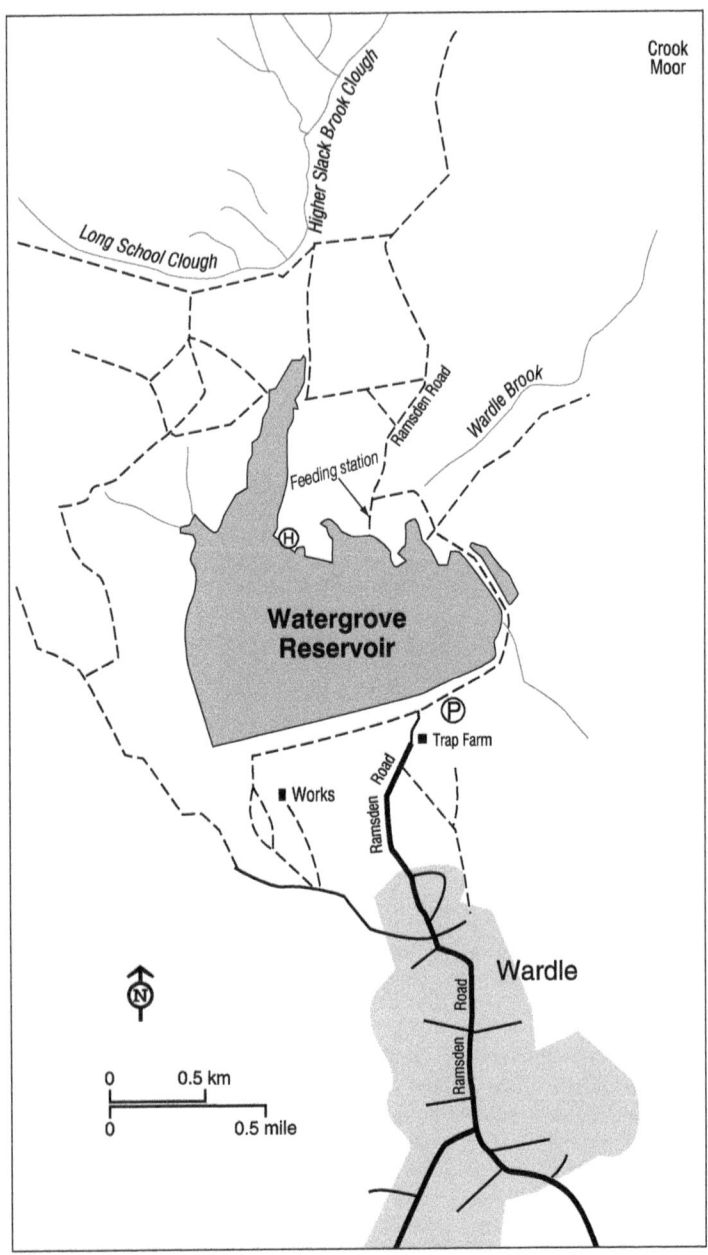

Woodpecker can be seen and heard around the plantations. Kingfisher can be seen around the reservoir, particularly on the hide pool.

Passage: searching the moorland edges in early spring can produce Ring Ouzel, and Common Scoter is fairly regular on the reservoir in both spring and autumn.

Exposed muddy edges on the shore attract passage waders such as Curlew, Dunlin, Ringed Plover, Redshank, Greenshank and Sanderling. Spotted Flycatchers pass through in early autumn, as do Wheatears in both passage periods and occasional Marsh and Hen Harriers.

Rarities and scarcities have included Red-throated Diver, Broad-billed Sandpiper, Gannet, Great Grey Shrike and Marsh Tit.

ACCESS
On the A58 Halifax Road between Rochdale and Littleborough, turn on to Wardle Road to Wardle Village (signposted Watergrove). Follow the road north through the village until it comes to a dead end in the car park below the dam at SD911176, shark.clear.bounty.

YOUR VISIT
There is a large free car park, toilets, bird hide and (separate) feeding station but there can be some disturbance on the reservoir from sailing boats and the site can be popular with walkers. Many of the paths, particularly around the dam and leading to the bird hide and feeding station, are suitable for wheelchair users but elsewhere they can be poorly surfaced, and gradients can be steep.

82 LIGHT HAZZLES, WHITEHOLME AND WARLAND RESERVOIRS

Light Hazzles, Whiteholme and Warland Reservoirs
Halifax Road, Littleborough, Rochdale, OL15 0LG
SD969179
palettes.flatten.weeds

Opening times: Access to all reservoirs is always available
Parking: Only possible in the small free car park just off Halifax Road, immediately west of The White House pub at milkman.cool.coil
OS Landranger Map 109, OS Explorer Map OL12

Light Hazzles Reservoir
SD962198
beans.dolly.cocktail

Whiteholme Reservoir
SD965196
newsreel.prove.manly

Warland Reservoir
SD957208
lightly.mentions.suiting

Peithorne Valley
Ogden Lane, Higher Ogden, Milnrow, Rochdale, OL16 3TQ
SD962123
companies.listings.grand

Opening times: Accessible at all times
Parking: Roadside parking possible along Ogden Lane
SBI

Ogden Reservoir
SD952122
encourage.porridge.reckoned

Kitcliffe Reservoir
SD962123
companies.listings.grand

Piethorne Reservoir
SD966123
sunset.sorters.drags

These moorland reservoirs are on the Pennine Way, 400m above sea level. Feeding stations were established for Twite at the first two waters but these no longer exist, although birds still breed in the vicinity. The feeding stations also used to attract small numbers of Snow Buntings.

SPECIES
Peregrine Falcon and Raven visit frequently while the reservoir edges attract passage waders such as Ringed Plover, Sanderling, Turnstone, Whimbrel and Dunlin. Whooper Swans can occasionally be found in winter and early spring/late autumn, as can Common Scoter. Scarcities and rarities have included Baird's Sandpiper, Long-tailed Duck, Caspian Gull and Lapland Bunting. Breeding birds in the area include Wheatear, Common Sandpiper and Red Grouse.

ACCESS
From the A58 Halifax Road, park at The White House pub and take the Pennine Way west by Blackstone Edge Reservoir.

Nearby site: Piethorne Valley
A series of three upland reservoirs (Ogden, Kitcliffe and Piethorne Reservoirs) in the Piethorne Valley. Access is along Ogden Lane from the A640 about 1km east of Newhey. Breeding birds include Stonechat, Common Sandpiper, Tufted Duck and Great Crested Grebe. Rarities have included Brent and White-fronted Geese, Honey Buzzard and Firecrest.

83 HOLLINGWORTH LAKE

Hollingworth Lake
Rakewood Road, Littleborough, OL15 0AQ
Website: https://www.visitrochdale.com/things-to-do/
hollingworth-lake-country-park-p85031
SD938151
wiser.heat.editor

Greater Manchester

Opening times: Access around the lake is always available
Parking: There is a large pay and display car park (free to blue badge holders) at finely.vibes.unique. There is a visitor centre also at the latter location which is open 10am to 3pm all days other than Wednesday and Thursday.
OS Landranger Map 109, OS Explorer Map 277
SBI

Originally built as a feeder for the Rochdale Canal, this large reservoir is ideally placed to attract cross-country migrants, particularly with its very close proximity to the Pennines. The nature reserve area at the south-east corner has a hide and some wooded areas. Beyond this to the south and east of the lake are more open areas with marshes, scrubland and mature hedgerows.

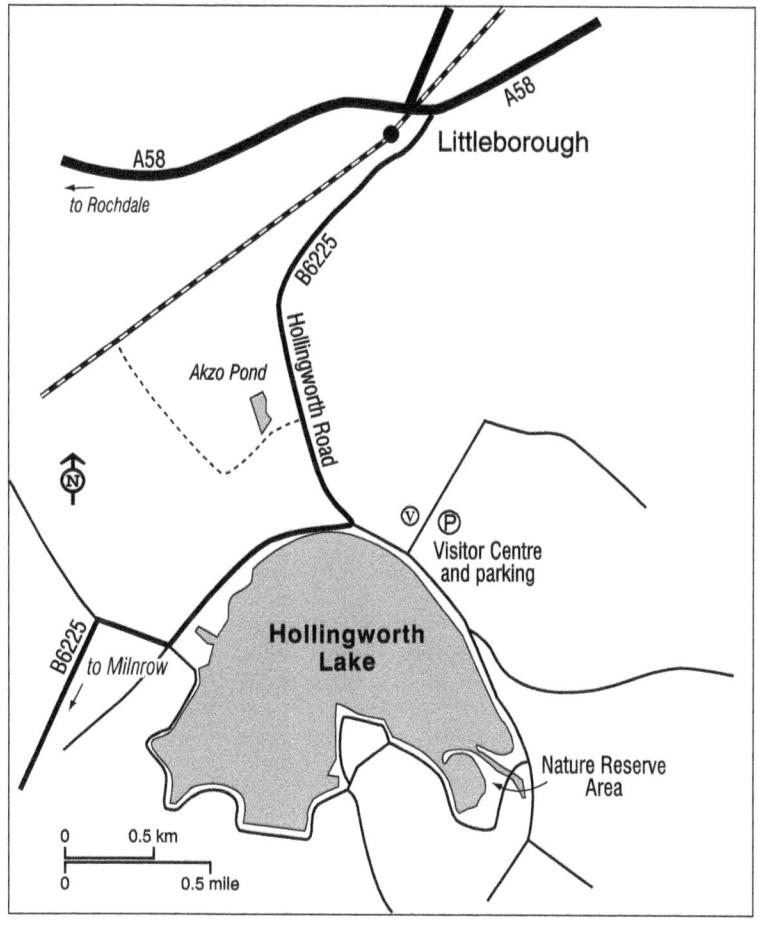

253

SPECIES

Winter: wildfowl such as Wigeon, Goosander and Goldeneye are regular as are Whooper Swans though their occurrence is much less predictable. The hedgerows hold winter thrushes, Kingfishers are often seen from the hide and Iceland Gulls occasionally drop in. Once, a very obliging juvenile Glaucous Gull frequented the shore by the Adventure and Water Activity Centre.

Passage: there can be some mud in the nature reserve area and along the exposed shoreline around the lake if water levels are low enough. In general, its fairly steep sides are not ideal for waders though both Collared Pratincole and Black-winged Stilt have occurred. Other species such as Sanderling and Turnstone take advantage when there is some shore. The passage periods see Common Scoter and Wigeon, terns (especially Sandwich and Arctic Terns) and annual Kittiwakes favouring early spring with the lake being one of the most productive sites in the county for the latter species. The woodlands and hedgerows can attract Redstarts and Spotted Flycatchers, particularly in late summer and early autumn, and in recent years late summer has also seen increasing records of Yellow-legged and Caspian Gulls.

Rarities and scarcities have included Pied-billed Grebe, Cattle Egret, Glossy Ibis, Collared Pratincole, Black-winged Stilt, Pectoral Sandpiper and Hawfinch.

ACCESS

From junction 21 of the M62, turn north following signs to Milnrow; turn right at the lights then left at a roundabout on to Kiln Lane (B6225). Look out for the sign to Hollingworth Lake (Wild House Lane) on the left after 300m. At a roundabout, turn right onto Lake Bank. Follow the edge of the lake, bearing right at the junction with Hollingworth Road and park at the visitor centre (SD939152, lined.turned.cove).

YOUR VISIT

There is a visitor centre with toilets and a large pay and display car park. All the paths are suitable for wheelchair users. The site is very popular on weekends and bank holidays and parking can often become an issue, as can disturbance on the water from water-sports enthusiasts.

84 STALYBRIDGE COUNTRY PARK, BRUSHES VALLEY

Stalybridge Country Park
Hartley Street, Brushes, Millbrook, SK15 3FH
SK002994
devours.equity.puns

Opening times: The entire area is accessible at all times
Parking: Limited roadside parking is available at frizz.income.feuds.
OS Landranger Map 109, OS Explorer Map OL1
SBI

Walkerwood Reservoir
SJ986991
query.cure.forge

Brushes Reservoir
SJ993991
ripe.chum.boasted

Lower Swineshaw Reservoir
SK002991
crew.magnum.laminate

Upper Swineshaw Reservoir
SK003995
applause.cascaded.blatantly

Set in the Brushes Valley, Stalybridge Country Park contains wooded valleys, four reservoirs and open, heather-covered moorland. There is a variety of footpaths and tracks within the country park, which means you can make routes as short or as long as you like. The views can be impressive, especially out on the open, expansive moorland.

SPECIES

The mill pond within the Lower Brushes Valley has Kingfishers but the walk along Brushes Road around Brushes Reservoir and beyond is most productive, particularly in spring and early summer when the area has a very healthy population of Willow Warblers, Redpolls and Bullfinches, a few Garden Warblers and sometimes Green Woodpecker and singing Wood Warbler. Early spring also brings Wheatears, which probably still breed out on the moorland, and occasional records of Ring Ouzel and Osprey; the latter have fished in the reservoirs. Curlews are present and breed, as do Common Sandpipers around Swineshaw Lower and Upper Reservoirs, Stonechats and Red Grouse out on the moors and Cuckoos in the upper valley, most often above the Gamekeeper's Cottage. Ravens patrol the moorland, as do occasional Hen Harriers and Merlins on passage, while the reservoirs can attract wildfowl, including Wigeon and Whooper Swan in late autumn and winter.

ACCESS

There is limited parking along Hartley Street, off the B6175 Huddersfield Road in Millbrook, Stalybridge (at SJ978994, frizz.income.feuds). There are many paths from here around the Lower Brushes Valley, all of which are wheelchair accessible. Follow the paths to Brushes Road, which continues east along the northern side of Walker Wood Reservoir and Brushes Reservoir and up the valley where the gradient can become fairly steep. Eventually the road turns northeast and passes Lower Swineshaw and then Upper Swineshaw Reservoirs and out onto open moorland.

85 DOVE STONE

Dove Stone
Holmfirth Road, Greenfield, Oldham, OL3 7NE
SE015032
jungle.conqueror.ringside

Website: www.rspb.org.uk/days-out/reserves/dove-stone
Phone: 01457 819880
Opening times: Access to all areas is always available
Parking: Always available. The large main pay and display car park is located at diamonds.giants.certainly and the smaller pay and display car park at Binn Green is located at powerful.televise.blink. This site is very popular at weekends and particularly bank holidays, when all car parks are often full.
OS Landranger Map 110, OS Explorer Map OL1

Yeoman Hey Reservoir
SE019045
envy.dreamers.fools

Chew Reservoir
SE035017
procured.claim.chicken

Set within the Peak District National Park, Dove Stone affords a dramatic and impressive landscape. Owned by United Utilities but managed by the RSPB, it is amongst the very best upland habitats in the county.

There is a 4km walk around the edge of the Dove Stone Reservoir, and a longer walk on rougher terrain can be found around Yeoman Hey and Greenfield Reservoirs. For truly spectacular views there is a 2.4km walk to Chew Reservoir. Above Dove Stone Reservoir there is internationally important blanket bog, and though most of the conifer plantations around Yeoman Hey Reservoir (once the best location in the county for Crossbills and boasting the only county record of Two-barred Crossbill) have now been felled, they have been replaced with deciduous plantations.

SPECIES

Winter: there are feeders close to the car park at Binn Green, attracting a good variety of woodland species including Siskin and Brambling. Stonechats remain in the area, Goosanders can be found on the reservoirs and Dippers on the smaller brooks, while Red Grouse are often found in good numbers, especially on the way up to Chew Reservoir. Crossbills still remain in fluctuating numbers and are best looked and listened for around Binn Green.

Passage and summer: spring sees the return of Wheatears and Ring Ouzels, both of which remain to breed in small numbers. The latter can often be found on the valley sides on the walk up to Chew Reservoir. Common Sandpipers are

Greater Manchester

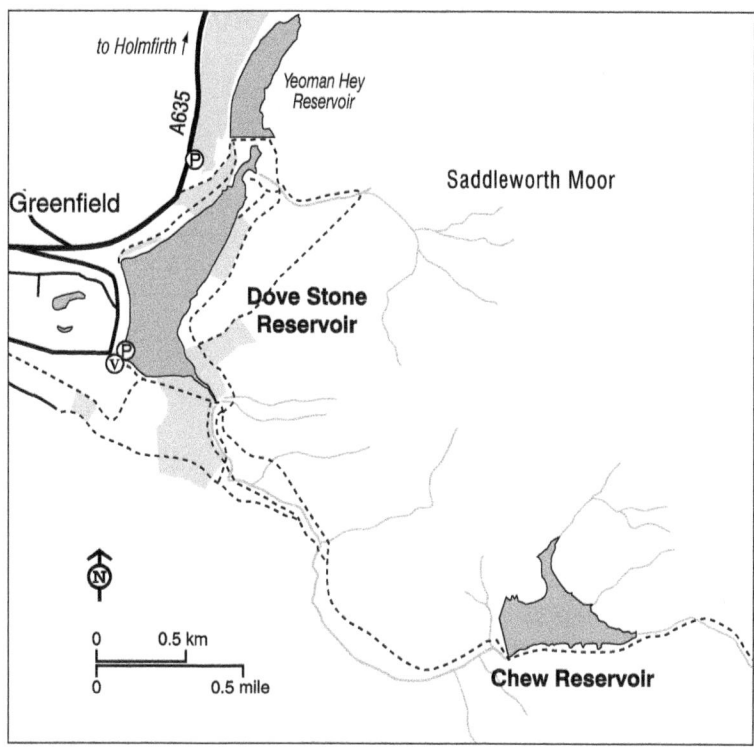

present around the shorelines and still probably breed. Willow Warbler and Chiffchaff inhabit the woodlands with occasional singing Wood Warblers. Redstart no longer breed but still occasionally visit the wider area. The blanket bog benefits Curlew, Golden Plover, Red Grouse and Dunlin as breeding birds, Red Kites are occasional visitors and Raven and Peregrine are both residents.

ACCESS
Drive up the A635 towards Holmfirth, following the brown signs for Dove Stone. The main car park is a little way up the hill on the right at SE013034, diamonds.giants.certainly, with the smaller car park at Binn Green further up the hill, also on the right at SE017044, powerful.televise.blink.

YOUR VISIT
The large pay and display car park is always open; RSPB members can park for free with their membership card on display in the windscreen. During weekends and bank holidays the area can be particularly busy, especially on warm, sunny days, so is best avoided then. There is also a smaller RSPB car park at Binn Green, off the A635 Greenfield to Holmfirth Road. The site has some tarmacked paths suitable for wheelchair users, but others are part gravel surfaced and have steep gradients in some places.

86 AUDENSHAW RESERVOIRS

Audenshaw Reservoirs
Audenshaw Road, Audenshaw, M34 5PH
SJ914964
whites.fuels.formed

Permits: Unfortunately, this site requires a permit and key to gain access, both of which are not readily available. Furthermore, no part of the reservoirs is viewable from outside the perimeter fence.
OS Landranger Map 109, OS Explorer Map 277
SBI

Redgate Recycling
Redgate Lane, Belle Vue, M12 4RY
SJ867965
bucket.prep.serve

Audenshaw comprises three late nineteenth-century stone-lined balancing reservoirs with a total area, including verges, of around 100ha, crossed by intersecting paths. It is easily the largest water area in Greater Manchester but is accessible by permit only, along with a key to the locked gate. Unfortunately, permits are limited, and the waiting list is very long. As the reservoirs are slightly elevated, they are not viewable from outside the substantial perimeter fence. The site is a great attraction for migrants crossing the Pennines and number one reservoir's low water levels and small islands are particularly attractive to waders and wildfowl.

SPECIES

The current site list stands at 224, including many remarkable rarities, and has proven that it is possible to find almost anything there. The major interest is its passage migrants, particularly waders and terns in the last fortnight in April and first two weeks of May, with return passage from July to September. Winter also holds good value, with the occasional diver or rarer grebe plus gulls (including a nightly roost), which regularly include Yellow-legged, Caspian and occasionally Iceland Gulls. The stone-lined edges of the reservoirs are the best location in the county for migrant Rock Pipits, primarily in March and October, whilst the change in water levels and vegetation of number three reservoir now attracts excellent wildfowl numbers. These often include good numbers of Wigeon and Pintail, the latter usually never easy birds to see in the county.

Rarities and scarcities have included Green-winged Teal, White-billed Diver, Kentish Plover, Pectoral Sandpiper, Red-necked Phalarope, Ring-billed and Sabine's Gulls; Alpine Swift, Wryneck, Shorelark, Iberian Wagtail, Red-rumped Swallow and Lapland Bunting.

NEARBY SITE: REDGATE RECYCLING

This relatively small recycling centre lies just south-east of Manchester City Centre in Gorton. It is located off the A57 Hyde Road along Redgate Lane (SJ867965, bucket.prep.serve) and has become a very productive and easy location to connect with gull species, particularly Yellow-legged, Caspian and, often, Iceland Gull. The site is a very busy area with many visiting refuse trucks so care should be taken when parking along Redgate Lane. Gulls can often be seen on the refuse within the yard itself or resting on the roofs of the many surrounding industrial units.

Yellow-legged Gulls can occur all year round, including summer, but for Caspian Gull, late autumn and winter are the most productive periods, winter being best for Iceland Gull.

87 ETHEROW COUNTRY PARK

George Street, Compstall, Stockport, SK6 5JD
SJ965907
newsreel.like.sooner

Website: www.stockport.gov.uk/etherow-country-park
Opening times: Access around the country park is always available
Parking: There is a large pay and display car park on site at asked.prop.magically.
OS Landranger Map 109, OS Explorer Map OL1
SBI

Keg Pool
SJ974916
barman.flattery.factories

Ernocroft Wood
SK974911
drifters.begun.hazel

Once part of the estate of George Andrew, a nineteenth-century cotton magnate, this local nature reserve consists of Compstall Lodges (once part of Compstall Mill), riverine woodland, the Keg Pool and a number of mature woods, some of which are amongst the best in the county. Paths are suitable for wheelchair users, and whilst the area around the lodges can be busy with visitors and water-sports enthusiasts at weekends in particular, the paths beyond the weir are usually quieter.

SPECIES

Mute Swan and Great Crested Grebe breed alongside introduced waterfowl, though Mandarin Duck has colonised naturally along the river valley and is present in good numbers. Beyond the lodges, a selection of paths leads into various mature woodlands alongside the River Etherow, where the large weir has breeding Dippers. The Keg Woodland was well known for breeding Pied Flycatcher; although these have declined dramatically in recent years, they can still be found. Ernocroft Wood often holds singing Wood Warbler and occasional Redstart, but it is likely neither breeds any more. Breeding birds include Woodcock, Spotted

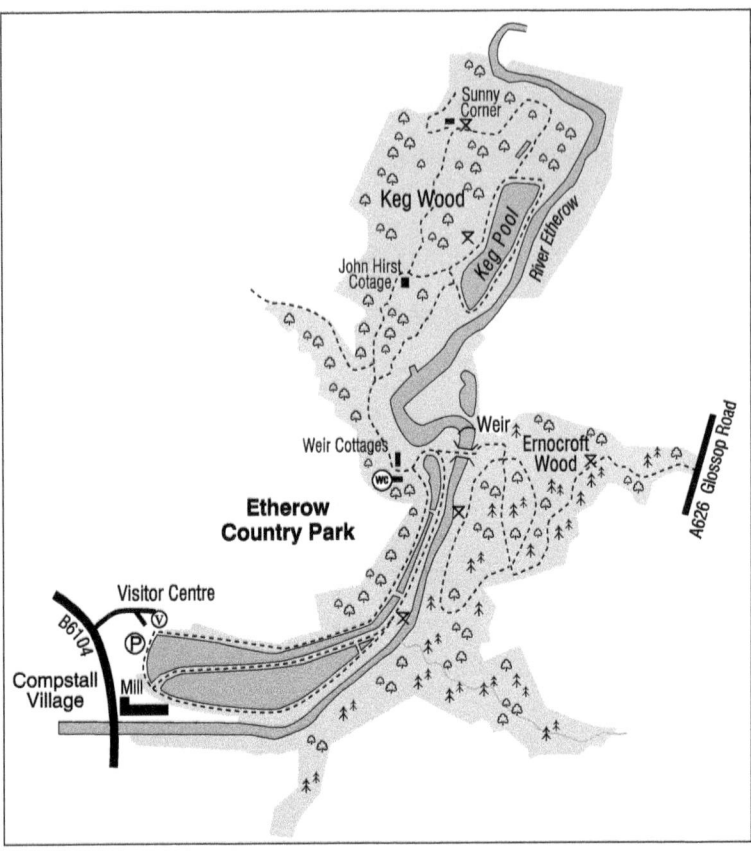

Flycatcher in some years, Tawny Owl and Garden Warbler among the more common woodland species. Kingfisher can frequently be seen from the hide by the river, just beyond the weir, whilst the Keg Pool attracts Goosander and Tufted Duck. Common Buzzard, Raven and Sparrowhawk are relatively common, and winter brings with it flocks of Siskin and Brambling.

ACCESS

Etherow Country Park is situated at Compstall on the B6104 between Romiley and Marple Bridge and is well signposted from the road. There is a large pay and display car park and visitor centre fronting on to Compstall Lodges at SJ965908, stole.redeeming.shorten. The visitor centre is open daily from 9am to 4.30pm and has toilets with wheelchair access and a café.

SEAWATCHING IN THE NORTH-WEST

The shallow waters of the Irish Sea mean that records of large shearwaters and scarcer tubenoses are much less frequent than in the south-west or even on the east coast. However, in the right conditions, our coastline can become the best place in the world to see Leach's Petrels from land, with birders travelling from all over the country and even further afield to see them. Very specific conditions bring petrels to Lancashire and Cheshire and their appearance can be predicted with a fair degree of accuracy in terms of timing and location.

The Leach's Petrels that appear on our coasts breed on remote islands in the North Atlantic, possibly from as far away as Newfoundland and Nova Scotia but more likely Iceland and the Faroes, or closer to home in Scotland. St Kilda, the Shiant Islands and North Rona all have significant colonies. From early September post-breeding birds migrate slowly south to tropical and subtropical waters of the Atlantic Ocean, particularly off the coasts of West Africa and South America, where they forage over the open ocean. Though perfectly adapted for a pelagic existence, they tend to drift according to the prevailing winds but also keep out of sight of land and diurnal avian predators such as large gulls.

Ordinarily most birds pass west of Ireland, but gales from anywhere in the westerly quadrant or just a sustained period of moderate westerlies can hold birds close to the Scottish coast and feed birds into the Irish Sea. Land sightings from the Outer Hebrides, and especially anywhere on the Ayrshire coast, are a sure sign that this is happening. Once in the Irish Sea, birds will continue south, just out of sight of land.

It is received wisdom that you need three days of north-westerly gales in September to produce Leach's Petrels, and whilst this will almost certainly result in petrels, the conditions are more nuanced than this. Once in the Irish Sea, birds tend to be closer to land when facing a headwind and this is almost as important as an onshore wind, so 45 degrees to the coastline is optimal. Very strong onshore winds will push birds against the shore, but bad visibility has a similar effect, and squally weather can be the equivalent of two or more extra points on the Beaufort scale. Similarly, the birds are closer to shore at night, so first light can be a good time to see birds. The peak time for Leach's Petrels is mid-September, but they can occur from late August to the start of November. In exceptional conditions, after sustained strong winds from the south, they can even be seen in winter. It usually takes two days from the event that first pushes petrels into the Irish Sea for them to reach far enough south to appear on our coasts.

Once birds are in the Irish Sea, the best location to see them from land will vary according to wind direction and visibility and, in areas with extensive mud or sand flats, the state of the tide. The best wind direction for the bulk of the Fylde is south-westerly, while for the Wirral coast it is westerlies that result in larger counts and closer sightings. Birds following the coast at night find themselves up the Wrye or Mersey rivers and be seen exiting the estuaries at Fleetwood or New Brighton respectively. If we can see a coastal promontory, so can the birds, and they will veer out to sea to avoid them. This means that it requires stronger winds to bring birds into the Sefton coast if Wirral is in view and to Wirral if, for example, North Wales is visible. In prolonged periods of very high winds birds will shelter in estuaries. Then, Fleetwood/Knott End and especially New Brighton/Crosby will come into their own.

On Wirral there are a number of locations that offer protection from the weather. New Brighton (site 24) has a number of shelters that become popular in good conditions as well as places from which scanning from a car pays dividends. Tide state is less important here than at sites along the rest of the north Wirral coast but unless the wind is force 7 or above, or the visibility is poor, most birds will 'cut the corner' and not pass close to the coast. Some shelter can be afforded behind the coastguards' hut at Harrison Drive, Wallasey, or from a car at the Derby Pool. If the tide is high enough the Gunsite (site 100) is probably the most popular location. It is possible to bring a car to the edge of the seawall and obtain very close views of passing petrels.

Dovepoint at Meols (site 93) is another place where seawatching from a car is possible. Petrels tend to be more distant there, but there is slightly more chance of seeing stronger-flying species which have a tendency to 'cut the corner'. Hilbre (site 20), though a long walk and a major time commitment if staying over the tide, is much less tide-dependent than other north Wirral sites.

In Lancashire the best sites depend on the precise wind direction. In sustained south-westerlies petrels can get pushed right into Morecambe Bay and will then struggle past the headlands of Jenny Brown's Point, Morecambe Stone Jetty and Heysham. The last is probably the most practical site in storms as viewing from a car on the North Harbour Wall is possible. On the Fylde coast Rossall Point is the best viewing point in the same conditions; when the gales are straight westerlies anywhere along the straight coastline can be productive and in Blackpool as well as Starr Gate the Middle Walk off the north Promenade gives observers more shelter. South of the Ribble, Formby Point can be productive in any onshore wind direction, but it is very exposed. Other options where watching from a vehicle may be an option to include Ainsdale. The Mersey estuary is shared with Cheshire so advice for watching from sites including Crosby and Seaforth is similar to the paragraphs above.

Other species are brought in by the same conditions as Leach's Petrels, notably Sabine's Gulls, terns, Grey Phalaropes and Long-tailed Skuas. Stronger flying species, including shearwaters, larger skuas, Kittiwakes, Fulmars and Gannets, are a little less likely to enter the estuaries unless it is very windy, very squally or dark. That said, skuas may appear to parasitise or predate petrels and wrecked auks.

From midsummer, westerly gales may bring Storm Petrels – which breed on the Welsh islands of Skomer, Skokholm and Bardsey, and the Treshnish Isles of Argyll and feed in the Irish Sea in small numbers – close to our coasts. 'Stormies' used to be a reasonably regular occurrence in the right conditions, with some lingering off the power station outfalls at Heysham, but they are an increasingly prized find in north-west England. Late summer and early autumn can also be productive for off-passage Arctic Skuas, with several birds sometimes lingering where they can harry feeding terns.

Spring seabird passage can also be rewarding with skuas and terns, especially Arctic Terns, passing north. Wirral tends to be in the shadow of the Lleyn Peninsula and Anglesey, but the north Lancashire coast can be excellent. Terns pass off the Fylde and sites in Morecambe Bay before being seen to gain height and head inland on a 'short cut' migration. This is rather less dependent on wind direction than autumn seabirds in the region and can actually be better after south-easterlies have pushed birds into the Irish Sea. Birds move more in the early morning and the light conditions are also best at this time. For Pomarine Skuas, experience suggests that they are less likely than other skuas to pass up the Fylde Coast, and most come into Morecambe Bay from offshore before gaining height like the terns and heading inland. Rossall Point and Heysham are therefore the best sites for a vigil to try and get a 'Pom' on your seawatching tally; sighting a brute of an adult or a small group resplendent with 'spoon' tail-feathers is always a highlight.

Whilst some conditions will more or less guarantee excellent seawatching off our coasts, things are not always so clear-cut. Small movements in wind direction can change which sites come to the fore, and the frequency of squally showers, for example, can influence how close to shore birds will come. It is therefore well worth being on local WhatsApp groups for bird news. For example, when Leach's Petrels are blown into Morecambe Bay regular news will be put out in real time on the Lancaster and Fylde news groups. This also worked in late March 2025 when Lancashire's third Forster's Tern was seen off Morecambe then twitched by many after it pitched up at Skippool Creek. Birds such as this, Fea's Petrel, Wilson's Petrel, Barolo Shearwater and Elegant Tern show just what can be seen in our area – but even without the 'megas' there is a good variety of seabirds to be enjoyed all year round.

TOP SITES FOR ACCESSIBILITY

CHESHIRE
1. Burton Mere: there is a wheelchair-friendly visitor centre and routes are suitable for visitors with limited mobility.
2. New Brighton: the mouth of the Mersey is viewable from the promenade, including excellent seawatching in strong autumn gales.
3. Parkgate Marsh: disabled parking is available and the marsh can be viewed from next to the car park.
4. West Kirby Marine Lake: wheelchair-friendly paths run around the lake, and beach wheelchair hire is available (booking in advance required).

LANCASHIRE
1. Grimsargh Reservoirs: viewing screens are designed to allow easy access for wheelchair users.
2. Martin Mere WWT: there is a wheelchair-friendly visitor centre and waterfowl gardens, the main paths are generally very accessible and hides have low or large windows for viewing.
3. Mere Sands Wood, Burscough: almost all the paths are wheelchair-friendly, there are six accessible hides, a viewing platform and visitor centre.
4. Rossall Point, Fleetwood: the off-road accessible promenade is close to wader roosts, and beach wheelchair hire is available.
5. Stocks Reservoir/Gisburn Forest, Slaidburn: an easy-access trail runs between both birdwatching sites and all-terrain mobile scooter hire is available.
6. Fairhaven Lake and Ribble Estuary: good viewing of the Ribble Estuary is possible from an elevated car park with disabled spaces, and there are accessible lakeside paths to view waterfowl and gulls on the marine lake.

GREATER MANCHESTER
1. Chorlton Water Park: well-surfaced paths are accessible for wheelchair users. A number of viewing points around the lake are suitable for those with limited mobility.
2. Pennington Flash: there are six hides, a wildlife-themed café and toilets, all of which are disabled accessible; a large play area includes specialist inclusive play equipment.

TOP SITES FOR PUBLIC TRANSPORT

CHESHIRE
1. Frodsham: Frodsham station is served by regular trains from Chester, Liverpool and Manchester and is a 1km walk to Frodsham Marsh.
2. Leasowe: Leasowe station is a 1km walk from the main birding sites.
3. New Brighton: the train station is a 500m walk from the main lake, breakwaters and views of the Mersey estuary mouth.
4. Tatton Park: a convenient walking distance from Knutsford town centre, which has a train station and numerous bus services.
5. West Kirby: the train station here is very convenient for the marine lake; from here it is also possible to walk out to Hilbre Island at the right state of the tide.

LANCASHIRE
1. Leighton Moss: Silverdale train station is adjacent, with reduced entry to the reserve for those arriving by train.
2. Crosby Marina and beach: a short walk from Waterloo rail station on the Liverpool-Southport line, with regular buses to the area.
3. Fleetwood area: the various sites around Fleetwood are generally easily accessed from several bus routes (particularly the 1 and 14 routes) and regular trams from Blackpool.
4. Starr Gate/St Anne's Beach: regular trams connect to a frequent rail service at Blackpool North Station; alternatively from the Blackpool South line, alight at Squires Gate station.
5. Marshside RSPB, Southport: regular bus services on several routes stop close by, including at Marshside Road and Crossens.

GREATER MANCHESTER
1. Pennington Flash: there is a designated Pennington Flash bus stop at Lowton Common, which is served by several routes.
2. Sale Water Park: served by the Sale Water Park tram stop on the Manchester Airport Line.

THIRTY SPECIES TO SEE IN NORTHWEST ENGLAND

STARLING
There are still spectacular murmurations to be enjoyed in the region, including at Blackpool North Pier, Leighton Moss, Martin Mere and Neumann's Flash. As well as being a stunning sight in their own right, these can also attract predators including Peregrines. They often also follow regular flightlines to these roost sites, and wave after wave of birds can be a good watch even without the swirling patterns of the evening roosts.

SNOW BUNTING
Snow Buntings are equally at home in winter on our sandy beaches and on upland fells. Many birders will be familiar with them feeding on seeds and the like on the shore at places including Carnforth, Hoylake, Red Rocks, Rossall Point, St Annes and West Kirby. Fewer birdwatchers will seek them out at upland sites where conditions are generally less hospitable, but getting a 'blizzard' of these charming birds to yourself is a treat.

WHOOPER SWAN
Alongside the Pink-footed Goose this is one of the most conspicuous winter wildfowl visitors to the region. Martin Mere has been a stronghold for many years with birds showing well at organised feeds, whilst other large flocks field feed in north Lancashire. Although inland wintering birds are now a thing of the past, birds can be seen in spring as they pass over east Lancashire and Greater Manchester en route to Iceland from the Ouse Washes.

HERRING GULL
Although other gulls can be seen in urban environments the Herring Gull is the archetypal bird of the seaside resorts of the north-west, increasingly nesting in these settings as well, even though numbers are declining and the species is on the Red List. The species does of course also occur inland in numbers and is the main species in the throngs at Whinney Hill, Accrington and roosts on nearby reservoirs, particularly Fishmoor and Lower Rivington.

MARSH HARRIER
The Marsh Harrier has colonised the area and goes from strength to strength. In the summer it is a familiar sight as it quarters the reedbeds at sites including Burton Mere, Leighton Moss and Marton Mere. In winter those that don't depart for warmer climes tend to be more coastal, occurring especially on estuarine saltmarshes, with several on the Ribble and Dee estuaries, sometimes in the same scope view as their Hen Harrier relatives.

COMMON SCOTER

Flocks upward of 10,000 birds are regularly seen in winter off both the Cheshire and Lancashire coastlines. Although many stay quite far offshore the large flocks in flight are still worth seeing. Birds are regular on overland passage at many freshwater bodies in east Lancashire and Greater Manchester. With the advent of 'noc-mig' recording even more birders inland are able to share in the experience of recording them as they migrate at night.

BLACK-TAILED GODWIT

The continental race *limosa* nests in very small numbers at the northern extremity of its range. More familiar is the *islandica* race which occurs in a range of coastal and inland habitats. Assemblies can be several hundred strong inland, particularly on floods on passage, whilst thousands may flock together on the Dee and Ribble estuaries. Checking them for colour rings can yield interesting life stories spanning several years.

PINK-FOOTED GOOSE

Lancashire in particular is one of the world's wintering strongholds for this species, but birds increasingly use sites in Cheshire also. Birds can be seen inland as they migrate to and from other winter staging posts in Norfolk and Lincolnshire. The flocks can often harbour other more scare species and a 'wild goose chase' can be very rewarding, but the spectacle of birds leaving nocturnal roost sites such as Martin Mere or Pilling Marsh is also something to behold.

LEACH'S PETREL

Leasowe and New Brighton are among the best places in Britain to see storm-driven Leach's Petrels as they struggle out of the Mersey mouth in north-westerly gales. Birds are also regular in strong onshore winds from Heysham down the Lancashire coast and off Hilbre island. In the strongest storms birds can pass over the beach rather than the surf within feet of birders and photographers, though they are very rare further inland.

WIGEON

The Ribble Estuary is the most favoured site for Wigeon in the United Kingdom, with often more than 50,000 birds in total and the marshes ringing out with the whistling calls of the drakes. Birds are also regular in good numbers on the Dee Estuary and in Morecambe Bay. Impressive as they are in their own right, these flocks often attract an American Wigeon, but beware of hybrids of the two species.

BITTERN

The secretive Bittern has nested at Leighton Moss for many years, but for many people their only experience is hearing male birds' 'booming' calls resonating over long distances. Recently birds also began nesting at RSPB Burton Mere. They also enliven winter birding at several other sites, including Martin Mere, Marton Mere and Pennington Flash. At some locations fortunate observers have seen them picking off Starlings coming in to roost in reedbeds.

RED-THROATED DIVER

This is the default north-west England diver species, with Great Northern fairly scarce and Black-throated usually only passing through in spring and autumn. Counts of over 300 have been made on seawatches, and though numbers have dropped a little, three-figure totals still do occur. They often lift their head when flying to confirm the ID. Occasionally birds will take refuge on marine lakes where they are a popular attraction for photographers.

RED GROUSE

The Red Grouse populations of Cheshire, Greater Manchester and Lancashire are not without controversy among birdwatchers, as management to generate sufficient numbers to shoot can include persecution of other species. That said, there is probably no sound more evocative of the upland heather moors than the *'go back go back go back'* calls as a grouse covey heads away from the birdwatcher or fellwalker to a place of perceived safety.

WILLOW TIT

Sadly, the Willow Tit is currently Britain's fastest declining resident bird, with the population having reduced by a sobering 95 per cent since the 1970s. There remain reasonable numbers in the Wigan flashes area, whilst the Cheshire strongholds including Woolston Eyes have recently received lottery funding to sustain the population there. Hopefully in a fourth edition of this book it will still be possible to give sites for this species in the north-west.

WATER PIPIT

Water Pipits were formerly associated with watercress beds and similar habitat in the south of England. They have adapted to a number of habitats in the north-west, including saltmarsh on several estuaries such as Marshside and Warton Bank on the Ribble, the Wyre Estuary and Neston Old Quay on the Dee. They occasionally occur in Greater Manchester, favouring sewage works filter beds including, particularly in recent times, Daisy Hill at Westhoughton.

WHIMBREL

Whimbrel don't breed in the area and only winter very rarely, but there are nationally important numbers on spring passage in Lancashire in particular. Flocks several hundred strong will roost at sites such as Barnacre Reservoir and Longton Marsh after feeding during the day in fields. Smaller numbers occur throughout the region, and the seven-note whistles of birds in a flock are familiar in a range of habitats and as birds pass close by offshore.

RING OUZEL

The 'upland Blackbird' is a treasured breeding species in Cheshire, Lancashire and Greater Manchester. It favours gullies and cloughs in our uplands but is declining in all three counties in line with a national trend that has seen them go on the Red List of birds of concern. On the coast it is a prized find among birders looking for migrants and tends to be more numerous (though still very scarce) in spring compared to autumn.

BEARDED TIT

Leighton Moss in Lancashire was until recently the only breeding site in the area for this species, and many birders would make an early autumn visit there to see them coming to the grit trays specifically provided for the species. Birds have also taken to reedbeds created at Burton Mere and Martin Mere, so the future is optimistic for them locally. In winter, irruptions can bring them to several other smaller reedbeds.

BLACK REDSTART

Black Redstarts nest at a small number of sites in central Manchester, reflecting their general preference for urban sites in the country following recolonisation after the Blitz. They also nest periodically in other towns and cities but are most familiar in the region as a passage migrant and winter visitor. They can be seen from rural locations such as Pendle Hill to coastal defences and breakwaters. Wintering birds may be very faithful to one site.

LITTLE GULL

The world's smallest gull used to winter in reasonable numbers in the Irish Sea off Cheshire and Lancashire, and three-figure flocks could be seen in storms. They also appeared in similar-sized groups at Seaforth Docks in spring. Although numbers have dropped, this species remains a potential winter seawatch highlight. In spring and autumn birds can drop in at reservoirs, particularly during showers, and they also linger at floods whilst resting on passage.

BARN OWL

Although the Barn Owl is a Schedule One breeding bird it occurs throughout the region in small numbers. Whilst the most reliable observation opportunities are on nature reserves including Burton Mere, Martin Mere and Marton Mere, they can also be observed hunting over less intensively farmed land including in particular Over Wyre and south-west Lancashire. Always observe birds away from nest-sites and take care not to disturb their feeding efforts.

GREY WAGTAIL

Really attractive birds with their subtle blends of yellows and blue-greys, Grey Wagtails have adapted somewhat to become denizens of urban areas with water as well as upland streams. They can therefore be seen more or less throughout the region where there is any kind of fresh water. Check any birds in autumn and winter for colour rings as there is a long-term monitoring programme at Heysham of birds moving south in the autumn.

COMMON TERN

There is a scatter of Common Tern colonies across the region. Their distribution is assisted by the presence of nesting rafts at several sites. Seaforth, Preston Dock and Pennington Flash are perhaps best known, but very good views can be had at Conder Pool from a lay-by off the Glasson Dock road. They are also a staple of seawatches from the Cheshire and Lancashire coast where care is needed to separate them from Arctic Terns.

KNOT

Although the Knot is generally scarce inland it is included in this Top 30 because of the spectacular flocks that gather on the coast. Thousands of birds form tight wheeling flocks, agile enough to be mistaken for Starling murmurations. Whilst they are predominantly winter visitors, non-breeding birds will sometimes summer on our estuaries. Morecambe Bay, the Dee, Alt and Ribble can all hold internationally important numbers of Knot.

STONECHAT

Stonechats have a preference for gorse but still nest widely in the area, with birds inland on hillsides and on the coast in dune edge. Some migrate for the winter and their return is one of the first signs of spring, as birds appear in small groups and sometimes impressive totals can be recorded. Stonechats are quite tolerant of people, so they are a staple of guided bird walks in coastal areas where their Robin-like coloration and chacking calls are striking.

PIED FLYCATCHER

Along with Redstart and Wood Warbler this is one of the special birds of the inland native woodlands of the region. All three are presently in decline, but hopefully steps can be taken to address this. Pied Flycatcher is also an annual but prized find among coastal passerine migrants. Favoured haunts on passage include Leasowe in Cheshire and the Fleetwood and Heysham areas of Lancashire.

AVOCET

When the first edition of this guide was produced it would have been fanciful to foresee the charismatic Avocet as anything other than a rare visitor. Colonisation has been speedy though, with major sites including Marshside, Burton Mere, Conder Pool, Martin Mere and the saltwater pools at Leighton Moss. Relatively large flocks can be seen on the estuaries before birds settle down to breed. They generally leave en masse in the winter months.

MEDITERRANEAN GULL

Another relatively recent coloniser, the Mediterranean Gull now nests among Black-headed Gull colonies at both coastal locations and inland (for example, at Belmont Reservoir on the West Pennine Moors). Relatively large assemblies occur at a number of sites, but particularly Heysham in Lancashire where high double figures can be present in late summer. Taking time to read the colour rings that many 'Meds' sport can often produce interesting life histories.

PEREGRINE FALCON

Peregrines nest in a number of very urban sites in the region, as well as still using more traditional quarries and cliffs. Feral pigeons form a significant part of the diet for town-dwelling Peregrines. They are a familiar sight as they terrorise waders on estuaries and even pick off exhausted Leach's Petrels in storms. Larger birds of potentially more northerly origin occur on occasion on coastal marshes.

BLACK-NECKED GREBE

North-west England is one of the UK strongholds for this attractive grebe, with sites in Greater Manchester and Woolston Eyes in Cheshire in particular holding around a quarter of all nesting pairs. They have also attempted to nest at Lunt Meadows in Lancashire. Outside the breeding season they are a prize find for local patch birders across the region, particularly as juveniles disperse from the natal sites to fend for themselves.

ONLINE RESOURCES AND CONTACTS

CHESHIRE

RECORDER
Jane Turner Recorder@Cawos.org

BIRD REPORT PRODUCTION
Cheshire and Wirral Ornithological Society: cawos.org/

WILDLIFE TRUST
cheshirewildlifetrust.org.uk/

BIRD CLUBS
Most, though not all, have sightings pages and produce local reports.
Wirral Bird Club: wirralbirdclub.com
South-East Cheshire Ornithological Society: secos.org.uk/
Mid-Cheshire Ornithological Society: midcheshireos.co.uk/
Knutsford Ornithological Society: 10x50.com/
Marbury report: foam.merseyforest.uk/

OTHER SIGHTINGS WEBSITES/BLOGS
Dee Estuary: up-to-date news at deeestuary.co.uk/

RESERVES
Woolston Eyes: woolstoneyes.com/
Risley Moss: rimag.org.uk/index.php/wildlife-resources/birds/
Burton Mere Wetlands RSPB: rspb.org.uk/days-out/reserves/dee-estuary-burton-mere-wetlands
Hilbre Bird Observatory: hilbrebirdobs.blogspot.com/

RSPB GROUPS
Chester: group.rspb.org.uk/chester
Macclesfield: group.rspb.org.uk/macclesfield
Wirral: group.rspb.org.uk/wirral/

LANCASHIRE

RECORDER
Steve White stevewhite102@btinternet.com

BIRD REPORT PRODUCTION
Lancashire and Cheshire Fauna Society: lacfs.org.uk

WILDLIFE TRUST
lancswt.org.uk

BIRD CLUBS
Most though not all have sightings pages and produce local reports.
Lancaster area: lancasterbirdwatching.org.uk
Blackpool and Fylde: fyldebirdclub.org
Preston: prestonsociety.co.uk
East Lancashire: eastlancsornithologists.org.uk
Chorley: chorleynats.org.uk/wpsite/
Blackburn: blackburnbirdclub.com

OTHER SIGHTINGS WEBSITES/BLOGS
Ainsdale and Southport: johndempseybirdblog.wordpress.com
Heysham Bird Observatory: heyshamobservatory.blogspot.com
Rossendale: rocforum.activeboard.com
Starr Gate: trektellen.nl/count/view/2486/20241116

RESERVES
Brockholes: lancswt.org.uk/nature-reserves/brockholes-nature-reserve
Leighton Moss: community.rspb.org.uk/placestovisit/leightonmoss/
Martin Mere: wwt.org.uk/wetland-centres/martin-mere
Mere Sands Wood: lancswt.org.uk/nature-reserves/mere-sands-wood-nature-reserve
Seaforth: lancswt.org.uk/nature-reserves/seaforth-nature-reserve

RSPB GROUPS
Lancaster: group.rspb.org.uk/lancaster/
Liverpool: group.rspb.org.uk/liverpool/
Southport: group.rspb.org.uk/southport/

GREATER MANCHESTER

RECORDER
Ian McKerchar ianmckerchar1@gmail.com

WILDLIFE TRUST
lancswt.org.uk

BIRD CLUBS
Leigh Ornithological Society: leighos.org.uk

OTHER SIGHTINGS WEBSITES/FORUMS
Manchester Birding: manchesterbirding.com/
Manchester Birding Forum: manchesterbirding.activeboard.com

RESERVES
Dove Stone: rspb.org.uk/days-out/reserves/dove-stone
Little Woolden Moss: lancswt.org.uk/nature-reserves/cadishead-little-woolden-moss
Pennington Flash: bewellwigan.org/location/attractions/pennington-flash/
Wigan Flashes: lancswt.org.uk/nature-reserves/wigan-flashes-local-nature-reserve

RSPB GROUPS
Bolton: group.rspb.org.uk/bolton/
High Peak: group.rspb.org.uk/highpeak/
Stockport: group.rspb.org.uk/stockport/
Wigan: group.rspb.org.uk/wigan/

BIBLIOGRAPHY

CHESHIRE

Conlin, A. & Williams, E. 2017. *Rare and Scarce Birds of Cheshire and Wirral.* Privately published.

Hardy, E. 1941. *The Birds of the Liverpool Area.* T Buncle & Co, Arbroath.

Norman, D. on behalf of CAWOS. 2008. *Birds in Cheshire and Wirral – A breeding and wintering atlas.* Liverpool University Press, Liverpool.

LANCASHIRE

Fylde Bird Club, McGough, M. and McGough, P. 2016. *Birds of Marton Mere (Second Edition).* Fylde Bird Club, Preston.

Lancashire and Cheshire Fauna Society, 2025. *Lancashire Bird Report 2024.* Lancashire and Cheshire Fauna Society, Rishton.

McCarthy, B. 2001. *Birds of Marshside.* Hobby Publications, Liverpool.

Oakes, C. 1953. *The Birds of Lancashire* Oliver and Boyd, Edinburgh.

Pyefinch, R. and Golborn, P. 2001. *Atlas of the Breeding Birds of Lancashire and North Merseyside.* Hobby Publications, Liverpool.

White, S., McCarthy, B. and Jones, M (Eds). 2008. *The Birds of Lancashire and North Merseyside.* Hobby Publications, Southport.

GREATER MANCHESTER

Holland, P., Spence, I. and Sutton, T. 1984. *Breeding Birds in Greater Manchester.* Manchester Ornithological Society.

REGIONAL

Conlin, A., Cullen, Dr J.P., Marsh, P., Reid, T., Sharpe, C., Smith, J. and Williams, S. 2008. *Where to Watch Birds – North West England & the Isle of Man.* Christopher Helm, London.

NORTHWEST ENGLAND BIRD LIST

Below is a list of the species to have been officially recorded in north-west England,* in at least one of Cheshire, Greater Manchester and Lancashire. The list is in the systematic order used by the British Ornithologists Union (bou.org.uk/british-list/), which maintains the list of wild birds to have occurred in Great Britain.

* In this context north-west England is used as a shorthand for the area covered by this book. Cumbria, which is covered in a separate guide in the Helm *Where To Watch Birds* series, is also in the north-west of England.

1. Brent Goose
2. Red-breasted Goose
3. Canada Goose
4. Barnacle Goose
5. Cackling Goose
6. Ross's Goose
7. Snow Goose
8. Greylag Goose
9. Taiga Bean Goose
10. Pink-footed Goose
11. Tundra Bean Goose
12. White-fronted Goose
13. Lesser White-fronted Goose
14. Mute Swan
15. Bewick's Swan
16. Whooper Swan
17. Egyptian Goose
18. Shelduck
19. Ruddy Shelduck
20. Mandarin Duck
21. Baikal Teal
22. Garganey
23. Blue-winged Teal
24. Shoveler
25. Gadwall
26. Wigeon
27. American Wigeon
28. Mallard
29. Pintail
30. Teal
31. Green-winged Teal
32. Red-crested Pochard
33. Canvasback
34. Pochard
35. Ferruginous Duck
36. Ring-necked Duck
37. Tufted Duck
38. Scaup
39. Lesser Scaup
40. Eider
41. Harlequin Duck
42. Surf Scoter
43. Velvet Scoter
44. Common Scoter
45. Black Scoter
46. Long-tailed Duck
47. Bufflehead
48. Goldeneye
49. Smew
50. Hooded Merganser
51. Goosander
52. Red-breasted Merganser
53. Ruddy Duck
54. Red Grouse
55. Black Grouse
56. Grey Partridge
57. Golden Pheasant
58. Lady Amherst's Pheasant
59. Pheasant
60. Quail
61. Red-legged Partridge
62. Common Nighthawk
63. Nightjar
64. Chimney Swift
65. Alpine Swift
66. Swift
67. Pallid Swift
68. Little Swift

69	Little Bustard		119	Broad-billed Sandpiper
70	Great Spotted Cuckoo		120	Sharp-tailed Sandpiper
71	Yellow-billed Cuckoo		121	Stilt Sandpiper
72	Black-billed Cuckoo		122	Curlew Sandpiper
73	Cuckoo		123	Temminck's Stint
74	Pallas's Sandgrouse		124	Sanderling
75	Rock Dove/Feral Pigeon		125	Dunlin
76	Stock Dove		126	Purple Sandpiper
77	Woodpigeon		127	Baird's Sandpiper
78	Turtle Dove		128	Little Stint
79	Collared Dove		129	White-rumped Sandpiper
80	Water Rail		130	Buff-breasted Sandpiper
81	Corncrake		131	Pectoral Sandpiper
82	Spotted Crake		132	Semipalmated Sandpiper
83	Moorhen		133	Western Sandpiper
84	Coot		134	Long-billed Dowitcher
85	Little Crake		135	Woodcock
86	Crane		136	Jack Snipe
87	Little Grebe		137	Great Snipe
88	Pied-billed Grebe		138	Snipe
89	Red-necked Grebe		139	Terek Sandpiper
90	Great Crested Grebe		140	Wilson's Phalarope
91	Slavonian Grebe		141	Red-necked Phalarope
92	Black-necked Grebe		142	Grey Phalarope
93	Stone-curlew		143	Common Sandpiper
94	Oystercatcher		144	Spotted Sandpiper
95	Black-winged Stilt		145	Green Sandpiper
96	Avocet		146	Solitary Sandpiper
97	Lapwing		147	Lesser Yellowlegs
98	Sociable Plover		148	Redshank
99	White-tailed Plover		149	Marsh Sandpiper
100	Golden Plover		150	Wood Sandpiper
101	Pacific Golden Plover		151	Spotted Redshank
102	American Golden Plover		152	Greenshank
103	Grey Plover		153	Collared Pratincole
104	Ringed Plover		154	Black-winged Pratincole
105	Little Ringed Plover		155	Kittiwake
106	Killdeer		156	Ivory Gull
107	Kentish Plover		157	Sabine's Gull
108	Dotterel		158	Bonaparte's Gull
109	Upland Sandpiper		159	Black-headed Gull
110	Whimbrel		160	Little Gull
111	Curlew		161	Ross's Gull
112	Bar-tailed Godwit		162	Laughing Gull
113	Black-tailed Godwit		163	Franklin's Gull
114	Hudsonian Godwit		164	Mediterranean Gull
115	Turnstone		165	Common Gull
116	Great Knot		166	Ring-billed Gull
117	Knot		167	Great Black-backed Gull
118	Ruff		168	Glaucous Gull

Northwest England Bird List

169	Iceland Gull		219	Little Bittern
170	Herring Gull		220	Night Heron
171	American Herring Gull		221	Squacco Heron
172	Caspian Gull		222	Cattle Egret
173	Yellow-legged Gull		223	Grey Heron
174	Lesser Black-backed Gull		224	Purple Heron
175	Gull-billed Tern		225	Great White Egret
176	Caspian Tern		226	Little Egret
177	Sandwich Tern		227	Osprey
178	Elegant Tern		228	Honey Buzzard
179	Little Tern		229	Golden Eagle
180	Roseate Tern		230	Sparrowhawk
181	Common Tern		231	Goshawk
182	Arctic Tern		232	Marsh Harrier
183	Forster's Tern		233	Hen Harrier
184	Whiskered Tern		234	Pallid Harrier
185	White-winged Black Tern		235	Montagu's Harrier
186	Black Tern		236	Red Kite
187	Great Skua		237	Black Kite
188	Pomarine Skua		238	White-tailed Eagle
189	Arctic Skua		239	Rough-legged Buzzard
190	Long-tailed Skua		240	Buzzard
191	Little Auk		241	Barn Owl
192	Common Guillemot		242	Little Owl
193	Razorbill		243	Scops Owl
194	Black Guillemot		244	Long-eared Owl
195	Puffin		245	Short-eared Owl
196	Red-throated Diver		246	Snowy Owl
197	Black-throated Diver		247	Tawny Owl
198	Great Northern Diver		248	Hoopoe
199	White-billed Diver		249	Roller
200	Wilson's Petrel		250	Kingfisher
201	Storm Petrel		251	Belted Kingfisher
202	Leach's Petrel		252	Bee-eater
203	Fulmar		253	Wryneck
204	Cory's Shearwater		254	Lesser Spotted Woodpecker
205	Sooty Shearwater		255	Great Spotted Woodpecker
206	Great Shearwater		256	Green Woodpecker
207	Manx Shearwater		257	Kestrel
208	Balearic Shearwater		258	Red-footed Falcon
209	Barolo Shearwater		259	Eleonora's Falcon
210	Black Stork		260	Merlin
211	White Stork		261	Hobby
212	Gannet		262	Gyr Falcon
213	Cormorant		263	Peregrine
214	Shag		264	Ring-necked Parakeet
215	Glossy Ibis		265	Great Grey Shrike
216	Spoonbill		266	Lesser Grey Shrike
217	Bittern		267	Woodchat Shrike
218	American Bittern		268	Daurian Shrike

269	Red-backed Shrike		319	River Warbler
270	Golden Oriole		320	Savi's Warbler
271	Jay		321	Grasshopper Warbler
272	Magpie		322	Blackcap
273	Nutcracker		323	Garden Warbler
274	Chough		324	Barred Warbler
275	Jackdaw		325	Lesser Whitethroat
276	Rook		326	Asian Desert Warbler
277	Carrion Crow		327	Sardinian Warbler
278	Hooded Crow		328	Western Subalpine Warbler
279	Raven		329	Whitethroat
280	Waxwing		330	Dartford Warbler
281	Coal Tit		331	Firecrest
282	Crested Tit		332	Goldcrest
283	Marsh Tit		333	Wren
284	Willow Tit		334	Nuthatch
285	Blue Tit		335	Wallcreeper
286	Great Tit		336	Treecreeper
287	Penduline Tit		337	Rose-coloured Starling
288	Bearded Tit		338	Starling
289	Woodlark		339	White's Thrush
290	Skylark		340	Song Thrush
291	Shore Lark		341	Mistle Thrush
292	Short-toed Lark		342	Redwing
293	Sand Martin		343	Blackbird
294	Swallow		344	Fieldfare
295	House Martin		345	Ring Ouzel
296	Red-rumped Swallow		346	Black-throated Thrush
297	Cetti's Warbler		347	Dusky Thrush
298	Long-tailed Tit		348	Spotted Flycatcher
299	Wood Warbler		349	Robin
300	Western Bonelli's Warbler		350	Bluethroat
301	Yellow-browed Warbler		351	Thrush Nightingale
302	Pallas's Warbler		352	Nightingale
303	Radde's Warbler		353	Red-flanked Bluetail
304	Dusky Warbler		354	Red-breasted Flycatcher
305	Willow Warbler		355	Pied Flycatcher
306	Chiffchaff		356	Black Redstart
307	Iberian Chiffchaff		357	Redstart
308	Greenish Warbler		358	Whinchat
309	Arctic Warbler		359	Stonechat
310	Great Reed Warbler		360	Siberian Stonechat
311	Aquatic Warbler		361	Wheatear
312	Sedge Warbler		362	Desert Wheatear
313	Paddyfield Warbler		363	Eastern Black-eared Wheatear
314	Blyth's Reed Warbler		364	Pied Wheatear
315	Reed Warbler		365	Dipper
316	Marsh Warbler		366	Tree Sparrow
317	Melodious Warbler		367	House Sparrow
318	Icterine Warbler		368	Dunnock

369	Yellow Wagtail		391	Crossbill
370	Citrine Wagtail		392	Two-barred Crossbill
371	Grey Wagtail		393	Goldfinch
372	Pied Wagtail		394	Serin
373	Richard's Pipit		395	Siskin
374	Tawny Pipit		396	Lapland Bunting
375	Meadow Pipit		397	Snow Bunting
376	Tree Pipit		398	Corn Bunting
377	Olive-backed Pipit		399	Yellowhammer
378	Red-throated Pipit		400	Ortolan Bunting
379	Buff-bellied Pipit		401	Cirl Bunting
380	Water Pipit		402	Little Bunting
381	Rock Pipit		403	Yellow-breasted Bunting
382	Chaffinch		404	Black-headed Bunting
383	Brambling		405	Black-faced Bunting
384	Hawfinch		406	Reed Bunting
385	Bullfinch		407	Dark-eyed Junco
386	Common Rosefinch		408	White-crowned Sparrow
387	Greenfinch		409	White-throated Sparrow
388	Twite		410	Song Sparrow
389	Linnet		411	Northern Parula
390	Redpoll		412	Blackpoll Warbler

INDEX TO SPECIES

Species index listed by site number

Auk, Little 20, 32, 56, 60, 70
Avocet 9, 12, 17, 28, 34, 52, 58, 59, 64, 69

Bee-eater 20, 21, 32, 60, 74
Bittern 1, 2, 5, 8, 9, 12, 13, 17, 18, 22, 27, 39, 50, 62, 64, 68, 69, 70, 73, 75
  American 39
  Little 39, 74
Blackbird 2, 6
Blackcap 1, 2, 3, 4, 6, 10, 14, 32, 50, 74
Bluetail, Red-flanked 20
Bluethroat 10, 20, 21, 23, 41, 55, 62
Brambling 2, 3, 4, 5, 6, 8, 12, 13, 14, 15, 18, 21, 27, 39, 64, 72, 74, 76, 81, 85, 87
Bullfinch 1, 3, 5, 8, 14, 15, 22, 70, 71, 74, 79, 81, 84
Bunting
  Black-faced 70
  Black-headed 64
  Cirl 21
  Corn 13, 14, 22, 35, 37, 59, 60, 62, 67, 76
  Lapland 8, 13, 18, 20, 21, 35, 38, 47, 56, 59, 60, 68, 72, 82, 86
  Little 8, 10, 20, 23, 55, 70
  Ortolan 37, 38
  Reed 2, 5, 8, 10, 11, 12, 13, 14, 17, 18, 19, 23, 25, 27, 32, 46, 52, 57, 61, 63, 64, 70, 71, 73, 76
  Snow 20, 21, 23, 24, 33, 35, 41, 47, 56, 58, 59, 60, 64, 72
  Yellow-breasted 20
Buzzard 1, 2, 3, 4, 5, 6, 7, 11, 12, 13, 14, 15, 16, 22, 43, 62, 64, 67, 72, 76, 77, 87
  Honey 6, 12, 13, 38, 44, 52, 61, 72, 82
  Rough-legged 2, 6, 43

Canvasback 70
Chaffinch 2, 3, 4, 6, 12, 14, 18, 20, 21, 27
Chiffchaff 1, 2, 3, 4, 8, 10, 14, 18, 20, 32, 50, 70, 74, 85

  Iberian 21, 26, 36, 55, 64
  Siberian 18, 23, 32, 60, 61, 62
Chough 19, 30, 39, 41, 59
Coot 1, 4, 6, 31, 50, 51, 52
Cormorant 1, 4, 5, 6, 11, 12, 13, 16, 20, 32, 41, 52, 70, 79
Corncrake 8, 13
Crake, Spotted 8, 12, 17, 58, 62, 78
Crane 2, 8, 13, 19, 37, 50, 59, 65, 74, 76
Crossbill 6, 14, 15, 26, 38, 43, 44, 60, 64, 72, 85
  Two-barred 85
Crow
  Carrion 1, 64
  Hooded 41, 49
Cuckoo 1, 5, 9, 12, 14, 15, 18, 23, 36, 37, 44, 64, 76, 84
  Black-billed 21
  Great Spotted 21
Curlew 1, 2, 5, 6, 7, 8, 9, 11, 12, 20, 25, 29, 34, 40, 43, 50, 52, 56, 60, 64, 68, 72, 74, 76, 80, 81, 84, 85

Dipper 6, 7, 44, 53, 54, 64, 66, 85, 87
Diver
  Black-throated 2, 20, 45, 46, 60
  Great Northern 2, 5, 8, 19, 20, 31, 32, 33, 41, 42, 43, 45, 46, 49, 60
  Red-throated 8, 20, 30, 31, 32, 33, 41, 42, 43, 45, 46, 60, 64, 66, 81
  White-billed 86
Dotterel 12, 21, 23, 35, 37, 44, 47, 49, 59, 64, 72
Dove
  Stock 1, 2, 11, 12, 14, 15, 16, 64, 67
  Turtle 8, 45, 46, 56
Dowitcher, Long-billed 9, 12, 17, 18, 28, 35, 55, 58, 59, 62
Duck
  Black-bellied Whistling 1, 2
  Ferruginous 1, 2, 10, 31, 38, 57, 59, 78
  Long-tailed 5, 10, 11, 12, 14, 20, 21, 28, 30, 31, 32, 33, 35, 36, 42, 43, 45, 46, 49, 55, 56, 59, 60, 63, 72, 78, 82

Index to species

Mandarin 1, 2, 5, 7, 16, 26, 46, 57, 65, 87
Ring-necked 5, 10, 12, 17, 27, 31, 40, 43, 45, 46, 48, 53, 55, 57, 63, 64, 65, 68, 70, 72, 73, 75, 78, 80
Tufted 1, 2, 4, 5, 6, 10, 11, 12, 14, 16, 17, 27, 31, 32, 39, 40, 41, 43, 49, 50, 52, 53, 54, 55, 58, 60, 63, 64, 70, 71, 72, 78, 80, 82, 87
White-faced Whistling 1
Dunlin 4, 5, 8, 9, 11, 12, 13, 19, 20, 21, 23, 32, 33, 34, 35, 36, 38, 40, 41, 43, 46, 58, 59, 60, 64, 67, 68, 70, 72, 73, 75, 76, 80, 81, 82, 85

Eagle
  Golden 44, 64
  White-tailed 6, 12, 13, 19, 44
Egret
  Cattle 1, 8, 9, 12, 13, 17, 22, 27, 34, 35, 36, 38, 43, 58, 59, 62, 63, 64, 83
  Great White 1, 5, 8, 9, 12, 13, 17, 18, 22, 27, 28, 30, 31, 34, 36, 38, 43, 45, 46, 49, 50, 52, 57, 58, 59, 62, 63, 70, 73, 76, 78, 79
  Little 1, 9, 11, 12, 13, 17, 18, 19, 22, 23, 27, 34, 39, 45, 46, 49, 50, 58, 59, 62, 63, 66, 67, 70, 74, 77, 79
Eider 8, 19, 20, 29, 32, 33, 34, 35, 38, 41, 42, 43, 45, 46, 59

Falcon, Red-footed 27, 39, 70, 72, 76
Fieldfare 6, 9, 11, 12, 32, 64, 72
Firecrest 6, 8, 26, 30, 36, 56, 57, 61, 64, 82
Flycatcher
  Pied 3, 6, 7, 8, 13, 14, 15, 20, 32, 39, 41, 44, 45, 58, 60, 64, 72, 87
  Red-breasted 2, 20, 21, 36, 60
  Spotted 3, 5, 7, 8, 13, 14, 15, 20, 39, 44, 58, 64, 70, 72, 73, 74, 81, 83, 87
Fulmar 1, 20, 35, 42, 56, 78, 79

Gadwall 1, 2, 4, 5, 9, 10, 11, 12, 13, 17, 27, 28, 31, 32, 35, 38, 39, 50, 53, 54, 57, 58, 63, 68, 70, 71, 75, 79
Gannet 8, 20, 24, 31, 42, 46, 51, 60, 79, 81

Garganey 4, 5, 8, 9, 10, 12, 13, 17, 27, 28, 35, 36, 38, 39, 43, 46, 49, 52, 54, 55, 57, 58, 59, 60, 62, 63, 70, 74, 75
Godwit
  Bar-tailed 12, 13, 23, 34, 35, 38, 43, 49, 54, 56, 58, 59, 60, 63, 64
  Black-tailed 5, 8, 9, 11, 12, 13, 17, 22, 25, 26, 28, 31, 34, 38, 40, 43, 46, 49, 54, 58, 59, 63, 64, 67, 68, 69, 70, 73, 75, 76, 80
  Hudsonian 13, 17
Goldcrest 1, 2, 3, 4, 6, 10, 14
Goldeneye 1, 2, 4, 5, 6, 9, 11, 12, 14, 16, 19, 27, 34, 39, 40, 41, 43, 50, 55, 56, 64, 70, 72, 74, 75, 78, 80, 81, 83
Goldfinch 1, 5, 6, 18, 20
Goosander 1, 2, 4, 5, 6, 7, 8, 9, 11, 16, 19, 20, 43, 49, 50, 51, 53, 54, 59, 63, 64, 65, 70, 71, 74, 78, 79, 80, 81, 83, 85, 87
Goose
  Bar-headed 1
  Barnacle 1, 2, 12, 36, 37, 39, 43, 58, 59
  Black Brant 36
  Brent 5, 12, 13, 20, 21, 32, 35, 36, 37, 43, 45, 58, 82
  Cackling 58
  Canada 1, 2, 4, 5, 6, 8, 9, 12, 13, 16, 17, 43, 46, 52, 57, 58, 59, 64
  Egyptian 1, 2, 5, 17, 49
  Grey-bellied Brent 58, 59
  Greylag 1, 4, 5, 8, 9, 16, 17, 28, 40, 43, 52, 53, 57, 64
  Lesser White-fronted 1, 17, 36, 37, 58
  Pink-footed 1, 5, 9, 10, 12, 16, 17, 18, 20, 22, 32, 34, 35, 36, 37, 38, 40, 41, 58, 59, 60, 62, 63, 67, 70, 72, 74, 76
  Red-breasted 5, 36, 37, 58
  Richardson's Cackling 59
  Ross's 36, 37, 58
  Snow 1, 17, 28, 36, 37, 58, 59, 60, 63
  Taiga Bean 36, 37, 55, 58, 59, 60
  Todd's Canada 37, 58, 59, 60
  Tundra Bean 5, 13, 17, 28, 35, 36, 37, 49, 58, 59, 62, 76

Index to species

White-fronted 5, 12, 17, 28, 35, 36, 37, 43, 45, 49, 58, 59, 62, 63, 76, 82
Goshawk 6, 15, 64
Grebe
    Black-necked 1, 2, 5, 10, 11, 12, 31, 32, 35, 43, 45, 46, 49, 50, 51, 53, 54, 55, 56, 59, 63, 70, 72, 75
    Great Crested 1, 2, 4, 5, 6, 8, 10, 11, 12, 14, 20, 21, 33, 41, 43, 46, 49, 50, 51, 52, 53, 54, 57, 63, 64, 65, 66, 70, 79, 82, 87
    Little 1, 2, 4, 5, 6, 8, 10, 11, 12, 14, 22, 31, 36, 39, 40, 41, 46, 49, 52, 53, 54, 57, 60, 63, 71, 72
    Pied-billed 27, 31, 53, 83
    Red-necked 1, 2, 8, 12, 38, 43, 45, 46, 48, 49, 50, 51, 53, 57
    Slavonian 1, 2, 5, 6, 8, 30, 31, 32, 33, 36, 45, 48, 49, 51, 52, 56, 65
Greenfinch 2, 8, 12, 18, 20, 27
Greenshank 4, 5, 8, 9, 12, 13, 17, 18, 23, 25, 28, 30, 34, 38, 40, 43, 46, 49, 58, 59, 64, 68, 69, 70, 73, 74, 75, 76, 80, 81
Grouse
    Black 6
    Red 7, 44, 47, 64, 72, 82, 84, 85
Guillemot
    Black 20, 32, 33, 41, 60
    Common 8, 20, 32, 60, 79
Gull
    American Herring 9, 55
    Black-headed 1, 5, 9, 10, 14, 17, 19, 28, 31, 32, 34, 40, 51, 52, 64, 68, 70
    Bonaparte's 13, 20, 24, 32, 40, 55, 80
    Caspian 1, 11, 13, 24, 36, 40, 43, 45, 48, 49, 51, 55, 56, 60, 65, 70, 72, 75, 80, 82, 83, 86
    Common 1, 56
    Franklin's 8, 9, 11, 12, 62
    Glaucous 11, 13, 24, 31, 33, 36, 40, 43, 45, 46, 48, 51, 55, 65, 70, 75, 80, 83
    Great Black-backed 80
    Herring 45, 80
    Iceland 11, 13, 24, 31, 33, 36, 39, 41, 43, 45, 46, 48, 49, 51, 55, 65, 70, 75, 80, 83, 86
    Ivory 38
    Kumlien's 13, 80
    Laughing 20, 24, 39, 50, 55
    Lesser Black-backed 5, 38, 45, 53
    Little 2, 5, 9, 10, 12, 13, 20, 23, 24, 27, 31, 32, 33, 37, 39, 40, 43, 45, 46, 48, 49, 50, 51, 52, 55, 56, 58, 59, 60, 62, 63, 65, 66, 70, 72, 74
    Mediterranean 1, 8, 10, 14, 17, 19, 21, 24, 25, 28, 31, 32, 33, 34, 35, 39, 40, 45, 48, 49, 50, 51, 52, 59, 64, 65, 70, 72
    Ring-billed 11, 20, 24, 34, 39, 40, 43, 48, 49, 51, 54, 55, 56, 62, 68, 70, 86
    Ross's 27, 39, 41, 55
    Sabine's 8, 13, 20, 23, 24, 27, 30, 32, 33, 35, 38, 41, 42, 56, 60, 70, 74, 86
    Yellow-legged 1, 5, 8, 13, 21, 33, 34, 36, 40, 43, 45, 48, 55, 56, 64, 70, 72, 75, 76, 80, 83, 86

Harrier
    Hen 6, 12, 13, 17, 18, 19, 22, 27, 33, 35, 37, 38, 43, 44, 47, 56, 58, 59, 62, 63, 64, 67, 72, 81, 84
    Marsh 4, 8, 9, 10, 12, 13, 17, 18, 19, 27, 28, 35, 37, 38, 39, 40, 43, 45, 56, 58, 59, 62, 64, 67, 68, 72, 76, 81
    Montagu's 12, 27, 36, 39, 47, 50, 62
    Pallid 18, 35, 44, 50
Hawfinch 8, 9, 13, 16, 27, 59, 64, 66, 72, 74, 83
Heron
    Grey 4, 5, 6, 9, 11, 12, 13, 16, 17, 19, 39, 52, 63, 66
    Night 8, 9, 17, 27, 39, 50, 53
    Purple 8, 27, 37, 50, 57, 62, 63, 69
    Squacco 27
Hobby 1, 3, 4, 5, 6, 8, 9, 10, 12, 13, 14, 15, 17, 27, 35, 37, 38, 39, 45, 46, 49, 50, 59, 61, 64, 76, 77
Hoopoe 8, 21, 22, 23, 24, 41, 46, 60

Ibis, Glossy 5, 8, 9, 10, 13, 17, 27, 36, 39, 59, 63, 70, 72, 83

Index to species

Jackdaw 1, 2, 6, 16, 64
Jay 1

Kestrel 1, 3, 6, 9, 12, 13, 14, 17, 18, 19, 22, 27, 37, 39, 40, 43, 50, 52, 58, 64, 67, 72
Killdeer 49
Kingfisher 1, 2, 4, 5, 7, 9, 10, 11, 16, 17, 26, 28, 34, 39, 50, 51, 52, 53, 54, 57, 62, 64, 66, 67, 70, 74, 79, 81, 83, 84, 87
  Belted 50
Kite
  Black 8
  Red 2, 4, 6, 7, 8, 9, 12, 44, 59, 64
Kittiwake 1, 9, 20, 24, 30, 32, 35, 37, 43, 45, 46, 48, 49, 51, 56, 64, 65, 83
Knot 9, 12, 13, 19, 20, 21, 23, 29, 32, 41, 43, 49, 55, 56, 58, 60, 64
  Great 40

Lapwing 1, 2, 3, 5, 8, 9, 10, 11, 12, 17, 18, 19, 20, 23, 34, 38, 40, 43, 46, 49, 50, 52, 58, 59, 60, 62, 63, 64, 67, 70, 77
  White-tailed 10
Lark, Short-toed 23, 41
Linnet 1, 3, 5, 11, 12, 14, 18, 20, 21, 23, 64, 76

Magpie 10, 64
Mallard 1, 2, 4, 5, 6, 9, 10, 11, 12, 22, 25, 29, 43, 52, 64
Martin
  House 2, 64
  Sand 2, 9, 10, 11, 31, 35, 39, 50, 54, 58
Merganser
  Hooded 1, 2
  Red-breasted 19, 20, 29, 41, 43, 45, 56, 59
Merlin 10, 12, 13, 17, 18, 19, 20, 38, 40, 43, 44, 47, 58, 59, 62, 64, 67, 72, 76, 81, 84
Moorhen 1, 4, 51

Nightingale 8, 20, 38
  Thrush 32
Nightjar 6, 14, 15, 26, 64

Nutcracker 2
Nuthatch 1, 2, 3, 14, 16, 21, 27, 53, 79

Oriole, Golden 8
Osprey 2, 4, 5, 6, 7, 8, 9, 12, 13, 18, 21, 24, 27, 32, 33, 39, 42, 43, 44, 45, 46, 49, 50, 52, 54, 56, 57, 59, 62, 63, 64, 65, 66, 72, 84
Ouzel, Ring 7, 12, 13, 15, 20, 22, 23, 32, 36, 41, 44, 47, 55, 58, 60, 61, 64, 72, 76, 77, 81, 84, 85
Owl
  Barn 1, 2, 5, 7, 8, 9, 13, 15, 17, 18, 22, 34, 37, 39, 49, 58, 61, 62, 64, 67
  Eagle 44
  Little 1, 2, 7, 8, 15, 39
  Long-eared 6, 8, 39, 64
  Short-eared 7, 10, 11, 12, 13, 17, 18, 19, 34, 35, 37, 38, 41, 43, 44, 49, 58, 59, 61, 62, 63, 64, 67, 76
  Snowy 64
  Tawny 2, 3, 5, 6, 7, 8, 9, 14, 15, 22, 26, 87
Oystercatcher 4, 5, 8, 9, 11, 12, 17, 18, 19, 20, 21, 23, 25, 29, 32, 33, 38, 40, 42, 43, 44, 49, 50, 52, 53, 55, 58, 59, 64, 67

Parakeet, Ring-necked 39, 74
Partridge
  Grey 1, 12, 13, 59, 63, 64, 75, 77
  Red-legged 12
Peregrine 1, 4, 5, 6, 7, 8, 9, 10, 11, 12, 13, 15, 16, 17, 18, 19, 20, 21, 22, 24, 27, 37, 39, 40, 41, 42, 43, 44, 50, 52, 56, 58, 59, 62, 64, 67, 74, 76, 81, 82, 85
Petrel
  Leach's 8, 20, 23, 24, 30, 32, 33, 35, 41, 42, 45, 56, 60
  Soft-Plumage 60
  Storm 20, 23, 24, 30, 31, 32, 35, 56, 60
  Wilson's 56
Phalarope
  Grey 2, 5, 8, 12, 20, 23, 24, 32, 33, 35, 38, 41, 42, 43, 46, 56, 58, 59, 60, 71, 72, 80
  Red-necked 8, 9, 12, 17, 19, 28, 38, 46, 49, 50, 56, 58, 62, 68, 69, 70, 86

Index to species

Wilson's 8, 10, 12, 17, 18, 28, 49, 55, 58, 62
Pintail 1, 8, 10, 12, 16, 17, 18, 25, 29, 36, 38, 43, 62, 86
Pipit
  Buff-bellied 18, 35
  Meadow 2, 5, 6, 8, 12, 20, 21, 23, 32, 41, 64, 70, 71, 72, 74, 76
  Red-throated 21, 23, 55
  Richard's 8, 18, 19, 21, 23, 36, 49, 56, 59, 60, 68, 70, 72, 73, 76
  Rock 2, 12, 18, 20, 21, 29, 32, 34, 35, 40, 41, 42, 45, 46, 49, 74, 86
  Tawny 21, 23, 32, 41, 70
  Tree 6, 7, 14, 15, 20, 21, 23, 37, 41, 44, 64, 70, 72, 74, 75
  Water 2, 12, 13, 17, 18, 27, 34, 35, 38, 40, 41, 56, 58, 64, 71
Plover
  American Golden 12, 21, 23, 35, 40, 48, 55, 58, 59, 63
  Golden 1, 5, 7, 11, 12, 13, 18, 19, 20, 34, 35, 40, 47, 48, 58, 63, 64, 72, 85
  Grey 12, 20, 35, 43, 49, 52, 60, 64
  Kentish 8, 35, 62, 70, 86
  Little Ringed 5, 8, 9, 10, 11, 12, 13, 31, 34, 43, 46, 49, 50, 52, 54, 58, 63, 64, 70, 73, 74, 75, 76
  Pacific Golden 36, 55
  Ringed 5, 8, 12, 13, 20, 23, 34, 35, 38, 40, 41, 43, 46, 58, 64, 70, 73, 74, 75, 76, 81, 82
  White-tailed 27, 28, 55
Pochard 1, 2, 4, 5, 6, 8, 10, 11, 12, 14, 16, 17, 27, 31, 41, 43, 50, 54, 55, 62, 64, 78
  Red-crested 10, 31, 49
Pratincole
  Black-winged 17, 28, 62
  Collared 12, 21, 23, 38, 39, 59, 62, 68, 83
Puffin 20, 32, 33, 41, 42, 60

Quail 13, 18, 35, 37, 59, 76, 77

Rail, Water 1, 2, 5, 8, 9, 10, 12, 17, 18, 21, 27, 38, 39, 57, 61, 63, 66, 68, 70, 74, 75

Raven 5, 6, 7, 12, 15, 16, 21, 22, 44, 64, 72, 76, 82, 85, 87
Razorbill 20, 60
Redpoll 1, 2, 3, 4, 5, 6, 8, 13, 14, 15, 16, 20, 21, 22, 23, 26, 32, 43, 44, 53, 57, 64, 68, 71, 75, 79, 81, 84
  "Arctic" 4
  "Mealy" 39
Redshank 4, 8, 9, 11, 12, 17, 18, 19, 20, 25, 29, 32, 34, 38, 40, 42, 43, 46, 49, 50, 52, 55, 56, 58, 59, 60, 62, 64, 68, 69, 70, 73, 76, 80, 81
  Spotted 12, 13, 17, 18, 28, 30, 34, 35, 37, 38, 40, 48, 49, 54, 55, 56, 58, 59, 60, 62
Redstart 3, 5, 6, 7, 8, 12, 13, 14, 15, 20, 21, 22, 23, 32, 39, 41, 44, 49, 58, 60, 64, 65, 70, 72, 73, 74, 75, 76, 77, 83, 85, 87
  Black 8, 12, 32, 33, 35, 41, 46, 47, 55, 56, 60, 64, 72
Redwing 2, 3, 6, 9, 10, 11, 12, 32, 64, 72
Roller 42
Rook 1
Rosefinch, Common 30, 60
Ruff 5, 8, 9, 11, 12, 13, 17, 28, 34, 35, 37, 38, 43, 46, 48, 49, 58, 59, 62, 63, 64, 67, 68

Sanderling 8, 9, 12, 13, 20, 23, 41, 43, 48, 49, 56, 60, 63, 64, 70, 74, 80, 81, 82, 83
Sandpiper
  Baird's 12, 21, 59, 82
  Broad-billed 12, 13, 17, 21, 34, 35, 58, 59, 69, 81
  Buff-breasted 10, 12, 36, 37, 55, 58, 59
  Common 1, 2, 4, 5, 6, 7, 8, 12, 16, 20, 22, 28, 34, 43, 44, 46, 50, 52, 55, 56, 64, 70, 72, 73, 74, 75, 79, 80, 82, 84, 85
  Curlew 5, 9, 12, 13, 17, 19, 21, 28, 30, 33, 34, 35, 36, 37, 40, 43, 46, 49, 55, 56, 58, 59, 60, 62, 63, 73
  Green 1, 4, 5, 8, 9, 10, 12, 16, 22, 28, 34, 43, 46, 49, 50, 52, 54, 58, 59, 63, 64, 67, 71, 73, 74, 75, 76, 77

Index to species

Marsh 8, 12, 17, 55
Pectoral 12, 13, 17, 28, 39, 43, 45, 49, 50, 52, 55, 58, 59, 62, 63, 70, 71, 73, 83, 86
Purple 5, 20, 24, 25, 33, 41, 42, 45, 80
Semipalmated 12, 21, 36, 40, 50
Sharp-tailed 12
Spotted 8, 43, 46, 50, 74
Stilt 8, 9, 12, 63
Terek 12, 17, 19, 55
Western 21
White-rumped 8, 13, 21, 28, 40, 55, 59
Wood 5, 8, 9, 10, 12, 13, 22, 28, 34, 43, 46, 49, 52, 54, 55, 56, 57, 58, 59, 60, 62, 63, 64, 69, 74, 76, 77
Scaup 5, 11, 12, 14, 17, 19, 28, 30, 31, 32, 33, 34, 35, 38, 39, 41, 43, 46, 47, 48, 49, 51, 54, 55, 56, 57, 58, 59, 60, 63, 64, 70, 72, 74, 78, 80
Lesser 8, 10, 12, 17, 27, 31, 39, 54, 58, 70, 74, 80
Scoter
 Black 21, 30
 Common 1, 2, 5, 9, 10, 14, 20, 31, 32, 33, 42, 43, 45, 46, 47, 48, 49, 50, 51, 52, 56, 60, 63, 64, 65, 70, 72, 74, 79, 80, 81, 82, 83
 Surf 21, 42, 57
 Velvet 20, 21, 30, 32, 33, 35, 42, 56, 60, 72
Serin 21, 23, 32, 55, 70
Shag 19, 31, 32, 46, 48, 51, 57, 78, 79
Shearwater
 Balearic 23, 42, 60
 Barolo 1, 23
 Cory's 60
 Manx 2, 20, 32, 35, 42, 46, 51, 60
 Sooty 23, 60
Shelduck 1, 9, 11, 12, 17, 18, 19, 25, 52, 62, 67, 69
 Ruddy 1, 17, 28, 38, 46, 49, 57, 60
 South African 1
Shorelark 35, 47, 56, 75, 86
Shoveler 1, 2, 4, 5, 9, 10, 11, 12, 13, 16, 17, 18, 27, 28, 31, 32, 35, 36, 39, 41, 50, 52, 53, 54, 57, 58, 63, 68, 69, 70, 71, 76, 79

Shrike
 Great Grey 6, 8, 11, 12, 22, 37, 38, 43, 45, 56, 58, 60, 64, 73, 76, 79, 81
 Isabelline/Daurian 43
 Red-backed 8, 12, 21, 28, 35, 41, 60
 Woodchat 12, 20, 27, 28, 38, 41, 58, 60
Siskin 1, 2, 3, 4, 5, 6, 8, 11, 13, 14, 15, 16, 20, 21, 22, 43, 44, 53, 64, 68, 70, 71, 75, 78, 79, 81, 85, 87
Skua
 Arctic 1, 20, 23, 30, 32, 35, 36, 45, 46, 50, 56, 60, 64, 72
 Great 1, 20, 23, 30, 32, 35, 36, 43, 45, 50, 51, 56, 60
 Long-tailed 20, 24, 30, 32, 36, 41, 42, 49, 56, 60
 Pomarine 20, 30, 36, 41, 46, 49, 50, 56, 60
Skylark 1, 5, 8, 9, 12, 17, 18, 19, 20, 23, 25, 29, 58, 60, 64, 67, 69, 70, 72, 75, 76, 77
Smew 1, 2, 5, 6, 10, 11, 12, 31, 39, 43, 45, 47, 48, 49, 50, 55, 56, 59, 63, 64, 68, 70
Snipe 2, 4, 5, 8, 9, 10, 12, 13, 14, 18, 21, 22, 23, 32, 34, 38, 39, 40, 41, 43, 49, 50, 52, 58, 60, 61, 62, 63, 64, 67, 68, 72, 76, 79
 Great 12
 Jack 2, 8, 9, 10, 12, 13, 14, 18, 21, 32, 34, 40, 41, 43, 49, 60, 61, 63, 67
Sparrow
 Song 55
 Tree 1, 2, 5, 8, 11, 37, 50, 57, 59, 62, 64, 67, 71, 76
 White-crowned 10, 55
 White-throated 32
Sparrowhawk 1, 5, 6, 8, 9, 12, 13, 14, 18, 27, 37, 39, 42, 43, 50, 52, 58, 62, 67, 70, 76, 87
Spoonbill 8, 9, 12, 13, 17, 18, 27, 28, 30, 38, 56, 58, 59, 62, 63
Starling 1, 6, 9, 10, 20, 21, 27, 39, 42, 57, 62
 Rose-coloured 12, 55, 74
Stilt, Black-winged 8, 9, 12, 17, 58, 62, 70, 76, 83

Index to species

Stint
  Little 5, 9, 12, 13, 17, 19, 21, 28, 34, 35, 36, 37, 40, 43, 46, 48, 49, 55, 56, 58, 59, 60, 62, 63
  Temminck's 9, 12, 13, 38, 43, 49, 50, 55, 58, 59, 63, 69, 71, 73, 79
Stone-curlew 23, 76
Stonechat 1, 5, 8, 9, 12, 13, 18, 20, 23, 44, 47, 50, 60, 64, 67, 70, 72, 74, 75, 76, 82, 85
  Siberian 9, 10
Stork
  Black 65, 72
  White 8, 49
Swallow 2, 10, 12, 17, 27, 31, 64
  Red-rumped 8, 20, 21, 31, 55, 61, 80, 86
Swan
  Bewick's 12, 13, 17, 35, 37, 43, 45, 46, 58, 59, 62, 64
  Black 1
  Mute 4, 8, 12, 17, 18, 35, 39, 41, 87
  Whooper 1, 5, 12, 13, 17, 18, 32, 35, 37, 41, 42, 43, 45, 46, 49, 52, 58, 59, 62, 64, 67, 68, 70, 72, 73, 74, 76, 81, 82, 83, 84
Swift 1, 2, 9, 10, 17, 70, 74, 76
  Alpine 8, 10, 17, 21, 70, 74, 86
  Chimney 10
  Little 24
  Pallid 21, 41, 55, 56

Teal 1, 2, 4, 5, 6, 8, 9, 10, 11, 12, 13, 14, 16, 17, 18, 19, 22, 25, 27, 28, 29, 32, 34, 35, 36, 39, 40, 41, 43, 46, 49, 50, 52, 53, 54, 55, 57, 59, 60, 62, 63, 64, 68, 71, 76, 79
  Baikal 58
  Blue-winged 9, 58, 70
  Green-winged 1, 8, 10, 12, 13, 17, 19, 27, 28, 34, 35, 38, 39, 43, 55, 57, 58, 59, 62, 63, 68, 69, 70, 86
Tern
  Arctic 1, 2, 5, 9, 12, 20, 21, 30, 31, 32, 35, 39, 41, 43, 45, 46, 49, 51, 55, 56, 59, 60, 64, 70, 72, 74, 83
  Black 1, 2, 4, 5, 9, 12, 20, 21, 24, 27, 30, 35, 39, 41, 43, 45, 46, 49, 50, 51, 52, 55, 56, 57, 58, 62, 63, 64, 70, 74

Caspian 5, 8, 9, 20, 21, 27, 28, 41, 79
Common 1, 2, 5, 8, 9, 12, 17, 18, 20, 21, 31, 32, 34, 39, 41, 43, 45, 46, 50, 51, 55, 56, 59, 60, 64, 68, 70, 74, 79
Elegant 41, 60
Forster's 55, 60
Gull-billed 8, 12, 17, 19, 21, 55, 62
Little 20, 21, 45, 46, 51, 56, 59, 60, 70
Roseate 21, 51, 55, 60
Sandwich 4, 20, 21, 31, 32, 41, 42, 46, 55, 59, 60, 64, 74, 83
Whiskered 8, 9, 12, 31, 64, 70
White-winged Black 8, 9, 12, 21, 27, 31, 34, 39, 45, 49, 55, 56, 59, 62, 63, 68, 70, 74, 80
Thrush, Song 6
Tit
  Bearded 1, 2, 10, 11, 17, 27, 39, 55, 62, 63
  Blue 3, 21
  Coal 1, 3, 6, 14, 21, 27
  Great 3
  Long-tailed 1, 6, 14
  Marsh 2, 15, 16, 27, 50, 53, 81
  Penduline 10, 27, 70
  Willow 1, 2, 8, 10, 11, 14, 57, 64, 66, 68, 69, 70, 71, 74, 75, 76, 79
Treecreeper 1, 2, 3, 11, 14, 27, 71, 79
Turnstone 9, 12, 20, 24, 32, 33, 41, 42, 43, 48, 49, 64, 70, 74, 80, 82, 83
Twite 7, 12, 18, 23, 29, 35, 36, 41, 49, 58, 59, 60, 64, 72

Wagtail
  Ashy-headed 50
  Black-headed 27
  Blue-headed 13, 36, 38, 49, 55, 56, 58, 59, 60
  Citrine 21, 38, 55
  Grey 1, 2, 6, 7, 9, 32, 39, 44, 51, 53, 64, 66, 71
  Grey-headed 55, 59
  Iberian 86
  Pied 2, 6, 31, 71
  White 2, 5, 6, 8, 13, 20, 21, 22, 23, 31, 49, 55, 58, 59, 76

287

Index to species

Yellow 2, 5, 6, 8, 9, 13, 20, 21, 22, 23, 31, 35, 36, 41, 55, 59, 60, 62, 67, 69, 74, 76, 77
Warbler
  Aquatic 12
  Arctic 21
  Barred 12, 17, 21, 23, 32, 38
  Blackpoll 55
  Blyth's Reed 10, 20, 21, 71
  Cetti's 1, 2, 8, 10, 12, 17, 21, 25, 27, 32, 39, 41, 50, 57, 58, 60, 61, 62, 63, 68, 69, 71, 79
  Dartford 20, 21, 64
  Desert 23
  Dusky 32, 38, 60
  Garden 2, 5, 6, 8, 9, 14, 15, 20, 26, 50, 84, 87
  Grasshopper 1, 2, 5, 8, 9, 10, 11, 12, 14, 18, 20, 23, 27, 32, 39, 50, 60, 61, 64, 66, 68, 69, 70, 74, 75, 77
  Great Reed 21, 27, 62, 70
  Greenish 21
  Icterine 21, 32, 41
  Marsh 21, 32, 70, 73
  Melodious 20, 21, 23, 32, 38
  Moltoni's 32
  Paddyfield 20
  Pallas's 19, 20, 21, 23, 26, 30, 41, 42
  Radde's 23
  Reed 1, 2, 5, 8, 9, 10, 11, 12, 17, 18, 21, 22, 23, 25, 27, 30, 36, 39, 41, 50, 52, 54, 57, 58, 61, 62, 63, 68, 69, 70, 71, 73, 74, 75
  River 68
  Savi's 12, 17, 21, 27, 50
  Sedge 1, 2, 5, 8, 9, 10, 11, 12, 17, 18, 20, 21, 22, 23, 25, 27, 39, 41, 50, 52, 54, 57, 58, 62, 63, 64, 68, 69, 70, 71, 73, 74, 75, 77
  Western Subalpine 18, 20, 21, 23, 32
  Willow 1, 2, 10, 11, 14, 20, 21, 70, 72, 74, 76, 84, 85
  Wood 3, 6, 14, 15, 17, 26, 39, 41, 43, 44, 45, 57, 58, 60, 64, 65, 84, 85, 87
  Yellow-browed 1, 8, 11, 13, 17, 20, 21, 22, 23, 24, 26, 27, 30, 32, 33, 35, 36, 38, 41, 42, 46, 55, 56, 58, 59, 60, 62, 66
Waxwing 57, 59
Wheatear 2, 5, 7, 8, 9, 11, 12, 13, 18, 20, 21, 22, 23, 25, 26, 32, 35, 36, 39, 41, 42, 44, 47, 55, 58, 59, 60, 61, 64, 65, 67, 72, 74, 75, 76, 77, 81, 82, 85
  Black-eared 74
  Desert 41, 42, 76
  Eastern Black-eared 36
  Pied 23, 36, 55
Whimbrel 5, 8, 9, 13, 19, 20, 23, 25, 40, 43, 49, 50, 52, 53, 64, 70, 74, 80, 82
Whinchat 8, 9, 12, 13, 18, 20, 21, 22, 23, 32, 35, 36, 39, 41, 44, 47, 49, 55, 56, 58, 59, 60, 61, 64, 67, 72, 74, 75, 76
Whitethroat 1, 2, 3, 9, 10, 11, 12, 14, 17, 20, 23, 25, 57, 61, 67, 71, 74, 77
  Lesser 1, 5, 8, 9, 17, 22, 25, 32, 54, 57, 69, 74
Wigeon 1, 2, 5, 6, 8, 9, 11, 12, 13, 16, 17, 18, 27, 28, 29, 31, 32, 34, 35, 36, 38, 39, 40, 41, 43, 46, 49, 50, 52, 54, 57, 59, 64, 68, 70, 74, 75, 79, 80, 81, 83, 84, 86
  American 1, 8, 12, 17, 27, 28, 35, 36, 38, 39, 40, 55, 58, 59, 62, 70
Woodcock 1, 2, 3, 6, 8, 9, 26, 39, 43, 64, 81, 87
Woodlark 32
Woodpecker
  Great Spotted 1, 3, 6, 9, 11, 14, 15, 53, 57, 66
  Green 2, 3, 4, 6, 9, 11, 14, 15, 26, 64, 81, 84
  Lesser Spotted 2, 6, 8, 9, 11, 14, 15, 16, 26, 64
Woodpigeon 1, 6, 8, 12, 13, 64, 70, 72, 74
Wryneck 6, 8, 17, 21, 23, 32, 35, 45, 55, 60, 65, 69, 72, 86

Yellowhammer 1, 11, 13, 14, 19, 37, 64, 67, 69, 76, 77
Yellowlegs, Lesser 8, 12, 17, 18, 28, 34, 37, 39, 50, 59